The Computer
from Pascal to
von Neumann

In September 1942, Lt. Herman H. Goldstine, a former mathematics professor, was placed in charge of the substation of the Ballistic Research Laboratory at the University of Pennsylvania's Moore School of Electrical Engineering. His mission was to develop new apparatus to expedite the production of firing tables. His achievement was to assist in the creation of the ENIAC, the first electronic digital computer. A team consisting of Lt. Goldstine, J. Presper Eckert, John W. Mauchly, and others produced a machine that operated at a speed several hundred times faster than the electromechanical devices previously in use.

The ENIAC was operational in December 1945, but even before it was completed ideas for its improvement were proliferating. The principal source of these ideas was John von Neumann, who became Goldstine's chief collaborator, and whose genius provided answers for numerous problems that many regarded as insurmountable. Before leaving the Moore School they developed the EDVAC, successor to the ENIAC. After World War II they worked together at the Institute for Advanced Study, where they built what was to become the prototype of the present-day computer.

Herman Goldstine combines history and scientific autobiography in his account of these momentous events. Part One provides a brief history of the development of the computer from early efforts in the seventeenth century to the

The Computer

from Pascal to von Neumann

Herman H. Goldstine

Princeton University Press

differential analyzers developed by Vannevar Bush and his associates in the 1930s. Part Two describes the wartime events at the Moore School, and Part Three discusses the postwar developments at the Institute for Advanced Study and the beginnings of the world-wide "Computer Revolution."

HERMAN H. GOLDSTINE began his scientific career as a mathematician and has had a lifelong interest in the interaction of mathematical ideas and technology. He received his Ph.D. in mathematics from the University of Chicago in 1936 and was an assistant professor at the University of Michigan when he entered the Army in 1941. After participating in the development of the first electronic computer, he left the Army in 1945, and from 1946 to 1957 he was a member of the Institute for Advanced Study, where he collaborated with John von Neumann in a series of scientific papers on subjects related to their work on the Institute computer. In 1958 he joined the IBM Corporation as a member of the research planning staff. He has been director of mathematical sciences at the Thomas J. Watson Research Center, 1958-1965; director of scientific development of the data processing division, 1965-1967; and consultant to the director of research, 1967-1969. Since 1969 he has been an IBM Fellow and is once again at the Institute for Advanced Study, where he is a member of the School of Natural Sciences.

This book has been composed in
Linofilm Caledonia

Printed in the
United States of America
by Princeton University Press
Princeton, New Jersey

Second printing, 1973

To Adele

But if the while I think on thee, dear friend,
All losses are restored, and sorrows end.

Preface

This work divides rather naturally into three parts. There is the pre–World War II era; the war period, particularly at the Moore School of Electrical Engineering, University of Pennsylvania; and the postwar years at the Institute for Advanced Study in Princeton, through 1957. I was concerned as a principal in the last two periods, and the reader will undoubtedly notice the stylistic differences that this has occasioned. It seemed to me in writing the text quite natural to record events as I viewed them then.

Of course, the writing of a history by a participant in the events is at best a tricky business since it may bring into the account some measure of personal bias. In mitigation it may be said, though, that it does provide a detailed, precise understanding of the people and of the events that actually took place. Such an understanding is very difficult for a non-participant. I therefore decided it was worthwhile to write this account with all the objectivity I could, with occasional warnings to the reader about possible traps awaiting him on his peregrination through the period.

There are, of course, at least two basic ways in which the history of apparatus can be written: by concentrating on the equipment or on the ideas and the people who conceived them. I have chosen rather arbitrarily to give the ideas and the people first place, perhaps because I find that approach more interesting personally. In any event, I have also tried to say enough about the apparatus to make it intelligible without becoming overly technical. The reader will find ever so often that the chronological narrative is interrupted by excursi in which I have attempted to explain some technical point so as not to leave the reader uninformed as to its nature. It is to be hoped that these interpolations do not unduly retard the account.

Rather fortuitously, I came out of the war with an extraordinarily large set of personal files covering the period 1942–1946, so that the documentation for these years is extremely complete. To a somewhat lesser extent, I kept files for the period 1946–1957 and have been fortunate in having had access to other files for the period at the Institute for Advanced Study. Thus the account given here is based soundly on the relevant documents of the times and does not depend merely on the memories of an individual. All this documentary material — letters, reports, etc. — is being turned over by me to the library of Hampshire College in Amherst, Massachusetts, so that future scholars may have full access to these papers.

In this connection I wish to acknowledge the many courtesies extended as well as documents furnished me by Prof. Arthur W. Burks of the University of Michigan, by Mr. Henry Halliday, Esq. of Minneapolis, and by Dr. Carl Kaysen, the Director of the Institute for Advanced Study, who made the Institute's files available to me. I also wish to acknowledge the patient, tireless, and good-natured help of my secretary, Mrs. Doris Crowell. I further wish to make suitable mention of all the help given me by Mr. Richard Luxner, reference librarian of the Thomas J. Watson Research Center, Yorktown Heights, New York. He was invaluable to me in procuring books, journal articles, etc. Finally, I wish most of all to thank Mr. Thomas J. Watson, Jr. for appointing me an IBM Fellow so that I have had the freedom to work on this history for the last several years.

I must acknowledge many valuable conversations with friends and with members of my family who patiently listened to me and gave freely of their advice and opinion. Equally, I wish to express my thanks to Princeton University Press for all its efforts in making this book possible and stylistically attractive.

H. H. G.

Yorktown Heights, N.Y.
January 1972

Contents

Illustrations
(following page 120)

1. Reconstruction of machine designed and built in 1623 by Wilhelm Schickard of Tübingen. (PHOT: IBM)
2. Calculating machine built by Blaise Pascal in 1642. (PHOT: IBM)
3. Calculator invented by the philosopher Leibniz in 1673. (PHOT: IBM)
4. Difference Engine of Charles Babbage. (PHOT: IBM)
5. Automated loom of Joseph Marie Jacquard. (PHOT: IBM)
6. Difference engine built in 1853 by Pehr Georg Scheutz of Stockholm. (PHOT: IBM)
7. Lord Kelvin's tide predictor. (PHOT: Reproduced by courtesy of the National Museum of Science and Industry, London)
8. Hollerith tabulating equipment used in the Eleventh Census of the United States in 1890. (PHOT: IBM)
9. The Michelson-Stratton harmonic analyzer. (PHOT: IBM)
10. The differential analyzer at the Moore School of Engineering, 1935. (PHOT: The Smithsonian Institution)
11. Harvard-IBM Mark I, the Automatic Sequence Controlled Calculator. (PHOT: IBM)
12. View of part of the ENIAC, the first electronic digital computer, operational in December 1945. (PHOT: The Smithsonian Institution)
13. Setting the function switches on the ENIAC. (PHOT: The Smithsonian Institution)
14. John von Neumann and J. Robert Oppenheimer at the dedication of the Institute for Advanced Study computer, 1952. (PHOT: Alan W. Richards)

Part One

The Historical Background
up to World War II

There is of course never an initial point for any history prior to which nothing of relevance happened and subsequent to which it did. It seems to be the nature of man's intellectual activity that in most fields one can always find by sufficiently diligent search a more or less unending regression back in time of early efforts to study a problem or at least to give it some very tentative dimensions. So it is with our field.

Since this is the case, I have chosen somewhat arbitrarily to make only passing references to the history of computers prior to 1600 and to say only the briefest word about the period before 1800. In fact, the only remarks I wish to make about this period are largely anticipatory ones to my main theme which concerns the early electronic digital calculators. The only excuse for this arbitrariness is that to say more on these earlier periods would add very little to our total knowledge of the electronic computer. I shall digress on a few occasions because of the colorfulness of one or another of the intellectual figures involved or because it seems desirable to establish in our minds some feeling for the intellectual, cultural, or social background of a given period.

Perhaps, however, this choice is not completely arbitrary. If we accept the quite reasonable point of view of scholars such as Needham,[1] we see that in a real sense the date of 1600 may be viewed as a watershed in scientific history. Prior to Galileo (1564–1642) there were of course intellectual giants, but his great contribution was to mathematicize the physical sciences. Many great scientists before him had investigated nature and made measurements, but the world needed Galileo to give these data "the magic touch of mathematical formulation."

It is worth recalling that prior to this time the state of mathematics in Europe was not substantially more advanced than that in the Arab world, based as it was on European and Chinese ideas and concepts. Then suddenly, as a result of a bringing together of

---

[1] Joseph Needham, *Science and Civilisation in China*, vol. 3 (Cambridge, 1959), pp. 155ff.

mathematics and physics, something happened in Europe that started science on the path that led from Galileo to Newton. This melding of practical and empirical knowledge with mathematics was the magic touchstone. In about 1580 Francois Vieta (1540–1603) in an earth-shaking discovery introduced the use of letters for unknowns or general parameters into mathematics. The subjects we now call algebra and arithmetic were called by him *logistica speciosa* and *logistica numerosa*,[2] respectively. He was followed, from our point of view, by John Napier, Laird of Merchiston (1550–1617), who in 1614 invented logarithms and who also was perhaps the first man to understand, in his *Rabdologia* in 1617, the use of the decimal point in arithmetical operations[3]; and by Edmund Gunter (1581–1626), who in 1620 invented a forerunner of the slide rule, which was actually invented by William Oughtred (1575–1660) in 1632 and independently by Richard Delamain, who also published the first account of the instrument in 1630.[4] The discovery of René Descartes (1596–1650) of analytical geometry in 1637 is perhaps the next great milestone on the road to the joint discovery of the calculus by Newton and Leibniz. The last stepping stone in that great chain lying between Descartes and Newton and Leibniz is, for our parochial purposes, Pascal's adding machine in 1642.

As Needham says in speaking of the great intellectual revolution that pushed Europe so far ahead of the Arabs, Indians, and Chinese, "No one has yet fully understood the inner mechanism of this development."[5] The mechanism that led to this forward thrust was the confluence of two formerly separate mathematical streams: of algebra, from the Indians and Chinese, and of geometry, from the Greeks. Again, according to Needham, "the marriage of the two, the application of algebraic methods to the geometric field, was the greatest single step ever made in the progress of the exact sciences."

In this great panoply of stars it was Galileo, as we have said, who produced the other great confluence of streams of ideas. He brought together the experimental and mathematical into a single stream "which led to all the developments of modern science and technology." These are the reasons why we have started our account when we did. It is the proper time in the intellectual history of our culture to do so.

[2] F. Cajori, *A History of Mathematics* (New York, 1893), pp. 138–139.
[3] *Ibid.*, p. 148.
[4] See D. E. Smith, *A Source Book of Mathematics*, vol. 1 (1959), pp. 156–164. Oughtred also invented the symbol × for multiplication.
[5] Needham, *op. cit.*, p. 156.

In starting at this place we are of course not doing justice to the interesting devices introduced by Moslem scientists very much earlier. The author is much indebted to Professor Otto Neugebauer, the historian of ancient astronomy and mathematics, for calling his attention to some special-purpose instruments invented by an Iranian astronomer and mathematician of the fifteenth century, Jamshīd ben Mas'ūd ben Mahmūd Ghīāth ed-Dīn al-Kāshī (1393–1449). He was the head of an astronomical observatory at Samarkand that was set up by Ulugh Beg, Tamerlane's grandson.[6]

Al-Kāshī was apparently a Moslem mathematician who made "contributions of a minor nature dealing with the summation of the fourth powers of the natural numbers, trigonometric computations, approximations. . . . To him is due the first use of a decimal fraction; . . . the value of 2π to sixteen decimal places." His instruments were aids to both astrologers and astronomers in simplifying their calculations. His "Plate of Conjunctions" was a means for finding when two planets would be in conjunction, i.e., when they have the same longitude. Such times were considered to be of peculiar importance by astrologers, and their days of occurrence could be determined from almanacs. Then al-Kāshī's plate was used to find the exact hour of occurrence.

His lunar eclipse computer was an ingenious device for simplifying the calculation of the important times associated with lunar eclipses. The method used gives an approximate solution which bears "a sufficiently close relation to reality to be useful."

Finally, his planetary computer was an instrument for determining the longitudes of the sun, the moon, and the visible planets. In Kennedy's words: "Al-Kāshī's instrument is now seen to be part of an extensive and more or less continuous development of mechanico-graphical scale models of the Ptolemaic system. This development was already well underway in classical times. Bronze fragments of what was probably a Greek planetarium of about 30 B.C. have been recovered from the Mediterranean. . . . The existence of a planetarium invented by Archimedes. . . . *The Hypotyposis Astronomicarum Positionum of Proclus Diadochus* (ca. A.D. 450) . . . description of a device for finding the longitude and equation of the sun. . . . In all this the work of al-Kāshī is of con-

[6] E. S. Kennedy, "Al-Kāshī's 'Plate of Conjunctions,' " *Isis*, vol. 38 (1947), pp. 56–59; "A Fifteenth-Century Planetary Computer: al-Kāshī's *Tabaq al-Manāteq*," parts I and II, *Isis*, vol. 41 (1950), pp. 180–183, and vol. 43 (1952), pp. 41–50; "A Fifteenth Century Lunar Eclipse Computer," *Scripta Mathematica*, vol. 17 (1951), pp. 91–97; and *The Planetary Equatorium of Jamshid Ghiath al-Din al-Kāshī* (Princeton, 1960).

siderable merit. His elegant constructions . . . carry the general
methods into branches of astronomical theory where they had not
previously been applied." [7]

Our story really opens during the Thirty Years War with Wilhelm
Schickard (1592–1635), who was a professor of astronomy, mathematics, and Hebrew in Tübingen. Some years ago (1957) Dr.
Franz Hammer, then assistant curator of Kepler's papers, discovered
some letters from Schickard to Kepler—both of whom were from
Würtemberg—containing sketches and descriptions of a machine
Schickard had designed and built in 1623 to do completely automatically the operations of addition and subtraction and, partially
automatically, multiplication and division. The first letter was
dated 20 September 1623, and a subsequent one 25 February 1624.
In the first one, Schickard wrote of the machine that it "immediately
computes the given numbers automatically, adds, subtracts, multiplies, and divides. Surely you will beam when you see how [it]
accumulates left carries of tens and hundreds by itself or while
subtracting takes something away from them. . . ." [8]

In his letter of 1624 he wrote: "I had placed an order with a local
man, Johann Pfister, for the construction of a machine for you; but
when half finished, this machine, together with some other things
of mine, especially several metal plates, fell victim to a fire which
broke out unseen during the night. . . . I take the loss very hard,
now especially, since the mechanic does not have time to produce
a replacement soon." [9]

No copy of the machine is extant but Professor Bruno, Baron von
Freytag-Löringhoff, with the help of a master mechanic, Erwin
Epple, as well as some others, reconstructed the instrument from
the information in the letters and a few working models were made.

The device is ingenious, and it is a great pity that its existence
was not known to the world of his day—unfortunately for the world
Schickard and all his family died in the plagues brought about by

[7] Kennedy, *Isis,* vol. 43, p. 50. See also D. J. deS. Price, "An Ancient Greek Computer," *Scientific American,* 200:6 (1959), pp. 60–67.

[8] Quoted in B. von Freytag-Löringhoff, "Wilhelm Schickards Tübinger Rechenmaschine von 1623 in Tübinger Rathaus," *Kleine Tübinger Schriften,* Heft 4, pp.
1–12. See also von Freytag, "Über der erste Rechenmaschine," *Physikalische
Blätter,* vol. 41 (1958), pp. 361–365. ". . . quae datos numeros statim automathos
computet, addat, subtrahat, multiplicet, dividatque. Rideres clare, si praesens
cerneres, quomodo sinistros denarium vel centenarium supergressos sua sponte
coacervet, aut inter subtrahendeum ab eis aliquid suffuretur. . . ."

[9] *Ibid.,* "Et curaveram tibi iam exemplar confieri apud Joh. Pfisterum nostratem,
sed illud semiperfectum, uno cum aliis quibusdam meis, praecipue aliquot tabellis
aeneis conflagravit ante triduum in incendio noctu et ex improvisibi coorto. . . ."

the Thirty Years War. It is interesting to speculate on how his invention might have influenced Pascal and Leibniz if war had not destroyed both Schickard and his machine. He must have been a man of many and great talents; Kepler said of him: "a fine mind and a great friend of mathematics; . . . he is a very diligent mechanic and at the same time an expert on oriental languages."

Our next great figure is Blaise Pascal (1623–1662) who, along with his many other acts of genius, had designed and built a small and simple machine in 1642–1644 when he was about twenty years old. His machine formed the prototype for several machines built in France, but all these represented devices of considerable simplicity in terms of their function, which was to effect by counting the fundamental operations of addition and subtraction. In fact the instrument was in some sense not as advanced as Schickard's in that it could not do the non-linear operations: multiplication and division. Apparently both he and his contemporaries viewed this machine as a most remarkable achievement. A version built in 1652 and signed by Pascal is in the Conservatoire des Arts et Métiers in Paris, and a copy of it is in the Science Museum in South Kensington, London. The machine was described in detail by Diderot in his famous *Encyclopédie*.

Some thirty years later Gottfried Wilhelm Leibniz (1646–1716), another of the great universalists of his or indeed of all time, invented a device now known as the *Leibniz wheel* and still in use in some machines. The mechanism enabled him to build a machine which surpassed Pascal's in that it could do not only addition and subtraction fully automatically but also multiplication and division. Leibniz said in comparing his device with Pascal's: "In the first place it should be understood that there are two parts of the machine, one designed for addition (subtraction) the other for multiplication (division) and that they should fit together. The adding (subtracting) machine coincides completely with the calculating box of Pascal." This device was viewed with the greatest interest both by the Académie des Sciences in Paris and by the Royal Society in London, to which Leibniz was elected a Fellow in 1673, the year his machine was completed and exhibited in London.[10]

Pascal is reputed to have built his machine as an aid to his father

[10] Leibniz's device enabled his machine to perform the operation of multiplication automatically by repeated additions. His idea was apparently re-invented in 1820 by Charles Xavier de Colmar of Alsace. Another interesting development was by a Frenchman, Léon Bollée, who built into his machine a device for storing the multiplication table and thereby obviating the need for repeated additions;

who was the discoverer of a famous curve known as the *limaçon* but who had need for help in computation.[11] In fact his father, Etienne, was a high official in Basse-Normandie, and following a revolt over taxes he reorganized the tax structure of the area.[12]

Curiously at this time the ability to do arithmetic was not generally to be found even among well-educated men. We thus find even such men as Pepys having, as a member of the Admiralty, to teach himself the multiplication tables. Perhaps therefore it was this general state of arithmetical knowledge rather than great filial piety that prompted Pascal to lighten his parent's burden.

In any case, it was Leibniz who summed up the situation very well indeed when he wrote: "Also the astronomers surely will not have to continue to exercise the patience which is required for computation. It is this that deters them from computing or correcing tables, from the construction of Ephemerides, from working on hypotheses, and from discussions of observations with each other. For it is unworthy of excellent men to lose hours like slaves in the labor of calculation which could safely be relegated to anyone else if machines were used." [13]

This notion which already had received such explicit formulation 300 years ago is in a very real sense to be the central theme of our story. It is fully in keeping with the genius of Leibniz that even when the field of computing was so very much in its infancy he already understood the point of the matter with such astonishing clarity. It is also of interest to realize that in the 1670s he had a third copy of his machine built for Peter the Great to send to the

---

this machine was first exhibited by Bollée in 1887 in Paris. He also played an important role in the development of the automobile.

Through the kindness of Dr. W. M. Turski of the Polish Academy of Sciences, I learned of still another interesting machine, or rather series of machines, developed by a Polish scholar, Abraham Stern (1769–1842). His machines handled the four arithmetic operations plus square root in a six-digit unit. The most advanced model was described in a public lecture at the Societas Scientiarum Varsoviensis on 30 April 1817 and was published in vol. VII of the annals of that society.

A number of other interesting machines followed after Leibniz's. The interested reader may consult R. Taton, *Le Calcul mécanique* (Paris, 1949); F. J. Murray, *The Theory of Mathematical Machines* (New York, 1960); M. d'Ocagne, *Le Calcul simplifié* (Paris, 1905); or W. de Beauclair, *Rechnen mit Maschinen, Eine Bildgeschichte der Rechentechnik* (Braunschweig, 1968). Another source containing English translations of Pascal's and Leibniz's papers is Smith, *op. cit.*, pp. 165–181.

[11] See R. C. Archibald, "Seventeenth Century Calculating Machines," *Mathematical Tables and Other Aids to Computation*, vol. 1 (1943), pp. 27–28.

[12] R. Taton, *op. cit.*, p. 19.

[13] Quoted in Smith, *op. cit.*

emperor of China to show the orientals the arts and industry of the occidentals and thereby increase commerce between the East and West.[14]

It is sad to realize that this great figure in our intellectual life should have had so difficult a time in his own period. Perhaps he was so far ahead of his contemporaries — excepting Newton — that none appreciated his work. On his death in 1716, his only mourner was his secretary, and an eye witness wrote: "He was buried more like a robber than what he really was, the ornament of his country." [15] In any case his little machine is still preserved in the State Library in Hanover, more as a curiosity than as a very early form of the modern computer. Leibniz's work did however stimulate a number of others to build improved machines, many of them ingenious variants of his.[16] Today these machines are still with us and play a significant role for minor calculations. Indeed during World War II they were of greatest importance.

It is also Leibniz, who in the grandeur of his genius realized, at least in principle, his Universal Mathematics. This was also to have greatest importance to our story — but much later. He wrote an essay in 1666 concerning Combinatorics, one of the great branches of mathematics, entitled *De Arte Combinatorica*. He described it as "a general method in which all truths of the reason would be reduced to a kind of calculation." This work was neglected in his time but was later to be picked up by George Boole and still later by Couturat, Russell, and other logicians.[17] Thus Leibniz contributed very profoundly to computers. Not only by his machine but also by his studies of what is now known as symbolic logic. We will return to this subject later in its appropriate place in our history. In closing our few paragraphs on Leibniz, we feel it is worth emphasizing his four great accomplishments to the field of computing: his initiation of the field of formal logics; his construction of a digital machine; his understanding of the inhuman quality of calculation and the desirability as well as the capability of automating this task; and, lastly, his very pregnant idea that the machine could be used for testing hypotheses. Even today there are only the beginnings of this type of calculation.

[14] L. Couturat, *Le Logique de Leibniz* (Paris, 1901), 295n, p. 527. Also Foucher de Careil, *Oeuvres de Leibniz*, vii (1859), 486–487, 498.

[15] *Encyclopaedia Britannica* (1948), s.v. "Leibnitz, Gottfried Wilhelm."

[16] See F. J. Murray, *op. cit.* His machine was basically an early version of the familiar desk calculator.

[17] E. T. Bell, *Men of Mathematics* (New York, 1937), pp. 117ff.

The theme of Leibniz—to free men from slavery by the automation of dull but simple tasks—was next taken up by one of the most unusual figures in modern intellectual history, Charles Babbage (1791–1871). As his biographer has so correctly said of Babbage everything about him was contentious including the date of his birth. According to her, he was born on 26 December 1791, in Devonshire, even though he stated that it was in 1792 in London. The title of her book *Irascible Genius* is another good clue to his character.[1]

In any case, he was born into an upper middle class English home and had the usual advantages, both intellectual and social, of such a background. During his college years he formed close associations with John Herschel, the son of the discoverer of the planet Uranus and himself one of the great astronomers, and George Peacock, who had three wonderful careers—as a mathematician, an astronomer, and later as a divine when he became Dean of Ely. The importance of these three men to mathematics is attested to by E. T. Bell who says: "The modern conception of algebra began with the British 'reformers,' Peacock, Herschel, de Morgan, Babbage, Gregory, and Boole."[2] Thus we see Babbage at the heart of the intellectual life of England. In fact, we see him as one of the founding members of the Royal Astronomical Society (12 January 1820) and the first recipient of its Gold Medal (13 July 1823) for his work on "Observations on the Application of Machinery to the Computation of Mathematical Tables."[3]

---

[1] Mabeth Moseley, *Irascible Genius, A Life of Charles Babbage, Inventor* (London, 1964).

[2] Bell, *Men of Mathematics*, p. 438.

[3] Moseley, p. 64 and p. 75, and Philip and Emily Morrison, *Charles Babbage and his Calculating Engines, Selected Writings by Charles Babbage and Others* (London, 1961). In particular see the excellent Introduction to that work by the Morrisons. Also see Lord Bowden, ed., *Faster than Thought* (London, 1953), chap. I. In each of these, various facets of this man's personality appear. See also Babbage's work, *Passages from the Life of a Philosopher* (London, 1864; reprinted in 1968). Here is a set of papers ranging from the sublime to the ridiculous, from profundities to nonsense in plain bad taste. Indeed much of Babbage's career is of this sort. It is a wonder he had as many good and loyal friends when his behavior was so peculiar.

There are, as the Morrisons point out, two different accounts of how young Babbage first hit upon his idea of automating computation. In his *Passages* Babbage says:

> The earliest idea that I can trace in my own mind of calculating arithmetical tables by machinery arose in this manner: —
>
> One evening I was sitting in the rooms of the Analytical Society at Cambridge, my head leaning forward on the table in a kind of dreamy mood, with a Table of logarithms lying open before me. Another member, coming into the room, and seeing me half asleep, called out, "Well, Babbage, what are you dreaming about?" to which I replied, "I am thinking that all these Tables (pointing to the logarithms) might be calculated by machinery."
>
> I am indebted to my friend, the Rev. Dr. Robinson, the Master of the Temple, for this anecdote. The event must have happened either in 1812 or 1813.[4]

The other version of the story is given by both Miss Moseley and the Morrisons, who state "the one written in 1822 seems more plausible than the other, which appeared in his autobiography."[5] The story is that Herschel and Babbage were once checking some astronomical calculations when Babbage in irritation is said to have remarked: "I wish to God these calculations had been executed by steam." To this Herschel replied: "It is quite possible."

It is quite interesting to note that Babbage thought instinctively in terms of the prime technology of his time: the steam engine. Here we see a manifestation of a phenomenon which is to be a dominant theme throughout our story: the development of radical new machines always comes about because some inspired person sees how to adapt a new technology to computers and thus to make a major advance in the state of the art. To be precise, it is usually the convergence of two very different concepts. One is the technology, but the other is the recognition of the importance and necessity for an advance. In a sense Babbage suffered because the need for his machines, his *Difference Engine* as well as his *Analytical Engine,* was not very great in his age. Indeed, Miss Moseley says: "A few trials had been made to show the rapidity with which it could calculate. The first was a table from the formula $x^2 + x + 41$. In the earlier numbers it was possible, writing quickly, to keep up with the engine, but when four figures were required the machine

---

[4] Babbage, *Passages*, p. 42.
[5] P. and E. Morrison, *op. cit.*, p. xiv.

was at least equal in speed to the writer. . . . As the machine could be made to move uniformly by a weight, the rate could be maintained for any length of time. Few writers would be found to copy with equal speed for many hours together." [6]

As we shall see later, this rate of operation was just not fast enough to revolutionize computation and was, at least in the author's opinion, one of the basic reasons why his work more or less disappeared from view. The other was that the technology needed to build and to machine gears, wheels, etc. to the tolerances he needed was not yet well enough developed. Indeed it is during this period that James Watt's steam-engine and George Stephenson's locomotive were invented. It was in 1830 that the Manchester and Liverpool Railway was inaugurated and thereby brought to England the great benefits of decent transportation; and in 1819–1825 Thomas Telford built an iron suspension bridge which was viewed as one of the wonders of the world. Thus it is not surprising that technology was not yet up to Babbage's needs.

It is interesting to note what Trevelyan has to say on the subject in one of his magnificent surveys of English history.

> A new civilisation, not only for England but ultimately for all mankind, was implicit in the substitution of scientific machinery for handwork. The Greek scientists of the Alexandrine age, the contemporaries of Archimedes and of Euclid, had failed to apply their discoveries to industry on the big scale. Perhaps this failure was due to the contempt with which the high-souled philosophy of Hellas regarded the industrial arts conducted by slave labour; perhaps the great change was prevented by the disturbed and war-like state of the Mediterranean world during those three hundred years between Alexander and Augustus, when Greek science was in its most advanced state. In any case it was left to the peaceful, cultivated but commercially minded England of the generation that followed Newton's death, to harness philosophic thought and experiment to the commonest needs of daily life. The great English inventors negotiated the alliance of science and business. [7]

It is also relevant to note what William Thomson (Lord Kelvin) and Tait wrote about Babbage's work in 1879. They refer in their

---

[6] Moseley, p. 69.

[7] G. M. Trevelyan, *British History in the Nineteenth Century and After, 1782–1919* (New York, 1966), p. 153. The reader will find much in this excellent book that bears on the period we are discussing.

classic work on natural philosophy to "the grand but partially realized conceptions of calculating machines by Babbage." [8] With this sentence they dismissed his work.

It is also quite significant that they as well as Thomson's brother James and James Clerk Maxwell worked on what are known as *continuous* or *analog* or *measurement* calculating machines and not on *digital* ones, as did Babbage and as do we today. We shall say more on this later. Suffice it to say at this point that the pre-eminent physicists of the nineteenth century turned away from Babbage's ideas and instead turned toward types of machines which were very much faster than his and were practically possible with the technologies of the period. It seems quite likely that they, as physicists, better understood the engineering problems than did the mathematician Babbage.

To return to our tale: In 1822 Babbage was writing a letter to Sir Humphry Davy, the then President of the Royal Society, about automating "the intolerable labor and fatiguing monotony" of calculating tables, writing a scientific paper entitled "On the Theoretical Principles of the Machinery for Calculating Tables," and reading a paper to the Royal Astronomical Society on this subject.[9]

Having gained considerable support from his colleagues, Babbage next applied formally to the British Government for funds to carry out his project. Rather soon after this a committee of the Royal Society, appointed at the request of the Lords of the Treasury to look into his application and chaired by Davy himself, returned its report stating *inter alia:* "It appears that Mr. Babbage has displayed great talents and ingenuity in the construction of his machine for computation which the committee thinks fully adequate to the attainment of the objects proposed by the inventor, and that they consider Mr. Babbage as highly deserving of public encouragement in the prosecution of his arduous undertaking." [10]

The Chancellor of the Exchequer accepted the report in 1823 and agreed in some general way to provide a governmental subsidy to build the Difference Engine since it would be of great value to the government—particularly the Navy—for making various nautical tables and was not salable profitably. Thus was the project

---

[8] Thomson and Tait, *Principles of Mechanics and Dynamics* (London, 1879), part 1, p. 458.

[9] Letter, Babbage to Davy, 3 July 1822, in *Brewster's Edinburgh Journal of Science*, vol. 8 (1822), p. 122.

[10] Moseley, p. 73.

launched. It is interesting to realize this must have been one of the very first arrangements between a scientist or engineer and a government. Today, this is all highly regularized and proceeds along clearly defined lines; then, it was all quite novel and as a result was to be a constant source of irritation both to the government and to Babbage.

Babbage certainly underestimated the scope of the task he set himself and had a breakdown in 1827. He went abroad to recover his health and while there was appointed to the chair at Cambridge once held by Newton; this appointment he viewed so lightly that during his entire tenure as Lucasian Professor of Mathematics (1827–1839) he did not live in at the university or teach there.

Upon his return from the continent, he received another grant from the Treasury and resumed his development of the Difference Engine, which, it has been said, was viewed favorably by the Duke of Wellington. This is a little hard to believe since Wellington had a life-long dislike for inventions. It was in fact due to his profound influence that the British Army in the Crimean War (1853) was outfitted with the same weapons it had used at Waterloo (1815). The event certainly showed that his innate dislike for change was almost catastrophic.

There were many crises and in 1833 Babbage stopped work on the machine. After many years of pressing on the government he at last was told in 1842 by Sir Robert Peel, the Prime Minister, and Henry Goulburn, the Chancellor of the Exchequer, that the project was dead as far as they were concerned. In part their decision was based upon the estimation of George Airy, the Astronomer Royal. He wrote in his diary: "I replied, entering fully into the matter, and giving my opinion that it was worthless." [11] Probably Airy was unjustified in making so harsh a judgment. He indeed had a certain record for overly conservative evaluations of this sort. Two famous ones are his advice against building the Crystal Palace for the International Exhibition of 1862 because he said the structure would collapse when salute guns were fired; and his advice to John C. Adams to proceed slowly and redo his calculations "in a leisurely and dignified manner." This resulted in Leverrier beating Adams to the draw as first discoverer of Neptune.[12] This was later corrected and both were viewed as co-discoverers.

It would carry us too far afield to describe in any detail the bitter

---

[11] Moseley, p. 145.
[12] See Lord Bowden, *Faster than Thought*, pp. 14ff.

quarrels that ensued. In any case, Peel remarked to a friend: "I should like a little previous consideration before I move in a thin house of country gentlemen a large vote for the creation of a wooden man to calculate tables for the formula $x^2 + x + 41$." [13] The interested reader may find Babbage's version of his interview in his *Passages*.[14] It should be remembered moreover that Peel was not uninformed on mathematics. Indeed he had taken a "double first" at Oxford—a first in mathematics and in classics—and was therefore quite capable of understanding Babbage's work.[15]

This machine, as far as Babbage got, and all drawings are now in the Science Museum in South Kensington. It was exhibited at the International Exposition of 1862 and is still in operating condition. In fact working copies of both the Difference Engine and parts of the Analytical Engine were made some years ago for IBM.[16]

A Swedish gentleman, Pehr Georg Scheutz (1785–1873), built a Difference Engine inspired by Babbage's ideas and displayed it in London in 1854 with considerable help from Babbage,[17] who wrote about the occasion:

> Mr. Scheutz, an eminent printer at Stockholm, had far greater difficulties to encounter. The construction of mechanism, as well as the mathematical part of the question was entirely new to him. He however undertook to make a machine having four differences and fourteen places of figures, and capable of printing its own Tables.
>
> After many years' indefatigable labor, and an almost ruinous expense, aided by grants from his Government, by the constant assistance of his son, and by the support of many enlightened members of the Swedish Academy, he completed his Difference Engine. It was brought to London, and some time afterwards exhibited at the Great Exhibition at Paris. It was then purchased for the Dudley Observatory at Albany (New York) by an enlightened and public spirited merchant of that city, John F. Rathbone, Esq.
>
> An exact copy of this machine was made by Messrs. Dunkin and Co., for the English Government, and is now in use in the Registrar-General's Department at Somerset House. It is very

[13] Moseley, p. 17.
[14] Babbage, pp. 68–96.
[15] Trevelyan, *op. cit.*, p. 267.
[16] Bowden, *op. cit.*, p. 10, and Morrison, p. xvii.
[17] Morrison, p. xvi.

much to be regretted that this specimen of English workmanship was not exhibited in the International Exhibition.[18]

The Scheutz machine remained for many years in Albany before it came into the possession of the Felt and Tarrant Co. of Chicago in 1924. It was later acquired by the Smithsonian Institution.[19]

Scheutz seems to have been an extraordinary man. He was first a practising lawyer, then part owner and co-editor of "Sweden's most important political newspaper in the 1820's." He translated into Swedish many of the literary masters including Boccaccio, Walter Scott, and especially Shakespeare; and he also published a number of technical and trade journals.

In 1834 he and his son Edvard read of Babbage's Difference Engine in an article in the *Edinburgh Review*. They became much interested and worked at building a considerably improved and modified version. In 1834 the machine was operable but they had no success in selling their ideas. Then in 1851 the Swedish Academy gave Georg Scheutz funds to build a larger and improved version, which was completed in 1853. The machine won a Gold Medal at the Exhibition in Paris in 1855. Scheutz was knighted by the King of Sweden in 1856 and made a member of the Swedish Academy. He received other substantial rewards from his government including an annuity and a prize for his translations of Shakespeare into Swedish. His citation, as quoted by Archibald, read in part: "Georg Scheutz, the first who successfully clothed Shakespeare in Swedish costume, and for whom literature, even though as author, an occupation, comprehensive and pursued to the evening of a long life, has in addition for a long period had a connection which has not been effaced by the fact that the man of letters has also acquired a respected name for himself in a field which lies beyond the boundaries of belles lettres."

It is curious that Scheutz was highly successful in building his two machines and Babbage was not. Neither one was a skilled

---

[18] Babbage, *Passages*, p. 48. The Registrar-General produced with his machine the life expectancy tables used for some years by English insurance companies. Then in 1914 "the machine was presented to the Science Museum, South Kensington, where it is still exhibited and occasionally operated for visiting experts."

[19] A short but excellent account of Scheutz' work may be found in a paper by R. C. Archibald, "P. G. Scheutz, Publicist, Author, Scientific Mechanician, and Edward Scheutz — Biography and Bibliography," *Mathematical Tables and Other Aids to Computation*, vol. 2 (1947), pp. 238–245. In the same issue Archibald has a short account of an improved version of the Scheutz machine designed by Martin Wiberg (1826–1905), also of Stockholm, in about 1874.

engineer yet Scheutz brought it off. Perhaps it was due to the fact that the Scheutz team started with a modest prototype, demonstrated its operability, and only then went on to a larger and more useful instrument. Perhaps there were also temperamental aspects.

Before continuing our brief survey of Babbage and his works perhaps we should now say a few more words about the *Difference Engine*. This is important because at a somewhat later period he began the development of a much more sophisticated machine that he called an *Analytical Engine* and which introduced a basically new idea into the field of computation.

To describe Babbage's Difference Engine we need first to embark on a brief discussion of tables — what they are and why they are or were important. Since the time of Leibniz and Newton mathematicians and, more generally, what were in those times known as natural philosophers were much concerned about the production of tables either by means of mathematical calculations as in the case of tables of multiplication, of logarithms, of sines and cosines, etc.; or by means of physical measurements or observations, as in the case of tables of precipitation in a given location, or of air density as a function of altitude at a particular time, of the gravity constant at different locations on the earth, etc. These tables are the means by which scientists very early in time learned how to record their experiences so that others could benefit.

The continued importance of tables is attested to by the establishment in 1943 of a journal called *Mathematical Tables and Other Aides to Computation* by Raymond C. Archibald of Brown University under the aegis of the National Research Council. The fact that so large a portion of that journal was devoted to listing errata in existent tables is proof of what an important problem Babbage was trying to solve.

When tables are produced by means of various mathematical principles as in the case, let us say of a table of logarithms, all experience is that humans make many mistakes, and this was the basis for Babbage's and Herschel's irritations with the people who had done their work. It was undoubtedly the idea of replacing "fallible" people by an "infallible" machine that appealed so strongly to Babbage at that moment. At any rate, it was this attempt which was in the end to wreck his life and embitter him with his fellow men.

To illustrate and hopefully to explicate the table problem, let us consider a very simple table in which we have recorded for the

consecutive whole numbers $N = 0, 1, 2, \ldots 9$ the values of the
expression $N^2 + N + 41$, an expression in which Babbage was interested. We also record two other columns which are headed D_1, D_2.

N	$N^2 + N + 41$	D_1	D_2
0	41		
1	43	2	
2	47	4	2
3	53	6	2
4	61	8	2
5	71	10	2
6	83	12	2
7	97	14	2
8	113	16	2
9	131	18	2

An inspection of the column D_1 shows it consists of the differences between the successive entries in the column headed
$N^2 + N + 41$; the column D_2, of differences between successive
entries in the column headed D_1. Since all entries in D_2 are the
same we can easily continue the table as far as we wish with almost
no labor. To see this let us see how to evaluate $N^2 + N + 41$ for
$N = 10$ without doing any multiplying. First we know that the entry
in the D_2 column will be 2 and thus the one in D_1 will be
$18 + 2 = 20$; hence, the one in the $N^2 + N + 41$ column will be
$131 + 20 = 151$, and we have found this without having to do a
single multiplication or division — only addition.[20]

There is a very remarkable and fundamental theorem in mathematics due to Karl W. T. Weierstrass (1815–1897) of Berlin, one of
the great nineteenth-century mathematicians, which asserts that
any smooth (continuous in technical parlance) function on an interval can be approximated as closely as desired by a polynomial.
A polynomial is an expression of the form $a + bx + cx^2 + \cdots + dx^n$
where n is some positive whole number. Thus most mathematical
or physical functions can be expressed approximately — but as
accurately as we wish — by such polynomials. Furthermore all
polynomials can be built up by difference tables as we saw above.

---

[20] Babbage apparently ran just such a table on his machine keeping 32 decimal
places in his numbers and found his machine could produce 33 per minute; he then
speeded it up to 44 per minute. These are rates faster than a human can achieve but
only barely so. This is perhaps why the machine was not of prime importance. See
the discussion earlier on this point.

Indeed in that example we had a polynomial of degree two and saw that its D_2 column – its second differences – were constant, and if we had written down its D_3 column we would have had only zeros in it. In general a polynomial of any degree has a difference column constant and it can be built up by successive additions. For a polynomial of degree three the D_3 column will be constant, of degree four the D_4 column will be constant, etc.

Babbage's Difference Engine – hopefully we have now a little better feel for the choice of words – was to be built to handle polynomials of degree six

$$a + bN + cN^2 + dN^3 + eN^4 + fN^5 + gN^6.$$

Thus his machine handled columns $D_1, D_2, \ldots, D_6$. Lady Lovelace (see below) says, "It can therefore tabulate *accurately,* and to an *unlimited extent,* all series whose general term is comprised in the above formula; it can also tabulate *approximately,* between *intervals of greater or less extent,* all other series which are capable of tabulation by the Method of Differences." (Her italics.) [21]

It is worth emphasizing at this point that this machine was very specialized in its function and could not in any sense be described as being a general purpose instrument. It had only one program or task it could carry out, albeit this was a complex one. In this sense it was a step back from Leibniz's machine which could only do the fundamental processes of arithmetic but which could be guided by a human to do any mathematical task. It may be viewed as an attempt, and indeed a brilliant one, to automate a specific process rather than as an attempt to automate numerical mathematics. (The idea was apparently first proposed by a J. H. Müller in 1786, but he did not build a prototype.[22])

In 1833 during a hiatus in his development of the Difference Engine, Babbage conceived his *chef d'oeuvre,* his Analytical Engine. This was to be in concept a general purpose computer, very nearly in the modern sense. It was very close in spirit to the Harvard-IBM machine, Mark I, about which we shall speak at a later point. This machine was to be the goal of his life, and he worked at it until his death in 1871; and after that his son H. P. Babbage carried

---

[21] Luigi F. Menabrea, *Sketch of the Analytical Engine Invented by Charles Babbage, Esq.,* Bibliothèque Universelle de Genève, No. 82, October 1842, translated into English with editorial notes by the translator, Augusta Ada, Countess of Lovelace, Taylor's Scientific Memoirs, vol. 111, Note A, p. 348. This work was reprinted by Lord Bowden and the Morrisons.

[22] d'Ocagne, *Le Calcul Simplifié,* p. 74.

on the work, building pieces of the device himself and presenting them to the Science Museum in London.[23]

The basic idea of this machine was totally different from that of his earlier Difference Engine. In this new one he saw rather clearly some of the ideas that characterized the modern computer. He got his idea from observing the Jacquard attachment to the loom, which revolutionized the textile industry in 1805.

Joseph Marie Jacquard was a product of the French Revolution. He invented a device as an attachment to a loom which essentially automated the process of weaving fabrics. In order to weave a pattern on a loom, the weaver must have a plan or *program* which tells him which threads of the warp he should go over and which under to create the pattern; he must also be told when to repeat or *iterate* the basic pattern. Without going into the details, we can see that this is a very intricate business and it was beyond the skill of humans to weave complex patterns without automation. Jacquard invented his famous device and almost at once changed the weaving business.

The key point in Jacquard's device is the use of a series of cards with holes in them arranged to depict the desired pattern. The holes allow hooks to come up and pull down threads of the warp so that when the shuttle passes through it goes over certain pre-determined threads and under others.[24] The device was an immediate success and by 1812 there were 11,000 Jacquard looms in use in France.[25]

It is perhaps of interest to read what Babbage said of Jacquard's instrument:

> It is known as a fact that the Jacquard loom is capable of weaving any design that the imagination of man may conceive. . . . holes [are punched] in a set of pasteboard cards in such a manner that when these cards are placed in a Jacquard loom, it will then weave . . . the exact pattern designed by the artist.

[23] Moseley, p. 260.

[24] It is interesting to note that the process of making the Jacquard cards was itself, up until recently an arduous task. The textile designer in fact drew on paper the threads of the warp and weft showing at each intersection their relative positions and from these the holes were punched. Then in the mid-1960s a very ingenious method was found by Miss Janice R. Lourie of IBM to automate the whole process with the help of the modern computer. This was displayed on a working loom at the HemisFair in San Antonio, Texas, in 1968. Thus the progenitor of the computer has become its child.

[25] *Encyclopaedia Britannica* (1948), s.v. "Jacquard, Joseph Marie."

Now the manufacturer may use, for the warp and weft of his work, threads that are all of the same colour; let us suppose them to be unbleached or white threads. In that case the cloth will be woven all in one colour; but there will be a damask pattern upon it such as the artist designed.

But the manufacturer might use the same card, and put into the warp threads of any other colour. Every thread might even be of a different colour, or of a different shade of colour; but in all these cases the *form* of the pattern will be precisely the same – the colours only will differ.

The analogy of the Analytical Engine with this well-known process is nearly perfect.

The Analytical Engine consists of two parts: –

1st. The store in which all the variables to be operated upon, as well as all those quantities which have arisen from the results of other operations, are placed.

2nd. The mill into which the quantities about to be operated upon are always brought.

Every formula which the Analytical Engine can be required to compute consists of certain algebraical operations to be performed upon given letters, and of certain other modifications depending on the numerical value assigned to those letters.

There are therefore two sets of cards, the first to direct the nature of the operations to be performed – these are called operation cards; the other to direct the particular variables on which those cards are required to operate – these latter are called variable cards. Now the symbol of each variable or constant, is placed at the top of a column capable of containing any required number of digits.

Under this arrangement, when any formula is required to be computed a set of operation cards must be strung together, which contain the series of operations in the order in which they occur. Another set of cards must then be strung together, to call in the variables into the mill, in the order in which they are required to be acted upon. Each operation will require three other cards, two to represent the variables and constants and their numerical values upon which the previous operation card is to act, and one to indicate the variable on which the arithmetical result of this operation is to be placed.

The Analytical Engine is therefore a machine of the most general nature. Whatever formula it is required to develop, the law of its development must be communicated to it by two sets of cards. When these have been placed, the engine is special for that particular formula.

Every set of cards made for any formula will at any future time, recalculate that formula with whatever constants may be required.

Thus the Analytical Engine will possess a library of its own. Every set of cards once made will at any future time reproduce the calculations for which it was first arranged. The numerical value of its constants may then be inserted.[26]

Much of this has a very modern ring. It is the introduction of strings of cards which cause the various operations to be performed automatically and in a given sequence that is the secret of the Analytical Engine. Lady Lovelace says: "We may say most aptly that the Analytical Engine weaves *algebraical patterns* [her italics] just as the Jacquard-loom weaves flowers and leaves. Here, it seems to us, resides much more of originality than the Difference Engine can be fairly entitled to claim." [27]

This is far enough to go at this time with a discussion of Babbage's machine. We shall see later how his ideas were to be carried out in the twentieth century and what the implications are. At this time, perhaps we need to say a few words more about Babbage and about Lady Lovelace.

In addition to conceiving of the computer, as he did, Babbage had a considerable spectrum of interests ranging from instruments such as dynamometer railway cars, diving bells, and ophthalmoscopes to operations research. In this latter connection he did a most interesting analysis of the costs in the British Post Office of handling mail. He was an early advocate of the "penny post," i.e., a uniform rate of postage. He formed this opinion by doing a study of the Bristol mail-bag for each night during one week. He also "recommended the enlargement of the duties of the Post-office by employing it for the conveyance of books and parcels." There is much of interest in his book *Economy of Manufactures and Machinery*.[28] Here he reveals his knowledge of and interest in a variety of topics

---

[26] Babbage, *Passages*, pp. 117ff.

[27] This is from Notes by the Translator, Note A, in Menabrea, *Sketch of the Analytical Engine Invented by Charles Babbage;* see Morrison, *op. cit.*, pp. 251–252. We shall say more later about Lady Lovelace. – The characteristics planned by Babbage for his Analytical Engine are not without interest. He conceived of his machine having a capacity for storing 1,000 numbers of 50 decimal digits, and he estimated a multiplying speed of one such operation per minute and an addition speed of one operation per second. We shall have occasion to refer to these again later.

[28] Published in London in 1832.

in political economy, including the postal ones mentioned above.

Babbage himself moved in the best social and intellectual circles of English and continental society, being friendly or acquainted with Prince Albert, the Duke of Wellington, Bessel, Browning, Carlyle, Darwin, Dickens, Fourier, Humboldt, Mill, and Thackeray — to mention just a few. Thus, we have the clear picture of Babbage as an upper-middle-class English gentleman of outstanding intellectual qualities.

In spite of all his manifold activities he seems to have been a most unhappy and maladjusted person. Miss Moseley quotes various incidents which suggest that Babbage had little ability to cope either with his intellectual peers or with ordinary people. For example, he published a work entitled *Reflections on the Decline of Science in England and on Some of its Causes*. The work is distinguished by its poor taste in the attacks it makes upon Davy, the late president of the Royal Society, and upon Davis Gilbert, the then president. Davy was accused of "lining his own pocket at the Society's expense." [29]

Not only was this accusation in bad taste, it was made even worse by the fact that Davy was already dead and therefore unable to respond. However, this sort of behavior was not atypical of Babbage in his paranoid moments. He could also not deal with ordinary people in a reasonable fashion, developing phobias against — of all things — Italian organ grinders. Miss Moseley tells of another incident: "He was driven on one occasion to ask Dr. Hooker, then President of the British Association, to ventilate his grievances at a B.A. Meeting, remarking that the previous Friday he and his manservant had gone out to find a constable. . . . They separated and Charles was followed through several streets, insulted and shouted at by hundreds of vagabonds, and pelted with mud. Finding a cab, he entered and was driven to a police station but received no assistance, and no one was taken into custody." [30]

It is not easy to form a clear picture of Babbage as an intellectual figure and to compare him to his contemporaries and immediate successors. Certainly his concept of the Analytical Engine is of breathtaking proportions; this idea is undoubtedly enough to put him well forward in the front ranks of intellectual figures. His other pursuits, although they are works of considerable virtuosity, seem much less in stature and also not connected. Apart from his two computers, he seems never to have come upon any topic which

[29] Moseley, p. 108.   [30] Moseley, p. 250.

provided him with real intellectual interest or excitement. Perhaps his inability to build his machines and use them destroyed his interest in anything else. Very likely all his serious efforts went into thinking about his machines. This leaves us with a very curious picture of this somehow incomplete intellectual giant thwarted in almost everything he held dear.

An insightful analysis of Babbage by an acquaintance is perhaps apropos at this point. Lord Moulton said of him in 1915:

> One of the sad memories of my life is a visit to the celebrated mathematician and inventor, Mr. Babbage. He was far advanced in age, but his mind was still as vigorous as ever. He took me through his workrooms. In the first room I saw the parts of the original Calculating Machine, which had been shown in an incomplete state many years before and had even been put to some use. I asked him about its present form. "I have not finished it because in working at it I came on the idea of my Analytical Machine, which would do all that it was capable of doing and much more. Indeed, the idea was so much simpler that it would have taken more work to complete the Calculating Machine than to design and construct the other in its entirety, so I turned my attention to the Analytical Machine." After a few minutes' talk we went into the next workroom, where he showed and explained to me the working of the elements of the Analytical Machine. I asked if I could see it. "I have never completed it," he said, "because I hit upon an idea of doing the same thing by a different and far more effective method, and this rendered it useless to proceed on the old lines." Then we went into the third room. There lay scattered bits of mechanism, but I saw no trace of any working machine. Very cautiously I approached the subject, and received the dreaded answer, "It is not constructed yet, but I am working at it, and it will take less time to construct it altogether than it would have taken to complete the Analytical Machine from the stage in which I left it." I took leave of the old man with a heavy heart. When he died a few years later, not only had he constructed no machine, but the verdict of a jury of kind and sympathetic scientific men who were deputed to pronounce upon what he had left behind him, either in papers or mechanism, was that everything was too incomplete to be capable of being put to any useful purpose.[31]

[31] Lord Moulton, "The Invention of Logarithms, Its Genesis and Growth," *Napier Tercentenary Memorial Volume*, ed. C. G. Knott (London, 1915), pp. 19–21.

Before leaving Charles Babbage and his engines, we must say a little about a colorful figure in his career, Augusta Ada Byron, later Countess of Lovelace and Baroness Wentworth, the only child of Lord Byron and his wife Annabella (née Milbanke). Ada's father and mother separated when she was but a month old, and she never after in her short life saw the poet. However, she evidently had great affection for her father and expressed the wish to be buried with him in the Byrons' vault in Nottinghamshire; father and daughter, each 36 years old at death, are buried side by side at Newstead. Byron also had great affection for her as indicated by these lines from "Childe Harold's Pilgrimage" (Third Canto):

> My daughter! with thy name this song begun —
> My daughter! with thy name thus much shall end —
> I see thee not, — I hear thee not, — but none
> Can be so wrapt in thee; thou art the friend
> To whom the shadows of far years extend:
> Albeit my brow thou never shouldst behold,
> My voice shall with thy future visions blend,
> And reach into thy heart, — when mine is cold, —
> A token and a tone even from thy father's mould.

He must have suffered greatly from his separation from her. Miss Moseley says: "As he lay dying he groaned: 'Oh, my poor dear child! — my dear Ada! my God, could I but have seen her!' and he instructed Fletcher, his valet, to give her his blessing." [32]

Ada was a most gifted young lady who displayed great abilities, both intellectual and artistic, and by the time she was fifteen she had already shown mathematical precocity. She was trained by a well-known mathematician of the period, Augustus de Morgan, cited earlier as one of the British "reformers" responsible for the modern conception of algebra.

Babbage and Ada became acquainted at this time, and in fact it was Mrs. de Morgan and Mary Somerville who escorted her frequently to see Babbage. We are indebted to the former for her statement after one of Ada's early visits, during which she observed Babbage's Analytical Engine: "While the rest of the party gazed at this beautiful instrument with the same sort of expression and feeling that some savages are said to have shown on first seeing a look-

---

John Fletcher Moulton (1844–1921) began his career in mathematics at Cambridge. He later resigned his fellowship in Christ's College and entered the legal profession, ultimately becoming a Lord of Appeal in Ordinary.

[32] Moseley, p. 156.

ing glass or hearing a gun, Miss Byron, young as she was, understood its working and saw the great beauty of the invention." [33]

Babbage had visited Turin at the invitation of a friend, Baron Gio-
vanni Plana (1781–1869), a famous astronomer and table compiler,
and had given a series of talks there to a distinguished group including General Luigi F. Menabrea (1809–1896), a leader in the
Risorgimento and a future Prime Minister of Italy. Menabrea was
so taken with Babbage's work that he wrote a short account of the
lectures which was published by the Bibliothèque Universelle de
Genève in 1842.[34] Lady Lovelace decided to translate this account
into English, and Babbage encouraged the young lady – she was
then 27 – to expand Menabrea's paper with extensive notes. In
fact, her notes are twice as long as the original paper.[35] They are
very well written and show that in addition to her mathematical
ability – which her mother had, too – she also possessed a literary
style – no doubt inherited from her father. Miss Moseley quotes
one of her letters to Babbage in which she took sharp issue with
Babbage over his making some emendation to her writings: "I am
much annoyed at your having altered my Note. You know I am
always willing to make *any* required alterations myself, but that I
cannot endure another person to meddle with my sentences. . . ." [36]
In spite of this, the notes constitute an excellent account of Babbage's work, and he clearly recognized their superior quality. They
were great friends and had a continuing close relationship until
her tragic and untimely death.

In 1835 Ada Byron married Lord King, who was to become the
Earl of Lovelace. While she is not the subject of this story, she is a
fascinating figure and deserves a place here because it was through
her efforts that Babbage's work was made accessible. Curiously
little has been written about her, and we are indebted to Lady
Wentworth, her granddaughter, for making her letters available to
Lord Bowden so that he could write his charming vignette of
her.[37] We are grateful, too, for Miss Moseley's somewhat longer por-
trait of Ada in her biography of Babbage.

[33] Quoted in Lord Bowden, *Faster than Thought,* p. 20.
[34] Cited above, n. 21.
[35] See Morrison, pp. 225–297; as printed in Morrison, the original text goes from
225 through 245, and Lady Lovelace's notes from 245 through 297.
[36] Moseley, p. 170.
[37] In his book *Faster than Thought.*

Until about the middle of the eighteenth century there was no satisfactory method available to a mariner for calculating his longitude. This, of course, constituted an extremely serious problem for sailors to navigate over great distances and arrive at predetermined ports with any accuracy.

In principle, it is not difficult to determine longitude given an *accurate* table of the moon's position as a function of the time. The problem lies in making such tables, because the moon's motion is very complex: it is compounded principally out of the mutual interactions of the moon with both the earth and the sun.

Because of the importance to navigation the study of the moon's motion has long enjoyed the scientific interest of man. As far back as the second century B.C. Hipparchus had described the moon's position. But it remained for Isaac Newton, in his *Principia,* to make the first great contribution to this very difficult subject. He wrote to Halley that the Lunar Theory "made his head ache and kept him awake so often that he would think of it no more." This is the genesis of what is known as the Three-Body Problem.

Virtually all the great applied mathematicians of the eighteenth century worked to improve Newton's theory. The names are a roster of the great: Euler, Clairaut, d'Alembert, Lagrange, and Laplace. During this period the importance of accurate tables of the moon's position was so great that a number of governments including that of Great Britain offered prizes for such tables. In this connection Euler first published some preliminary and unsatisfactory tables in 1746, and in the following year both Clairaut and d'Alembert presented papers on lunar theory to the Paris Academy on the same day. Both presentations contained difficulties, but Clairaut cleared them up in 1749. He won the prize offered by the St. Petersburg Academy in 1752, and both he and d'Alembert published lunar tables as well as theories in 1754. Meanwhile, in 1753 Euler brought out his corrected *Lunar Theory,* and Johann Tobias Mayer of Göttingen "compared Euler's tables with observations and corrected them so successfully that he and Euler were each granted a reward of £3,000 by the English Government." [1]

In 1755 Mayer sent his tables to the British Admiralty which turned them over to James Bradley, third Astronomer Royal, and one of the great astronomers. Bradley saw the value of the work and recommended the use of these tables as a means for calculating longitude. In fact he estimated they could give it to within a half a degree. [2]

After Bradley's death, Nevil Maskelyne, the fifth Astronomer Royal, saw Mayer's tables and proceeded to create the *British Nautical Almanac and Astronomical Ephemeris for the Meridian of the Royal Observatory at Greenwich.* Using Mayer's tables, Maskelyne's people produced a useful ephemeris—a table of the moon's position—for every midnight and noon. Maskelyne wrote in the preface to this first ephemeris in 1767:

The Commissioners of Longitude, in pursuance of the Powers vested in them by a late Act of Parliament, present the Publick with the NAUTICAL ALMANAC and ASTRONOMICAL EPHEMERIS for the year 1767, to be continued annually; a Work which must greatly contribute to the Improvement of Astronomy, Geography, and Navigation. This EPHEMERIS contains every Thing essential to general Use that is to be found in any Ephemeris hitherto published, with many other useful and interesting Particulars never yet offered to the Publick in any Work of this Kind. The Tables of the Moon had been brought by the late Professor MAYER of Göttingen to a sufficient Exactness to determine the Longitude at Sea, within a Degree, as appeared by the Trials of several Persons who made Use of them. The Difficulty and Length of the necessary Calculations seemed the only Obstacles to hinder them from becoming of general Use: To remove which this EPHEMERIS was made; the Mariner being hereby relieved from this Necessity of calculating the Moon's Place from the Tables, and afterwards computing the Distance to Seconds by Logarithms, which are the principal and only very delicate Part of the Calculus; so that

[1] F. R. Moulton, *An Introduction to Celestial Mechanics* (New York, 1931), p. 364. The account of this by E. T. Bell is slightly different in detail but not in substance; cf. Bell, *Men of Mathematics*, p. 143. Still another account is given by Arthur Berry in *A Short History of Astronomy* (London, 1898), p. 283. According to Berry, the money was paid to Mayer's widow in 1765 and his theory as well as his tables were published in 1770 at the expense of the British Board of Longitude. This is apparently the correct version.

[2] Berry, p. 283.

the finding the Longitude by the Help of the EPHEMERIS is now in a Manner reduced to the Computation of the Time, and Operation. . . .

All the Calculations of the EPHEMERIS relating to the Sun and Moon were made from Mr. MAYER'S last manuscript Tables, received by the Board of Longitude after his Decease, which have been printed under my Inspection, and will be published shortly. . . .[3]

Maskelyne died in 1811, and subsequent almanacs were full of errors and highly unreliable—so much so in fact that in 1830 the British Admiralty requested the Royal Astronomical Society to rectify the situation. This was done and the *Nautical Almanac* of 1834 was a greatly improved and accurate ephemeris. Probably it is this situation that made it possible for Babbage to secure the help of the British government. It is also very possible that Babbage's failure to produce a workable machine just when it was so badly needed caused the astronomer Airy to make so harsh a judgment as he did.[4] In any case it is so that Babbage was working during a period in which celestial mechanics and astronomy in general were evolving rapidly and many theories of the moon's orbit and tables of its position were being developed. While this is not the place to recount the history of lunar theory, it is interesting to note what Moulton says on the state of the subject during the period of Babbage's life:

Damoiseau [sic] (1768–1846) carried out Laplace's method to a high degree of approximation in 1824–28, and the tables he constructed were used quite generally until Hansen's tables were constructed in 1857. Plana (1781–1869) published a theory in 1832, similar in most respects to that of Laplace. An incomplete theory was worked out by Lubbock (1803–1865) in 1830–34. A great advance along new lines was made by Hansen (1795–1874) in 1838, and again in 1862–64. His tables published in 1857 were very generally adopted for Nautical Almanacs. DePontécoulant (1795–1874) [sic] published his *Théorie Analytique du Système du Monde* in 1846. . . . A new theory of great mathematical elegance, and carried out to a very high degree

[3] Explanatory Supplement to *The Astronomical Ephemeris and The American Ephemeris and Nautical Almanac* (London, 1961), p. 3.
[4] See above, p. 14.

of approximation, was published by Delaunay (1816–1872) in
1860 and 1878.[5]

The period of Babbage's life saw other great calculations. Two
of the most notable of these were to result in the discovery of the
planet Neptune. Shortly after the discovery of Uranus by William
Herschel, the father of Babbage's friend John, it was seen that the
orbit of the planet was not what calculations showed it should be.
In 1845 John Couch Adams in England and independently Urbain
Jean Joseph Leverrier in France postulated the existence of still
another planet whose presence was perturbing the orbit of Uranus.
They calculated the size and orbit of this unknown planet so that
it would exactly explain the anomalous behavior of Uranus. Then
on 23 September 1846 Galle at the Berlin Observatory, upon a
request by Leverrier, found this planet: Neptune.

[5] Moulton, *op. cit.*, p. 364. Danoiseau's name is misspelled in the passage quoted,
as are the dates of dePontécoulant (1764–1853).

Chapter 4

The Universities:

Maxwell and Boole

Thus we observe in Babbage's period a great flowering of astronomy and celestial mechanics. It was still too soon however to see a similar situation obtain in other branches of natural philosophy. As a consequence, great calculations were not attempted or, indeed, needed in other branches of physics. In fact, we may say as a general principle that large calculations will not be attempted in a given field until its practitioners can write down in mathematical form an unambiguous description of the phenomena in question.

This was not possible in Babbage's time, for example, in the field of electricity, where Michael Faraday (1791–1867) was just then performing his beautiful experiments, which were to form the basis for Maxwell's wonderful work. Moreover, it was not until the year of Babbage's death that a chair of experimental physics was established at Cambridge. In 1871 James Clerk Maxwell (1831–1879) was called to the chair which was intended "especially for the cultivation and teaching of the subjects of Heat, Electricity, and Magnetism." This was the first time that instruction in physics was provided at Cambridge. Prior to this physical research was done outside of the English university structure. In 1874 the Cavendish Laboratory was presented to the University by its Chancellor, the Duke of Devonshire, in honor of his kinsman, Henry Cavendish (1730–1810), the distinguished physicist who anticipated, among others, Coulomb and Faraday. Upon Maxwell's untimely death in 1879, Lord Rayleigh succeeded him.

The university world of Babbage's time was a very different one from our own, and it is a mistake to extrapolate back from our times to his without realizing the differences. Although Oxford and Cambridge were the only universities in England and Wales, until 1827 Roman Catholics, Protestant Dissenters, and Jews were excluded from them by an act of Parliament that was not revoked until 1856. Moreover, it was not until the year of Babbage's death that restrictions were removed making it possible for such people to hold lay offices in either school. Thus both these schools ex-

cluded some of the most able people from the scientific and political fields. During the eighteenth century annual admissions at
Cambridge were under 200; in the 1820s they rose to 400, in the
1870s to 600, and in the 1890s to 900. One should not however
imagine that Oxford and Cambridge had completely changed.
A. J. P. Taylor says that as late as the 1920s "the two systems
of education catered for different classes and provided education,
different in quality and content, for rulers and ruled." [1] To validate
this point he states that "only one in a hundred [students] came
from a working-class family."

This exclusion was going on just at the time when the Industrial
Revolution was making education ever more essential for all members of society. In 1823 George Birkbeck (1776–1841) founded his
first Mechanics' Institute in Scotland, and similar institutes spread
into England under the patronage of Henry Brougham (1778–1868).
These brought to the workingman the advantages of a technological
training just when it was most needed in England. These schools
are the place in which the engineers and mechanics learned their
business—for an annual fee of one guinea. Most of these men were
not middle-class; for example, Stephenson, the inventor of the locomotive, was a poor boy who taught himself to read when he was
seventeen.

In 1827 Brougham founded University College in London, where
those excluded from Oxford and Cambridge could obtain a real
education without religious tests for students or faculty and without any theology in the curriculum. Science received an important
role in this school. Then in competition the Church of England
founded King's College in London in 1828, and in 1836 these two
London colleges were joined into the University of London. This
new University admitted women for degrees in 1878 and has
served as the model for all the universities founded in England
since its establishment.

It is worthy of note that as late as 1860 Thomas Huxley (1825–
1895) had to defend at Oxford Darwin's and, more generally, science's right to do research and to profess the results of this research
without reference to the theological implications of these results.
Trevelyan says:

> Science was not yet a part of the established order. In the past,
> most investigation had been done by individuals like Priestley or

[1] A. J. P. Taylor, *English History, 1914–1945*, Oxford History of England (New York and Oxford, 1965), p. 171.

Darwin working at their own expense and on their own account, and this era of private initiation was only gradually passing into the new era of endowed and organized research. In the fifties the Natural Science Tripos had been set up at Cambridge, largely owing to the intelligent patronage of the Chancellor of the University, the Prince Consort, whose royal presence was able to charm away opposition to new-fangled learning in just those academic quarters which were most obscurantist. In the sixties science was making itself felt as a power in the land. And so long as it was still struggling for freedom and recognition, with the word "evolution" inscribed on its banner militant, it could not fail to exert an influence favourable in a broad sense to liberal reform.[2]

Curiously, the Presbyterian Church in Scotland did not suppress free thought and was eager to provide education to all. Thus the university scene in Scotland was entirely different from that in England and Wales. In 1904 Kelvin, upon his installation as Chancellor of the University of Glasgow, was able to say: "The University of Adam Smith, James Watt, Thomas Reid was never stagnant. For two centuries and a quarter it has been very progressive. Nearly two centuries ago it had a laboratory of human anatomy. Seventy-five years ago it had the first chemical students' laboratory. Sixty-five years ago it had the first Professorship of engineering in the British empire. Fifty years ago it had the first physical students' laboratory—a deserted wine cellar of an old professorial house, enlarged a few years later by the annexation of a deserted examination room. Thirty-four years ago . . . it acquired laboratories of physiology and zoology."[3]

Clearly the creation of a chair in experimental physics and the calling of Maxwell to it was a most signal event in English education. It is not possible to conceive of a more remarkable figure than Maxwell to hold the Cavendish Chair. (He even edited Cavendish's researches in electricity, which had been largely unpublished since doing this was of no interest to Cavendish.) We say more about him below. It was he who fitted up the Cavendish Laboratory and got experimental physics started in England. It was his successor, Lord Rayleigh (John William Strutt, 3d Baron Rayleigh, 1842–1919), who undertook to establish physics instruction in the thoroughgoing manner characteristic of all his activities. It was due to Rayleigh and his assistants that modern laboratory courses in heat,

[2] Trevelyan, *British History in the Nineteenth Century*, p. 342.
[3] Kelvin, *Mathematical and Physical Papers*, vol. VI (Cambridge, 1911), p. 374.

electricity, magnetism, matter, light, and sound were established on sound footings.

Neither Maxwell nor Rayleigh regarded the chair in physics as a sinecure as did Babbage his in mathematics. The holder of the professorship was expected to be in residence eighteen weeks during each academic year and to give at least forty lectures during this period.[4] Contrast this with Babbage, who was never in residence and never lectured at the university.

Thus we see at least faintly the differences between the traditions in astronomy and in mathematical physics. In astronomy there was a longstanding tradition of large-scale calculation of very extensive tables to high accuracy and with great precision. In physics there was as yet little tradition for calculation, and in the future as it developed the need arose for rather smaller and more ad hoc—what are now called in slang "quick and dirty"—calculations. The physicist, to make a too-sweeping generalization, generally did not need the very large tables of the astronomer. Instead he needed much less accurate solutions of the differential equations of motion which described various phenomena. These differences were to cause the physicists to go off on an entirely different tack from the astronomers. The two groups did not come together until modern times with the advent of the modern, general-purpose computer.

In this connection it is interesting to read Maxwell's views on calculation. He said in an address to the British Association:

> I do not here refer to the fact that all quantities, as such, are subject to the rules of arithmetic and algebra, and are therefore capable of being submitted to those dry calculations which represent, to so many minds their only idea of mathematics.
>
> The human mind is seldom satisfied, and is certainly never exercising its highest functions, when it is doing the work of a calculating machine. What the man of science, whether he is a mathematician or a physical inquirer, aims at is to acquire and develope clear ideas of the things he deals with. For this purpose he is willing to enter on long calculations, and to be for a season a calculating machine, if he can only at last make his ideas clearer.[5]

---

[4] See the Introduction by R. Lindsay, p. viii, to Lord Rayleigh, *The Theory of Sound* (New York, 1945).

[5] J. C. Maxwell, *Address to the Mathematical and Physical Sections of the British Association*, British Association Report, vol. XL (1870), in Collected Works, II, 215.

Before ending our brief description of the English university scene perhaps we can illustrate the problem by saying some words about George Boole (1815–1864), who was one generation after Babbage. Boole was born in Lincoln of lower middle-class parents and was in a stratum of society from which children did not go to university in spite of their religion. Instead, he, like Abraham Lincoln, six years his senior, educated himself. He trained himself in Latin and Greek, and his father, remarkably enough, was able to start him in mathematics. At age sixteen he needed to go to work to support his parents and took a job as an usher (an assistant teacher) in a school; in fact he taught in two schools for four years. During this period he taught himself French, German, and Italian in preparation for a career in the Church. Bell amusingly says: "In spite of all that has been said for and against God, it must be admitted even by his severest critics that he has a sense of humor. Seeing the ridiculousness of George Boole's ever becoming a clergyman, he skilfully turned the young man's eager ambition into less preposterous channels." [6]

When he was twenty Boole opened his own school and taught himself mathematics — indeed, higher mathematics. He managed to read and to digest all the great masters of his time. This lonely study produced great results, which came about through the good offices of a Scot, D. F. Gregory, who was the editor of the *Cambridge Mathematical Journal* and was therefore able to bring Boole's work before the mathematical world. The magnitude of his total contribution may be judged by Bertrand Russell's summation: "Pure mathematics was discovered by George Boole in his work published in 1854."

One of his important contributions, from our point of view, was to the so-called calculus of finite differences. This apparatus is the basic tool of the numerical analyst, and Boole's accomplishments in helping to build it are certainly far from negligible.

But most important is his contribution to formal logics. In 1848 he published a little book entitled *The Mathematical Analysis of Logic*, which was to be the prelude to his great work in 1854: *An Investigation of the Laws of Thought, on which are founded the Mathematical Theories of Logic and Probabilities.*[7] In these works he rendered logics into a precise and mathematical form. He set down

---

[6] Bell, *Men of Mathematics*, p. 436.
[7] Reprint edition, *Laws of Thought* (New York, 1953), the edition quoted here.

postulates or axioms for logics for the first time—just as Euclid
and others had done for geometry. Further, he gave to the whole
subject an algebraic treatment. He said in his first chapter:

1. The design of the following treatise is to investigate the
fundamental laws of those operations of the mind by which reasoning is performed: to give expression to them in the symboli-
cal language of a Calculus, and upon this foundation to establish
the science of Logic and instruct its method; . . . and finally to
collect from the various elements of truth brought to view in the
course of these inquiries some probable intimations concerning
the nature and constitution of the human mind.

. . . .

3. But although certain parts of the design of this work have
been entertained by others its general conception, its method,
and to a considerable extent, its results, are believed to be original. For this reason I shall offer, in the present chapter, some
preparatory statements and explanations, in order that the real
aim of this treatise may be understood, and the treatment of its
subject facilitated.

It is designed, in the first place, to investigate the fundamental laws of those operations of the mind by which reasoning is
performed. It is unnecessary to enter here into any argument to
prove that the operations of the mind are in a certain real sense
subject to laws, and that a science of the mind is therefore *possible*. If these are questions which admit of doubt, that doubt is
not to be met by an endeavour to settle the point of dispute *a
priori*, but by directing the attention of the objector to the evidence
of actual laws, by referring him to an actual science. And thus the
solution of that doubt would belong not to the introduction to
this treatise, but to the treatise itself. . . .

4. Like all other sciences, that of the intellectual operations
must primarily rest upon observations,—the subject of such observation being the very operations and processes of which we
desire to determine the laws. . . .

. . . .

6. . . . There is not only a close analogy between the operations of the mind in general reasoning and its operations in the
particular science of Algebra, but there is to a considerable extent
an exact agreement in the laws by which the two classes of operations are conducted. . . .

Now the actual investigations of the following pages exhibit

Logic, in its practical aspect, as a system of processes carried on by the aid of symbols having a definite interpretation, and subject to laws founded upon that interpretation alone. But at the same time they exhibit those laws as identical in form with the general symbols of algebra, with this single addition, viz., that the symbols of Logic are further subject to a special law (Chapter II), to which the symbols of quantity, as such, are not subject. . . .[8]

The monumental work of Boole was to remain a curiosity for many years, and it was not until Whitehead and Russell wrote their great *Principia Mathematica* in 1910–13 that serious mathematicians took up formal logics. Since then the field has flowered into the stupendous achievements of Gödel and Cohen in our era.

In any case Boole's contribution to logics made possible the works of subsequent logicians including Turing and von Neumann. As we shall see later, the work of von Neumann was essential to the modern computer. Even Babbage depended a great deal on Boole's ideas—as well as those of de Morgan, Herschel, and Peacock—for his understanding of what mathematical operations really are. We mentioned before how Babbage understood the notion of a mathematical operation and the quantities upon which it operated. This was made possible for the first time in this period by this group of English algebraists.

Since Boole showed that logics can be reduced to very simple algebraic systems—known today as Boolean Algebras—it was possible for Babbage and his successors to design organs for a computer that could perform the necessary logical tasks. Thus our debt to this simple, quiet man, George Boole, is extraordinarily great and probably not adequately repaid. However, it is nice to know that he was befriended by de Morgan and that in 1849 he was called to a chair in mathematics of the newly-founded Queen's College in Cork, Ireland.

His remark about a "special law to which the symbols of quantity are not subject" is very important: this law in effect is that $x^2 = x$ for every x in his system. Now in numerical terms this equation or law has as its only solution 0 and 1. This is why the binary system plays so vital a role in modern computers: their logical parts are in effect carrying out binary operations.

In Boole's system 1 denotes the entire realm of discourse, the set of all objects being discussed, and 0 the empty set. There are two operations in this system which we may call $+$ and $\times$, or we may say

[8] Boole, *Laws of Thought*, chap. 1.

or and *and*. It is most fortunate for us that all logics can be comprehended in so simple a system, since otherwise the automation of computation would probably not have occurred — or at least not when it did. We need to say much more on this point and will do so a little later when it fits in more naturally against a larger background of understanding.

To describe the next major advance in the field of computing and to set the stage for our real topic, the early history of the electronic digital computer, we need to embark on an excursus. It is desirable at this point to acquaint those of our readers who are not already familiar with them with the differences between digital and analog or continuous computing instruments. These are the two broad categories into which we can divide all forms of computing equipment or processes.

Every computing instrument or machine of whatever sort has certain fundamental operations which it performs and out of which are formed its overall operations. Thus Pascal's machine had addition and subtraction as its operations; Leibniz's had addition, subtraction, multiplication, and division as its; Babbage's Difference Engine had addition and subtraction.

In these three cases the fundamental operations were performed by the well-known techniques of counting, that is, they did their fundamental operations by a mechanization of the methods that we as humans employ. Machines of this class are generically described as *digital* or *arithmetical*. The former name clearly calls attention to the quantities employed and the latter to the processes performed on these quantities. The premier device in this category is the abacus, the simplest form of digital computer still used in many places throughout the world.

Analog machines are very different. They are often described as being *continuous* or *measurement*. They are rather difficult to explain and describe. Here the former name again calls attention to the quantities employed and the latter to the processes performed on them. In all cases analog machines depend upon the representation of numbers as physical quantities such as lengths of rods, direct current voltages, etc. Usually they are developed for a fairly specific purpose.

During the latter part of the nineteenth century physicists had developed sufficient mathematical sophistication to describe in mathematical equations the operation of fairly complex mechanisms. They were also able to perform the converse operation: Given a set of equations, to conceive a machine or apparatus whose motion is in accord with those equations.[1] This is why these machines are called analog. The designer of an analog device decides what operations he wishes to perform and then seeks a physical apparatus whose laws of operation are *analogous* to those he wishes to carry out. He next builds the apparatus and solves his problem by *measuring* the physical, and hence continuous, quantities involved in the apparatus. A good example of an analog device is the slide rule. As is well-known, the slide rule consists of two sticks graduated according to the logarithms of the numbers, and permitted to slide relative to each other. Numbers are represented as lengths of the sticks and the physical operation that can be performed is the addition of two lengths. But it is well-known that the logarithm of a product of two numbers is the sum of the logarithms of the numbers. Thus the slide rule by forming the sum of two lengths can be used to perform multiplication and certain related operations.

Another old but important analog device is the planimeter, a device for measuring the area bounded by a simply closed curve (Fig. 1).

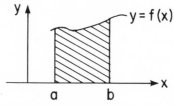

Fig. 1

George Stibitz in an article on mathematical instruments attributes the first such device to a German engineer, J. H. Hermann, in 1814.[2] In any case various improved versions appeared both on the Continent and in Britain. The most famous names connected with these are James Clerk Maxwell, who invented one in 1855, and James Thomson (1822–1892), who produced one in the early 1860s based on Maxwell's. Thomson presented his idea to the Royal Society in 1876. The delay was occasioned by the fact that Thomson saw no use for the instrument until his brother, Lord

[1] S. Lilley, "Machinery in Mathematics, A Historical Survey of Calculating Machines," *Discovery*, vol. 6 (1945), pp. 150–182. See also, Lilley, "Mathematical Machines," *Nature*, vol. 149 (1942), pp. 462–465.

[2] *Encyclopaedia Britannica* (1948), s.v. "mathematical instruments."

Kelvin, discussed with him an idea for a tide-calculating machine. At that point it became clear that his integrator could provide exactly what was needed. Thomson wrote thus: "The idea of using pure rolling instead of combined rolling and slipping was communicated to me by Prof. Maxwell, when I had the pleasure of learning from himself some particulars as to the nature of his contrivance. . . . I succeeded in devising for the desired object a new kinematic method. . . . Now, within the last few days, this principle, on being suggested to my brother as perhaps capable of being usefully employed towards the development of tide-calculating machines which he had been devising, has been found by him to be capable of being introduced and combined in several ways to produce important results." [3]

In all these analog devices the fundamental operation is that of integration, i.e. they all produce as an output $\int_a^b f(x)dx$ given $f(x)$ as an input. (The reader will recall this is the same as the area marked in by slanted lines in Fig. 1 above.)

If one were asked to name the great British physicists of the nineteenth century, or indeed of all time, James Clerk Maxwell's name would very likely be next to that of Newton's. It is not our place here to describe either his life or work but we must at least indicate his most famous and wonderful accomplishment: a complete treatment of the relation between light and electromagnetism, by means of a mathematical theory resting on two fundamental equations. His work is basic to all our understanding of electricity, magnetism, light, and other forms of electromagnetic radiation. In his monumental work, *A Treatise on Electricity and Magnetism*, he not only expressed in his equations the beautiful experimental findings of Michael Faraday (1791–1867) but also predicted the existence of electromagnetic waves, which we will discuss briefly later.

For the benefit of those readers who do not know how integrators and planimeters work, we quote Maxwell's own description since it so well describes the situation:

In considering the principle of instruments of this kind, it will be most convenient to suppose the area of the figure measured by an imaginary straight line, which, by moving parallel to itself, and at the same time altering in length to suit the form of the area, accurately sweeps it out.

[3] Thomson and Tait, *Principles of Mechanics and Dynamics*, Appendix B′: III, "An Integrating Machine having a new Kinematic Principle," pp. 488–490.

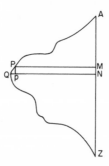

Let AZ be a fixed vertical line, APQZ the boundary of the area, and let a variable horizontal line move parallel to itself from A to Z, so as to have its extremities, P and M, in the curve and in the fixed straight line. Now, suppose the horizontal line (which we shall call the generating line) to move from the position PM to QN, MN being some small quantity, say one inch for distinctness. During this movement, the generating line will have swept out the narrow strip of the surface, PMNQ, which exceeds the portion PMNp by the small triangle PQp.

But since MN, the breadth of the strip, is one inch, the strip will contain as many square inches as PM is inches long; so that, when the generating line descends one inch, it sweeps out a number of square inches equal to the number of linear inches in its length.

Therefore, if we have a machine with an index of any kind, which, while the generating line moves one inch downwards, moves forward as many degrees as the generating line is inches long, and if the generating line be alternately moved an inch and altered in length, the index will mark the number of square inches swept over during the whole operation. By the ordinary method of limits, it may be shown that, if these changes be made continuous instead of sudden, the index will still measure the area of the curve traced by the extremity of the generating line.

. . . .

We have next to consider the various methods of communicating the required motion to the index. The first is by means of two discs, the first having a flat horizontal rough surface, turning on a vertical axis, OQ, and the second vertical, with its circumference resting on the flat surface of the first at P, so as to be driven round by the motion of the first disc. The velocity of the second disc will depend on OP, the distance of the point of contact from the centre of the first disc; so that if OP be made always equal to the generating line, the conditions of the instrument will be fulfilled.

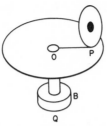

This is accomplished by causing the index-disc to slip along the radius of the horizontal disc; so that in working the instrument, the motion of the index-disc is compounded of a rolling motion due to the rotation of the first disc, and a slipping motion due to the variation of the generating line.[4]

It remained for Lord Kelvin—then Sir William Thomson—to make fundamental use of integrators to build some machines which could really do work transcending what a human could. Kelvin in his long life was enormously broad in his interests and worked in virtually every aspect of the physics of his day. One subject seemed to have great attraction for him: the motion of the tides. To gain an understanding of tidal motion, he invented several instruments which played an important part in this task. They are a tide gauge, a tidal harmonic analyser, and a tide predictor.[5]

The first of these instruments does not concern us except in a passing way. It is intended to record "by a curve traced on paper, the height of the sea level at every instant above or below some assumed datum line." The second is much more germane to our discussion. Kelvin, writing about his tidal harmonic analyser, said: "The object of this machine is to substitute brass for brain in the great mechanical labour of calculating the elementary constituents of the whole tidal rise and fall. . . . The machine consists of an application of Professor James Thomson's Disk-Globe-and-Cylinder Integrator to the evaluation of the integrals required for the harmonic analysis. . . . It remains now to describe and explain the ac-

---

[4] *The Scientific Papers of James Clerk Maxwell* (Dover edition, New York, 1965), pp. 230–232; the paper itself is entitled "Description of a New Form of the Platometer, an Instrument for Measuring the Areas of Plane Figures drawn on Paper," and first appeared in *Transactions of the Royal Scottish Society of Arts*, vol. IV (1855).

[5] Kelvin, *Mathematical and Physical Papers*, vol. VI, pp. 272ff.

tual machine . . . which is the only Tidal Harmonic Analyser hitherto made." [6]

Let us now turn attention to a few of the details of his harmonic analyser and tide predictor. To do this it is convenient first to discuss very briefly what is meant by simple harmonic motion.

Perhaps the easiest example of simple harmonic motion is afforded by musical instruments and in particular by musical notes, as distinguished from noise. Such notes correspond to *periodic* vibrations, i.e. after a certain interval of time, called the *period*, those vibrations repeat themselves with perfect regularity. The number of vibrations made in a given time is referred to as the *frequency*. Finally, the notes whose frequencies are multiples of that of a given one, are called its *harmonics*.

However, musical notes, while periodic, are frequently made up of simpler vibrations called *tones* or *simple harmonic motions*. It has been shown that these tones are mathematically representable by the so-called circular functions, the sines and cosines of trigonometry. What are these functions or simple motions? Let us imagine a point P which is moving uniformly — at constant speed — around a circle and let us drop a perpendicular from this point P to the diameter as in Fig. 2 below. The perpendicular cuts the diame-

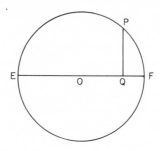

Fig. 2

ter at a point Q which moves back and forth along the diameter. The motion of Q is what is called simple harmonic motion.

Before going into more details we should remark that the reader who finds mathematical formulas unfamiliar or even frightening can safely skip over these in the text. They are introduced for possibly pedantic reasons by the author and may be safely ignored.

Now in Fig. 3 the *amplitude* of the motion is the distance $r =$ OE $=$ OF; the *period* is the time it takes Q to go from its present

[6] Kelvin, *Papers*, vol. VI, p. 280.

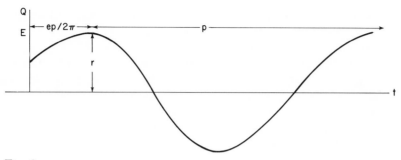

Fig. 3

position — whatever that may be — back to the same position and moving in the same direction as it was. This is represented mathematically as

$$x = OQ = r \cos\left(\frac{2\pi t}{p} - e\right)$$

where r is the radius of the circle or the amplitude, p is the period, t is the time, and e is the *phase* or *phase angle*. This phase measures essentially when we start reckoning time, i.e., at $t = ep/2\pi$ we have $x = r$. Different simple harmonic motions may, or course, start at different times. They are thus out of phase with each other.

These simple facts about musical tones or simple harmonic motion apply equally to all types of periodic motions. Such motions are particularly important in physics since very many phenomena are periodic — indeed, our lives are constantly regulated by such motions, as for example the rotation of the earth about its axis on a daily basis or its revolution about the sun on a yearly one. Many such examples can be cited, but that is not the point. The study of simple harmonic motions probably sounds to the uninformed reader as possibly interesting but certainly highly special.

That this is not the case was first shown by a most remarkable man, Jean-Baptiste-Joseph Fourier (1768–1830), the orphaned son of a tailor in Auxerre, France. According to Bell:

> By the age of twelve he was writing magnificent sermons for the leading church dignitaries of Paris to palm off as their own. At thirteen he was a problem child, wayward, petulant, and full of the devil generally. Then, at his first encounter with mathematics, he changed as if by magic. He knew what ailed him and cured himself. To provide light for his mathematical studies after he was supposed to be asleep he collected candle-ends in the

kitchen and wherever he could find them in the college. His secret study was an inglenook behind a screen. . . .

To check the mere blood-letting Napoleon ordered or encouraged the creation of schools. But there were no teachers. All the brains that might have been pressed into immediate service had long since fallen into the buckets. It became imperative to train a new teaching corps of fifteen hundred, and for this purpose the École Normale was created in 1794. As a reward for his recruiting in Auxerre Fourier was called to the chair of mathematics.[7]

In 1807 he wrote a treatise on the mathematical theory of heat, *Théorie analytique de la chaleur,* that is now world famous. In it is contained a theorem, now named after him, which is described by Thomson and Tait as follows: "Fourier's theorem . . . is not only one of the most beautiful results of modern analysis, but may be said to furnish an indispensable instrument in the treatment of nearly every recondite question in modern physics." [8]

This result may be described roughly in this way: An extremely wide class of periodic phenomena are describable as a sum of a simple harmonic motion and of its harmonics with suitably chosen amplitudes. Thus the physicist may resolve virtually any complex periodic vibration or motion into simple ones based on the familiar sines and cosines of trigonometry. In mathematical language we may write a periodic function f as

$$f(x) = a_0 + a_1 \cos \left(\frac{2\pi t}{p} - e_1 \right)$$
$$+ a_2 \cos \left(\frac{4\pi t}{p} - e_2 \right) + a_3 \cos \left(\frac{6\pi t}{p} - e_3 \right) + \cdots,$$

where the number of terms in the sum may be finite or infinite. If we permit not only cosines but also sines, the phase angles may be eliminated, and we write

$$f(x) = b_0 + b_1 \cos \frac{2\pi t}{p} + b_2 \cos \frac{4\pi t}{p} + b_3 \cos \frac{6\pi t}{p} + \cdots$$
$$+ c_1 \sin \frac{2\pi t}{p} + c_2 \sin \frac{4\pi t}{p} + c_3 \sin \frac{6\pi t}{p} + \cdots.$$

Even the case where there are only two periodic terms in the series can be quite interesting. If one considers the lunar and solar

[7] Bell, *Men of Mathematics,* chap. 12.
[8] Thomson and Tait, *op. cit.,* part 1, p. 54.

tides, they are each found to be very nearly simple harmonic
motions — except in certain cases such as tidal rivers or long channels or deep bays. Thus in many cases one can as a first approximation consider tidal motion as the sum of just these two simple
harmonic ones. (The amplitude of the lunar tide is about twice
that of the solar one.) The interested reader is referred to Thomson
and Tait for the details.

We are now in a much better position to describe what a harmonic analysis is and how a harmonic analyser does it. First of all,
given a periodic function or phenomenon, then a harmonic analysis
of it is a determination of the amplitudes of the fundamental and
its overtones or harmonics. That is, a harmonic analysis of a given
periodic phenomenon is intended to find the coefficients b_0, b_1, etc.
and c_1, c_2, etc. in our formula for $f(x)$.

It may be shown by the mathematically inclined reader that the
coefficients are expressible as

$$b_0 = \frac{1}{p} \int_0^p f(t)dt,$$

$$b_n = \frac{2}{p} \int_0^p f(t) \cos \frac{2\pi nt}{p} \, dt,$$

$$c_n = \frac{2}{p} \int_0^p f(t) \sin \frac{2\pi nt}{p} \, dt \ (n = 1, 2, \ldots).$$

Thus a harmonic analysis consists of forming a number of integrals of the general form

$$\int_0^p f(t)g(t)dt,$$

where g is a sine or cosine function. The evaluation of integrals of
this sort is what Kelvin succeeded in doing by an ingenious adaptation of his brother's disk-globe-and-cylinder integrator. In fact he
says: ". . . and I believe that in the application of it to the tidal
harmonic analysis he will be able in an hour or two to find by aid
of the machine any one of the simple harmonic elements of a result
which hitherto has required not less than twenty hours of calculation by skilled arithmeticians." [9] We need not go now into the
details of how he carried out the adaptation.

Here we see for the first time an example of a device which can
speed up a human process by a very large factor, as Kelvin asserts.

[9] Sir W. Thomson, *Proc. Roy. Soc.*, vol. 24 (1876), p. 266; see also Thomson and
Tait, p. 496.

That is why Kelvin's tidal harmonic analyser was important and Babbage's difference engine was not.

Maxwell in an aricle on harmonic analysis stated:

> Sir W. Thomson, availing himself of the disk, globe, and cylinder integrating machine invented by his brother, Professor James Thomson, has constructed a machine by which eight of the integrals required for the expression of Fourier's series can be obtained simultaneously from the recorded track of any periodically variable quantity, such as the height of the tide, the temperature of the pressure of the atmosphere, or the intensity of the different components of terrestrial magnetism. If it were not on account of the waste of time, instead of having a curve drawn by the action of the tide, and the curve afterwards acted on by the machine, the time axis of the machine itself might be driven by a clock, and the tide itself might work the second variable of the machine, but this would involve the constant presence of an expensive machine at every tidal station.
>
> It would not be devoid of interest, had we opportunity for it, to trace the analogy between these mathematical and mechanical methods of harmonic analysis and the dynamical processes which go on when a compound ray of light is analyzed into its simple vibrations by a prism, when a particular overtone is selected from a complex tune by a resonator, and when the enormously complicated sound-wave of an orchestra, or even the discordant clamours of a crowd, are interpreted into intelligible music or language by the attentive listener, armed with the harp of three thousand strings, the resonance of which, as it hangs in the gateway of his ear, discriminates the multifold components of the waves of the aerial ocean.[10]

Another interesting use for a harmonic analyser may be noted. Kelvin also built a small harmonic analyser which could be used to handle the analysis needed to adjust the compasses in iron ships — a process that was still in use in World War II. The instrument was capable of handling harmonic functions of the form

$$A + B \sin \Theta + C \cos \Theta + D \sin 2\Theta + E \cos 2\Theta.$$

This is precisely the form the deviation of the compass needle from the true reading takes as a function of the heading angle Θ of the ship; once the coefficients $A, B, \ldots, E$ are known small

---

[10] J. C. Maxwell, "Harmonic Analysis," Collected Works, II, 797–801.

blocks of iron are inserted into the compass housing to compensate for the deviation.

The tide predictor of Kelvin is an interesting and important instrument based on his harmonic analyser. In his words: "After having worked for six years at the Tidal Harmonic Analysis the Author designed an instrument for performing the mechanical work of adding together the heights (positive or negative) above the mean level, due to the several simple harmonic constituents, determined by analysis, from observation or from the curves of a self-recording tide gauge for any particular part, so as to predict for the same port for future years, not merely the tides of high and low water, but the position of the water-level at any instant of any day of the year." [11]

The last of Kelvin's inventions that is relevant to our history at this point is his work on what is now called a differential analyzer. In a most interesting preliminary paper he set himself the task of mechanically integrating the most general linear differential equation of the second order. This is a very important equation to the mathematical physicist because it describes, *inter alia*, "vibrations of a non-uniform stretched cord, of a hanging chain, of water in a canal of non-uniform breadth and depth, of air in a pipe of non-uniform sectional area, conduction of heat along a bar of non-uniform section or non-uniform conductivity, Laplace's differential equation of the tides, etc., etc." It is also of great importance to electrical engineers since many circuits can be described by means of it.

Kelvin put his differential equation into the form

$$\frac{d}{dx}\left(\frac{1}{p}\frac{du}{dx}\right) = u,$$

where p is a given function of x. He then set up an iterative scheme whereby he made an initial guess for u, say u_1. He then envisioned a device containing two integrators so that he could form by the first integrator

$$g_1(x) = \int_0^x u_1 dx$$

and then by the second one

$$u_2(x) = \int_0^x p(x)g_1(x)dx.$$

[11] Kelvin, *Papers*, vol. VI, p. 285.

He then took his result u_2 and put it into the first integrator in place of u_1 and repeated his procedure, finding u_3, etc. It is known that this procedure will in general converge. In Kelvin's words: "After thus altering, as it were, u_1 into u_2 by passing it through the machine, then u_2 into u_3, by a second passage through the machine, and so on, the thing will, as it were, become refined into a solution which will be more and more nearly rigorously correct the oftener we pass it through the machine. If u_{i+1} does not sensibly differ from u_i, then each is sensibly a solution." [12]

This point of view, while interesting, still lacked something which apparently came to Kelvin in 1876. He wrote: "So far I had gone and was satisfied, feeling I had done what I wished to do for many years. But then came a pleasing surprise. Compel agreement between the function fed into the double machine and that given out by it. . . . Thus I was led to a conclusion which was quite unexpected; and it seems to me very remarkable that the general differential equation of the second order with variable coefficients may be rigorously, continuously, and in a single process solved by a machine." [13]

To describe exactly what Kelvin meant we need now to pause for a moment. Observe that in Kelvin's second integral appears the function $g_1(x)$ calculated in the first one. What he now noticed was that if the result of the second integration is continuously fed into the first integral, it forces $u = u_1$ to be u_2 and in one iteration the problem is solved. In differentials we have

$$dg_1 = udx, \; du = pg_1dx$$

under this scheme. Now if we eliminate g_1 from these relations, we have

$$\frac{d}{dx}\left(\frac{1}{p}\frac{du}{dx}\right) = u.$$

In other words, u is the solution of the differential equation.

This technique can of course be generalized, and Kelvin wrote another paper entitled *Mechanical Integration of the General Linear Differential Equation of Any Order with Variable Coefficients*.[14] This paper contains basically the idea which Vannevar Bush was to rediscover when he developed and constructed the first successful machine to do what Kelvin proposed.

[12] Thomson and Tait, part 1, p. 498.
[13] *Ibid.*
[14] Thomson and Tait, p. 500.

Principals of the Electronic Computer Project of the Institute for Advanced Study: Julian H. Bigelow, chief engineer; Herman H. Goldstine, assistant project director; J. Robert Oppenheimer, director of the Institute; and John von Neumann, project director. The IAS, Princeton, or von Neumann machine, as it was variously called, was begun in 1946 and was operational in 1952, the year the above photograph was taken. (PHOT: Alan W. Richards)

The difficulty with Kelvin's scheme was technological and he never built a machine for solving differential equations. The problem arises from the need to take out the output of one integrator and use it as the input to another. Why is this a difficulty? The answer lies in the fact that, in Maxwell's explanation, the output is measured by the rotation of a shaft attached to a wheel that is turned by pressing on a rotating disk. The so-called torque of this shaft—its ability to turn another shaft—is very slight and hence it cannot in fact form the input to another integrator. This was to be a barrier to progress for fifty years.

Chapter 6   Michelson,

Fourier Coefficients, and the

Gibbs Phenomenon

There was one substantial difficulty with Kelvin's Harmonic Analyzer: it was not possible to increase by very much the number of terms in the series since his device for adding the terms together led to an accumulation of errors. If the number of terms was fairly large, this accumulation could vitiate the results completely. In fact, Albert A. Michelson (1852–1931) and Samuel W. Stratton (1861–1931) were to write in 1898:

> The principal difficulty in the realization of such a machine lies in the accumulation of errors involved in the process of addition. The only practical instrument which has yet been devised for effecting this addition is that of Lord Kelvin. In this instrument a flexible cord passes over a number of fixed and movable pulleys. If one end of the cord is fixed, the motion of the other end is equal to twice the sum of the motions of the movable pulleys. The range of the machine is however limited to a small number of elements on account of the stretch of the cord and its imperfect flexibility, so that with a considerable increase in the number of elements the accumulated errors due to these causes would soon neutralize the advantages of the increased number of terms in the series.[1]

Michelson was one of the great physicists of our time. Born in Germany, he was a graduate of the United States Naval Academy where he taught for several years. He then went to the Continent and was a graduate student both in Germany and France. He became the first head of the physics department at the newly-created University of Chicago in 1892. There he set an intellectual tone and standard which was to be responsible for the great tradition in physics at the University and in the United States ever since. In 1907 he received the Nobel Prize in physics, the first American to be so distinguished, in recognition of "the methods which you have discovered for exactness of measurements. . . ." His work on

[1] A. A. Michelson and S. W. Stratton, "A New Harmonic Analyzer," *American Journal of Science*, 4th ser., vol. v (1898), pp. 1–13.

the velocity of light and the non-existence of ether was to lay the foundation for Einstein's special theory of relativity. (After Michelson there was not another award of the Nobel Prize in physics to an American until Robert Millikan, who worked in Michelson's department for 25 years, received one in 1923. A number of other great American scientists worked under him including Arthur H. Compton, a Nobel Laureate, Edwin P. Hubble, and George E. Hale, to mention a few.)

The things that Michelson did are only very peripherally related to our history but serve to show his great ability to detect minute differences. One such observation served, as we shall see shortly, to introduce a most remarkable phenomenon into pure mathematics. Before describing it however we should say a word about his greatest accomplishment. It had been believed for some time that absolute space was pervaded by a medium—the ether—which served as the carrier for electromagnetic phenomena. In moving in its orbit around the sun the earth therefore must move through this substance. If the velocity of light relative to the ether is c—about 300,000,000 meters per second $= 3 \times 10^8$ m/s—and if an object is moving with velocity v through the ether, then $c - v$ is the velocity of light relative to the object. In this expression we understand that if the object is moving in the same direction as the light, their velocities subtract, whereas if they are approaching each other head-on they add, and an appropriate correction is made for oblique motions.

Now the earth moves in an orbit around the sun with a velocity of about 30 km/sec $= 30,000$ meters per second. The great Michelson-Morley experiment of 1887 was to see whether it was possible to detect any difference in the velocity of light relative to an observer on earth depending on the direction of motion of the light. In this experiment a time difference of 10^{-15} sec had to be detected. Michelson invented a very precise tool for measuring the interference of light waves slightly out of phase with each other. This is his famous interferometer; it was built in 1881 through the financial generosity of Alexander Graham Bell. With this device Michelson made his first experiments on the ether in the laboratory of the superb German physicist Hermann von Helmholtz in Berlin. Then came the great and careful measurement of 1887. (Incidentally his measurement also resulted in a very precise determination of the velocity of light.) Michelson and Morley showed that the velocity of light did not depend on the direction of motion. Thus they completely overturned the notion, originally postulated by Newton, of a stationary medium pervading space. Newton had said: "Absolute space, in its own nature, without relation to anything external, remains always similar and immovable." [2]

But now let us return to our story. Michelson apparently some years before the paper cited above had considered other devices for carrying out the addition process — with special reference to the harmonic analyzer — without the ruinous loss of precision experienced by Kelvin. He decided that the simplest way was by the addition of the forces of spiral springs. Using this idea in 1897, he and Stratton built an analyzer which could handle Fourier Series of 20 terms. This worked so well they obtained backing from the Alexander Dallas Bache Fund of the National Academy of Sciences and built a machine which handled 80 terms, i.e., it could take account of terms corresponding to $\cos x$, $\cos 2x$, . . . , $\cos 80x$. It could also handle series consisting of sines.

Their machine could be fed the Fourier coefficients of a trigonometric series and would then produce a graph of the sum function, and it could also perform the inverse process: given the function, the machine could produce the Fourier coefficients. Michelson and Stratton calibrated their machine against two examples to measure the accuracy with which it would evaluate $\int \phi(x) \cos kx dx$. In the case where $\phi(x)$ is a constant on the interval 0 to 4 and zero elsewhere they calculated the first twenty Fourier coefficients with an average error of 0.65 percent of the greatest term; for $\phi(x) = e^{-.01x^2}$ they evaluated the first twelve coefficients and found a comparable error. They summed up by saying this:

> It appears, therefore, that the machine is capable of effecting the integration $\int \phi(x) \cos kx dx$ with an accuracy comparable with that of other integrating machines; and while it is scarcely hoped that it will be used for this purpose where great accuracy is required it certainly saves an enormous amount of labor in cases where an error of one or two percent is unimportant.
>
> The experience gained in the construction of the present machine shows that it would be quite feasible to increase the number of elements to several hundred or even to a thousand with a proportional increase in the accuracy of the integrations.
>
> Finally it is well to note that the principle of summation here employed is so general that it may be used for series of any

[2] Isaac Newton, *Mathematical Principles of Natural Philosophy and His System of the World*, English translation by A. Motte, revised by F. Cajori (Berkeley, 1934), p. 6.

function by giving to the points (p) the motions corresponding to the required functions, instead of the simple harmonic motion. . . . A simple method of effecting this change would be to cut metal templates of the required forms mounting them on a common axis. In fact the harmonic motion of the original machine was thus produced.[3]

From our point of view the importance of this work is two-fold: first, it was a continuation of the tradition of analog computation which owed so much to Kelvin; and second, the junior member of the team of Michelson and Stratton was to leave Chicago, first to found the National Bureau of Standards and then to become President of the Massachusetts Institute of Technology.

These moves were to be important because Stratton established at the National Bureau a tradition for computing which was to persist there through the present, and at MIT he was to be a sympathetic and understanding colleague of Vannevar Bush, who was to carry out Kelvin's ideas on solving differential equations in a practical way. It was Bush's great influence which was to make MIT one of the greatest centers for innovation in the field of computing.

In our next chapter we will again find Stratton. The Bureau of the Census set up "the Census Machine Shop . . . in limited quarters furnished by the Bureau of Standards, under the general supervision of S. W. Stratton, Chief of that bureau. . . ." [4] This laboratory was commissioned to continue the developments of the machine originally started by Hollerith — more on this later.

It is of some interest to note that President William Rainey Harper, the great first president of the University of Chicago, succeeded in persuading Michelson to send an interferometer, the echelon spectroscope, and his new harmonic analyzer to the World's Fair in Paris in 1900. These instruments won the University a Grand Prize, and Michelson wrote Harper saying: "I thank you sincerely for your own share in the work and appreciate the wisdom of your course in the matter of sending in the apparatus in the face of many obstacles — my own objections included." [5]

In 1948 the U.S. Navy built the Albert Abraham Michelson Laboratory in his honor at the U.S. Naval Ordnance Test Station,

[3] Michelson and Stratton, "A New Harmonic Analyzer," pp. 12–13.
[4] Leon E. Truesdell, *The Development of Punch Card Tabulation in the Bureau of the Census, 1890–1940* (U.S. Government Printing Office, 1965), p. 121.
[5] B. Jaffe, *Michelson and the Speed of Light* (Garden City, 1960), p. 130.

Inyokern, China Lake, California, in the Mojave Desert. The harmonic analyzer together with various of his other instruments and memorabilia were on exhibit in the main lobby until they were moved to the Smithsonian Institution in Washington, D.C.

There is yet another importance to Michelson's work which is almost unique: he discovered by numerical calculations a phenomenon which he described in a paper and whose true explanation was then worked out by Josiah Willard Gibbs (1839–1903), the greatest American mathematical physicist of the nineteenth century, and perhaps one of the greatest in the world. This is to this day known as the "Gibbs phenomenon." [6] Although the phenomenon is only peripheral to our discussion, it may nonetheless be of real interest to some readers so we digress long enough to describe the situation which is mathematical in nature and may be ignored by the non-technical reader.

Let $f(x)$ be a function whose graph is piece-wise smooth and let

$$f_n(x) = \frac{1}{2} a_0 + \sum_{i=1}^{n} (a_i \cos ix + b_i \sin ix)$$

be the nth partial sum of the Fourier series for $f(x)$. We now ask how the graphs of $f_n(x)$ approach the graph of $f(x)$. It is well known that in each interval where there are no discontinuities all is well: the graphs of the f_n approach that of f without difficulty. But in the neighborhood of a discontinuity of f the graphs of the f_n contain oscillations which do not die down in amplitude as n increases. To make this precise let us take the example of Michelson and Gibbs. They considered the function defined by the Fourier series

$$f(x) = 2 \left(\sin x - \frac{1}{2} \sin 2x + \frac{1}{3} \sin 3x - \frac{1}{4} \sin 4x + \cdots \right)$$

$$= 2 \sum_{m=1}^{\infty} \frac{(-1)^{m+1}}{m} \sin mx,$$

[6] There is a considerable and somewhat confused literature on this subject published in *Nature* during 1898 and 1899, starting with a paper by Michelson and culminating in Gibbs' explication of the apparent paradox and by an appeal from Michelson to Poincaré with his response. The reader is referred to vol. LVIII, pp. 544, 569; LIX, 200, 271, 314, 606; and LX, 52, 100. Although there was much attention given the subject at the time, it seems to have disappeared from view until Bôcher in 1906 discussed the topic in general terms. Curiously, in 1848 the phenomenon was first noticed and fully explained by a little known but evidently excellent mathematician at Trinity College, Cambridge, Henry Wilbraham, B.A., and published in the *Cambridge and Dublin Mathematical Journal*, vol. 3 (1848), p. 198, under the title "On a Certain Periodic Function."

and the partial sums

$$f_n(x) = 2 \sum_{m=1}^{n} \frac{(-1)^{m+1}}{m} \sin mx.$$

It is not difficult to show that $f(x)$ represents the function $y = x$ on the interval $-\pi < x < \pi$ and extended periodically over congruent intervals as shown in Fig. 4.

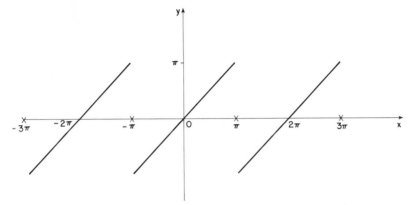

Fig. 4

We see from this graph that f represents $y = x$ on the interval $-\pi \leq x \leq +\pi$ except at $x = \pm\pi$ where the series sum is zero.

When Michelson and Stratton evaluated f_n for $n = 80$, they discovered the graph of this function behaved most peculiarly near $x = \pi$. Here it oscillated with a small but substantial amplitude; in the figures below there are several graphs showing this behavior near to π. There are also included two of the figures from the Michelson-Stratton paper showing the "Gibbs phenomenon" as first observed.

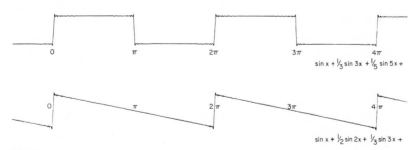

Fig. 5

Very likely only so great an observer as Michelson—recall the exceedingly small putative difference in velocities he proposed to detect—would have been able to note the minute oscillation shown in Fig. 5. In any case he did and this clearly troubled him to the point where he described the situation in *Nature* in October of 1898 as mentioned earlier.

It may be shown that the limiting graph will not be as seen in Fig. 4 above but instead will be as in Fig. 6 below.

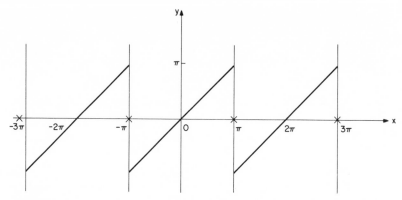

Fig. 6

The overshoot indicated in the graph is quite remarkable. The total length of the vertical line is four times the definite integral

$$\int_0^\pi \frac{\sin u}{u}\, du.[7]$$

The value of this integral is about 1.8519, and so the vertical line is 7.4076 instead of the $2\pi = 6.2832$ one might have reasonably but erroneously conjectured. In Figs. 7 and 8 below are given two enlargements of the overshoot to show what happens as the number of terms in the series increases. The first graph shows the neighborhood of $x = \pi$ when 50 terms are summed, and the second when 200 terms are. In both cases $f_n(x)/2$ is summed. It is clear that the amplitude of the overshoot is unchanged but the frequency of the oscillation becomes higher.

[7] Gibbs, *Nature*, vol. LIX (1899), p. 606.

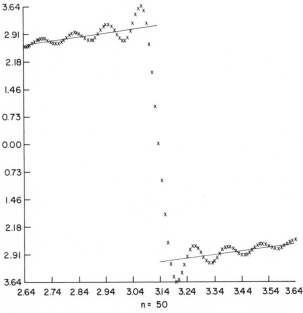

Fig. 7

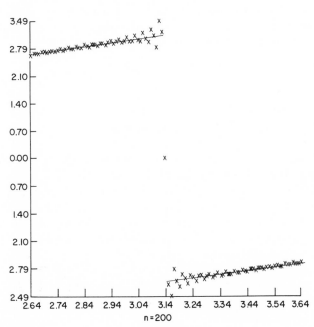

Fig. 8

Chapter 7   Boolean Algebra:

$$x^2 = xx = x$$

Earlier we discussed very briefly George Boole's contribution to logics and hinted at what he had done. Perhaps now it is worth our while to describe in somewhat more detail the nature of Boolean algebra, since we can do this in not too technical terms and since it has such relevance to digital computers. As Boole said, his aim was to "give expression . . . to the fundamental laws of reasoning in the symbolic language of a Calculus." He did this by a most ingenious reduction of logic to algebraic form. It is worth our while to try to understand a little of what he did.

In his first proposition he supposes that he is dealing with a system of signs or arbitrary symbols such as x, y, or z that stand for the things—whatever they may be—to be discussed; with two operations called $+$ and $\cdot$ that operate upon the signs giving still other signs—thus if x and y are signs, $x + y$ and $x \cdot y$ are also signs; with an identity operation $=$; and with various rules governing the behavior of the operations such as

$$x + y = y + x, \quad x \cdot y = y \cdot x,$$
$$x \cdot (y + z) = x \cdot y + x \cdot z.$$

In fact, the rules are, with one exception, those of the algebra we all learned in school. This exception is that

$$x^2 = x \cdot x = x$$

for every x. We discuss this in a few moments.

Now let us return to Boole's signs. What can they be? A simple answer is things or groups or classes. What about the operations of $+$ and $\cdot$? Consider first the latter, $\cdot$. If x stands for "black things" and y for "cows," then $x \cdot y$ or xy stands for "black cows." In general xy represents that thing having both properties x and y. Thus if x stands for "horned things," y for "cows," and z for "black things," then xyz stands for "horned, black cows." Now if both x and y are the same thing—say, "cows"—then xy expresses no more than x. In other words

$$x^2 = xx = x,$$

which is Boole's special law that sets logic apart from ordinary algebra.

Let us now ask about the operation of +. This is used to relate totally disparate things and in fact to collect them together into a new thing or class. Thus $x + y$ is a thing made up of x and/or y. Suppose x stands for "men" and y for "women"; then $x + y$ stands for "men and women"; i.e., if x is the class of men and y the class of women, then $x + y$ is the class of both men and women. Moreover if z stands for "American," then $z(x + y)$ represents the class of American men and American women.

Boole recognized that the operation of collecting together into a whole must in general have an inverse operation of separating a whole into parts. To illustrate this, let x represent the stars, y the suns, and z the planets; then in symbols

$$x = y + z.$$

It seems reasonable then to be able to solve this equation and write

$$z = x - y;$$

i.e., the "stars, except the planets, are suns."

Boole had a very clear idea of the relation of his logical system to algebra. He said: "Let us conceive, then, of an Algebra in which the symbols x, y, z, &c. admit indifferently of the values 0 and 1, and of these values alone. The laws, the axioms, and the processes, of such an Algebra will be identical in their whole extent with the laws, the axioms, and the processes of an Algebra of Logic. Difference of interpretation will alone divide them. Upon this principle the method of the following work is established." [1] The reason for the 0, 1, of course, is that these are the only quantities satisfying the equation $x^2 = x$. But what do they mean in Logic, in Boole's system?

First, let us address 0. How must it behave? Well, in an algebra we have $0 \cdot y = 0$ for every y. Boole therefore concludes that "a little consideration will show that this condition is satisfied if the symbol 0 represents Nothing. In accordance with a previous definition we term Nothing a class. In fact, Nothing and Universe are the two limits of class extension, for they are the limits of the possible interpretations of general names, none of which can relate to fewer individuals than are comprised in Nothing, or to more than are comprised in the Universe." [2] Furthermore the symbol 1 in

[1] Boole, *Laws of Thought*, pp. 37–38.
[2] Boole, p. 47.

algebra is characterized by the relation $1y = y$ for every y. "A little consideration will here show that the class represented by 1 must be 'the Universe' since this is the only class in which are found *all* the individuals that exist in *any* class. Hence the respective interpretations of the symbols 0 and 1 in the system of Logic are *Nothing* and *Universe*." [3]

We see from this that if x is a class of objects, then $1 - x$ is the complementary class of all objects not in x. Thus if the equation $x = x^2$ can be written as $x(1 - x) = 0$, we see that it asserts the following: a class and its complement can have nothing in common. If x is the set of all dogs, $1 - x$ is the set of all non-dogs — things that are not dogs — and no thing can be both a dog and a non-dog.

Moreover the law $x^2 = x$ enabled Boole to reduce virtually any expression $f(x)$ in a sign x to the extremely simple form

$$ax + b(1 - x).$$

To see this we recall that a wide class of important functions $f(x)$ in algebra are expressible as polynomials

$$f(x) = a_0 + a_1 x + a_2 x^2 + \cdots + a_n x^n.$$

But since $x^2 = x$, $x^3 = x^2 = x$, $x^4 = x^3 = x$, ..., $x^n = x^{n-1} = x$, we can write

$$f(x) = a_0 + (a_1 + a_2 + \cdots + a_n)x = ax + b(1 - x),$$

where $b = a_0$, $a = a_0 + a_1 + \cdots + a_n$. Now if

$$f(x) = ax + b(1 - x),$$

then

$$f(0) = a \cdot 0 + b(1 - 0) = b, f(1) = a \cdot 1 + b(1 - 1) = a;$$

thus

$$f(x) = f(1)x + f(0)(1 - x).$$

Of course in this formula, $f(1)$ means the value of the logical expression $f(x)$ when x is the Universe 1 and $f(0)$ the value when x is Nothing 0. An exactly similar situation obtains for functions of several variables. Thus, for example,

(a) $f(x, y) = f(1, 1)xy + f(1, 0)x(1 - y) + f(0, 1)(1 - x)y$

$$+ f(0, 0)(1 - x)(1 - y).$$

By way of illustration of these ideas let us consider the opera-

[3] Boole, p. 48.

tions $\oplus$ and $\otimes$, the former being the true sum of binary digits modulo 2, i.e., $x \oplus y$ being the units digit of the arithmetical sum of x and y; the latter, the true arithmetical product of the digits. Thus

$$x \oplus y = \begin{cases} 0 \text{ if } x = 0, \quad y = 0, \\ 1 \text{ if } x = 1, \quad y = 0 \text{ or } x = 0, \quad y = 1, \\ 0 \text{ if } x = 1, \quad y = 1 \end{cases}$$

$$x \otimes y = \begin{cases} 0 \text{ if } x = 0, \quad y = 0 \text{ or } 1 \text{ or } x = 0 \text{ or } 1 \text{ and } y = 0 \\ 1 \text{ if } x = 1, \quad y = 1. \end{cases}$$

Then we see with the help of formula (a) that

$$x \oplus y = x(1-y) + y(1-x), \quad x \otimes y = xy.$$

Thus the arithmetical product is the same as the logical product— this is why binary multiplication is so easy to execute in circuitry, and the arithmetical sum is not much harder.

Generally, logical propositions are of the form $f = 0$ where f may be a function of one or more variables. To illustrate consider with Boole the Mosaic proposition "Clean beasts are those which divide the hoof and chew the cud." Let x be clean beasts, y beasts dividing the hoof, and z beasts chewing the cud. Then the proposition in question is in symbols

(1) $x = yz,$

which is expressible as $f(x, y, z) = x - yz = 0$. This may be written as

(2) $0 \cdot xyz + xy(1-z) + x(1-y)z + x(1-y)(1-z) - (1-x)yz$

$$+ 0 \cdot (1-x)y(1-z) + 0 \cdot (1-x)(1-y)z$$

$$+ 0(1-x)(1-y)(1-z) = 0.$$

We proceed now to simplify this equation; to this end let us first operate upon both sides by $(1-x)$ and recall that for every variable u, $u^2 = u$ or $u(1-u) = 0$. Thus

(3) $(1-x)yz = 0;$

similarly operating with $(1-y)z$, we find

(4) $x(1-y)z = 0$

with $(1-z)y$

(5) $$xy(1-z)=0;$$

and with $(1-y)(1-z)$,

(6) $$x(1-y)(1-z)=0.$$

These four trivial consequences (3), (4), (5), (6) of the relation (1) state the following propositions:

(a)   There are no unclean beasts that divide the hoof and chew the cud.
(b)   There are no clean beasts that do not divide the hoof and chew the cud.
(c)   There are no clean beasts that divide the hoof but do not chew the cud.
(d)   There are no clean beasts that neither divide the hoof nor chew the cud.

In fact these four denials are equivalent to the original proposition (1).

Using (3), (4), (5) in (2), we find the proposition

$$0 \cdot xyz + 0 \cdot (1-x)y(1-z) + 0 \cdot (1-x)(1-y)z$$
$$+ 0 \cdot (1-x)(1-y)(1-z) = 0$$

and by negation

$$xyz + (1-x)(1-z)y + (1-x)(1-y)z + (1-x)(1-y)(1-z) = 1.$$

This proposition asserts that all living beasts are in one of the following classes:

(a)   Clean beasts that divide the hoof and chew the cud.
(b)   Unclean beasts that divide the hoof but do not chew the cud.
(c)   Unclean beasts that do not divide the hoof but do chew the cud.
(d)   Unclean beasts that neither divide the hoof nor chew the cud.

Chapter 8   Billings, Hollerith, and

the Census

At this point we take up consideration of digital machines that are basically founded on Boole's ideas. Here we find an anomalous development. Up till now we have dealt with scientific and engineering usages and needs for computation. Now all of a sudden we find a totally new situation confronting us: needs of the Census Office of the United States Department of Interior are such in the 1880s that a system of machines is developed to automate at least partially the procedure. We have here then for the first time a statistical and a commercial need for computation. A very complete account of the history of the punch card in the Census Bureau is given in the excellent book by Leon Truesdell, cited above (p. 55, n. 4); in it the interested reader may find an authoritative account of the period 1890–1940. As A. Ross Eckler, the Director of the Bureau of the Census, wrote in his Foreword to the book: "Dr. Truesdell is in a unique position to relate the history of the period of the gradual development of punch card machinery and the techniques of the attendant procedures. He was intimately connected with four censuses during the period of rapid development, 1910 to 1940, and is also familiar with the methods used in 1950 and 1960."

From our somewhat parochial point of view the key census is that of 1890, and the two names which stand out are those of Herman Hollerith (1860–1929) and John Shaw Billings (1839–1913). (We must also mention James Powers.) Like Michelson, Hollerith was the son of parents who left Germany as a result of the political disturbances in the historic year of 1848. Unlike Michelson however he was born in Buffalo and went to the School of Mines at Columbia University. His doctoral dissertation was a paper on his own tabulating system. Upon graduation from Columbia he was befriended by a well-known engineer-educator of that period, William P. Trowbridge (1828–1892), who placed him in a position in the Census Office set up for the 1880 census. (Recall that the Census Bureau as a permanent organization was not established until 1902 and the following year it was transferred from the Department of Interior to the newly-created Department of Commerce and Labor. It is amusing to notice that the first five censuses were conducted under the direct supervision of the then Secretaries of State, and it was not until the 1840 census that a lesser official was appointed to supervise the work.)

Hollerith was associated with the Census Office from October 1879 to August 1883, at which point he went to the Patent Office for about a year. After he left that post he worked on the development of his machine for tabulating population statistics and received patents on this equipment in 1889. This was the system that was to be used in the census of 1890, the first one that was ever automated in any significant part. It is also significant that he spent the academic year 1882–83 at the Massachusetts Institute of Technology as an instructor in mechanical engineering.

Out of this work was to grow a commercial organization, Tabulating Machine Company. In 1911 this company became the Computer-Tabulating-Recording Company, which Thomas J. Watson, Sr., joined in 1914. This in turn (1924) became the International Business Machines Corporation under Watson's masterful leadership. How all this came about, while fascinating, is not germane to our story except in part and is, in any case, already well told.[1] We will make many references to IBM and its equipment in our story, but we will in general not treat the development of IBM or any other company in the detail it deserves since such matters are peripheral to our story.

The other figure of note is John Shaw Billings. In 1870 the then Superintendent of the Census asked the Surgeon General of the U.S. Army for help in connection with a biostatistical question of procedure for the 1870 census. The latter gentleman assigned two Assistant Surgeons to the task, one of whom was Billings. This work apparently was intensely interesting to Billings, and in some fashion he made himself an important member of the Census Office staff, though he was never on the census payroll. At this point he was a major and had the rank of Surgeon in the Army; he became Deputy Surgeon General in 1894. (He saw service during the Civil War and later was medical inspector for the Army of the Potomac.) But his chief interest was closely related to the Surgeon General's library and the statistical and public health consequences that could be deduced from these records.

Clearly Billings had great administrative abilities and was sought

[1] T. G. and M. R. Belden, *The Lengthening Shadow, The Life of Thomas J. Watson* (Boston, 1962).

after by a number of institutions in addition to the Census Office. Thus in 1876 he was made Medical Advisor to the trustees of the Johns Hopkins Fund. In this capacity he had a most important hand in the planning of the hospital as well as other medical and public health problems. Again in 1891 he was lecturing on hygiene and vital statistics at the University of Pennsylvania and was made Director of the University Hospital there in 1893. He retired from the Army in 1895 and became Professor of Hygiene at Pennsylvania. Finally, in 1896, he left to found, in its present form, the New York Public Library.

To return to our story: Billings was in charge of the work on vital statistics both for the 1880 and 1890 census. It was he who was in charge of the collection and tabulation of the data. The picture that emerges is of a man who not only was in nominal charge of an operation but was the true intellectual leader and pioneer. He must have been a man of tremendous vigor and breadth of interest.

This vital man was most fortunate in having young Hollerith as a member of the staff, and he acknowledged this in the Letter of Transmittal of the 1880 report on *Mortality and Vital Statistics.* At this point the story becomes complex and contradictory. The details may be read either in Truesdell's work or in a thesis by John H. Blodgett.[2] We recall that there were two accounts of how Babbage first got his idea for a Difference Engine, so are there two for Hollerith. Truesdell regards as most reliable the account written by Dr. Walter F. Willcox, who worked in the Census Office in 1900 and wrote an account of the genesis of the punch card for the census as follows:

> While the returns of the Tenth (1880) Census were being tabulated at Washington, Billings was walking with a companion through the office in which hundreds of clerks were engaged in laboriously transferring items of information from the schedules to the record sheets by the slow and heartbreaking method of hand tallying. As they were watching the clerks he said to his companion, "There ought to be some mechanical way of doing this job, something on the principle of the Jacquard loom, whereby holes in a card regulate the pattern to be woven." The seed fell on good ground. His companion was a talented young engineer in the office who first convinced himself that the idea was practicable and then that Billings had no desire to claim or use it.[3]

[2] Truesdell, *op. cit.,* or J. H. Blodgett, "Herman Hollerith, Data Processing Pioneer," master's thesis, Drexel Institute of Technology, 1968.
[3] Truesdell, pp. 30–31.

Willcox also stated that "This statement is based on my memory of a conversation with Mr. Hollerith." On the other side is the recollection by Hollerith in 1919, who wrote as follows:

> One Sunday evening at Dr. Billings' tea table, he said to me there ought to be a machine for doing the purely mechanical work of tabulating population and similar statistics. We talked the matter over and I remember . . . he thought of using cards with the description of the individual shown by notches punched in the edge of the card. . . . After studying the problem I went back to Dr. Billings and said that I thought I could work out a solution for the problem and asked him if he would go in with me. The Doctor said he was not interested any further than to see some solution of the problem worked out.[4]

In reading over various speeches and papers by Billings on the subject it seems to the author clear that Hollerith was the real implementor of Billings' basic idea. Dr. Raymond Pearl, who was for many years one of the world leaders in biology and biostatistics and who was professor of biology and public hygiene at Johns Hopkins, analyzed the situation in 1938 and concluded that "In all essentials the case seems clear. Billings was the originator, the *discoverer*, who contributed that which lies at the core of every scientific discovery, namely, an original idea that proved in the trial to be sound and good; Hollerith built a machine that implemented the idea in practical performance, the accomplishment here, as always of the successful *inventor*." [5]

It is difficult to say more about the allocation of credit than Pearl did. We will have occasion to see other examples of the difficulties of allocating credit among men all of whom are deeply immersed in a common project. Let us then agree with Pearl that Billings had the basic idea and that Hollerith implemented it.

More importantly let us consider what the system was and what use was made of it. Hollerith, proceeding on Billings' suggestions, used a system of holes in a punch card to represent various characteristics such as male or female, black or white, native or foreign-born, age, etc. He first designed his system using a continuous roll of paper instead of individual cards. The card or roll of paper then ran under a set of contact brushes which completed an electrical circuit if and only if a hole was present. The completed circuits activated counters which advanced one position for each

[4] Truesdell, p. 31.   [5] Truesdell, p. 33.

hole counted—the counters were operated by electromechanical relays. For the 1890 census, cards were used and the roll of paper abandoned. Hollerith realized that cards could be prepared by different people in different locations and at different times and then assembled in one large deck for subsequent tabulation; he also saw that cards could be sorted according to a given characteristic. This last point was to be of crucial importance. Thus in the analysis of population statistics one could in a few sorts determine out of a given population how many people had characteristics A, B, C, for example, and how many did not.

Hollerith devised a card $6^5/_8$ by $3^1/_4$ inches in size containing 288 locations at which holes could be made. (He chose the size because this was the size of the dollar bill and saved him building some additional equipment.) He built a machine for punching these cards and a simple sorter. This sorter had a box containing 24 bins each with a lid held closed by an electromagnetic latch working against a spring. Normally all lids were closed, but when a hole was sensed the electric current that flowed as a result turned off the latch and the spring opened the lid. The card was then dropped into the open bin by hand. It was to be much later before this part of the operation was automated.

In 1889 the Superintendent of the 1890 census, Robert P. Porter, set up a committee of three to decide how to tabulate the forthcoming data. The committee contained, among others, Billings and William C. Hunt. It first considered the Hollerith system and a so-called chip system proposed by Charles F. Pidgin. Then Hunt stepped down from the committee to propose a system somewhat analogous to Pidgin's. For a test a part of the 1880 data of St. Louis was used; it contained information on about 10,000 people. Hollerith's system was about twice as good as his nearest rival in total time spent; i.e., in transcribing onto cards and tabulating. The chief advantage was in tabulation where Hollerith's apparatus was about eight times faster.

The 1890 census was tabulated on Hollerith's system with great success. Just a month after all returns arrived in Washington Porter was able to announce the total population count.[6] In a paper on the subject Porter stated in 1891:

> The Eleventh Census handled the records of 63,000,000 people and 150,000 minor civil divisions. One detail (characteristic) alone required the punching of one billion holes. Because

[6] Blodgett, *op. cit.*, p. 54.

the electrical tabulating system of Mr. Hollerith permitted easy counting, certain questions were asked for the first time. Examples of these were:

>   Number of children born
>   Number of children living
>   Number of family speaking English

By use of the electric tabulating machine it became possible to aggregate from the schedules all the information which appears in any way possible. Heretofore such aggregations had been limited. With the machines, complex aggregations can be evolved at no more expense than the simple ones.[7]

After this success Hollerith set up his Tabulating Machine Company in 1896 and made both machines and cards. The results of the American experiment excited the census takers of western Europe and of Canada, and his business flourished. His equipment, with various improvements, was again used in the Twelfth (1900) Census. This time it was used on a rental basis by the Census Bureau.

It was in this period that the U.S. Congress passed the Permanent Census Bureau Act and S. N. D. North became the first Director of the Census in 1903. Very soon thereafter he and Hollerith disagreed about the rental charges for the use of the Hollerith machines. North therefore requested and received an appropriation from Congress of $40,000 to do—in modern parlance—research and development work on tabulating equipment. It was in this connection that we mentioned earlier S. W. Stratton's part in the Census work.[8]

Out of this laboratory were to come several refinements of Hollerith's tabulating machine. Among other things, the new device had printing counters which automatically recorded the tallies, thereby obviating the necessity for manual reading of the old dial faces. This was done in 1906–07. In the latter year James Powers, one of the engineers in the Census shop was commissioned to develop an automatic card-punching machine. He did this in the next two years.

Somewhat later (1911) Powers was to form the Powers Tabulating Machine Co., and this was to be Hollerith's principal competitor for

---

[7] Robert P. Porter, "The Eleventh Census," *Proceedings of the American Statistical Association*, no. 15 (1891), p. 321; Blodgett, pp. 54–55.
[8] See above, p. 55.

many years. The Powers Company merged with Remington-Rand in 1927, and this in turn with Sperry Gyroscope in 1955. Evidently the Powers system was a good one. Belden and Belden say: "Powers had a machine that automatically printed results as opposed to hand-posting in the Hollerith system, an electrical punch instead of Hollerith's clumsy hand-operated one, and, to get cards in order for tabulation, a horizontal sorter in place of the backbreaking vertical sorter Hollerith had devised for crowded railroad offices. Besides, Powers rented machines for only $100 a month whereas the inferior CTR equipment cost $150." [9]

It is interesting to note that the thing which gave IBM its edge over Powers came in 1914 with the creation by Watson of a development group under E. A. Ford, one of Hollerith's people. An engineer in Ford's group, Clair D. Lake, then invented a superior printer-lister which saved the Tabulating Machine Company from extinction.

It is a noteworthy feature of our American system that much of the computer field owes its existence to the generosity of our government in giving to its employees and university contractors the rights to inventions made with government funds.

[9] T. G. and M. R. Belden, *The Lengthening Years*, pp. 111–112.

Chapter 9

Ballistics and the Rise of

                          the Great Mathematicians

Let us now put to one side the development of punch card equipment and return to discussions of scientific and engineering needs
and advancements in the field of computing. In particular, let us
examine what World War I did in these respects.

In a one-sentence oversimplification, that war was, from a scientific point of view, dominated by chemistry and did much to further
the field in the United States. At the beginning, the United States
was extremely dependent on Europe for potash, nitrates, and dyes.
To overcome this dependence the government did a number of
things, including the construction of the plant at Muscle Shoals in
the Tennessee Valley to spur the manufacture of nitrogen. Out of
these developments was to come the whole synthetics field: for
example, between 1914 and 1931 the output of rayon increased 69
times.[1]

But this is not the point of this history. We will interest ourselves
in the development of the science and practice of ballistics and primarily exterior ballistics. This is the subject in which one studies
the motion of a projectile from the moment it emerges from the
muzzle of a gun until it reaches its target. Clearly this is a branch of
the dynamics of rigid bodies moving under gravitational and aerodynamical forces. The reason why we wish to discuss this recondite
and perhaps uninteresting branch of mechanics is because it was
to have a vital impact on our subject. We shall see how the ballisti-
cal needs of the United States were to be a primary incentive to the
development of the modern computer.

Just as virtually all aspects of classical mechanics can be traced
back to Newton, so can ballistics. The concept is of course as old
as man and his warlike proclivities. Even the word ballistics itself
is derived from the Latin *ballista,* a missile-throwing engine and
the Latin word in turn is derived from a much earlier Greek word
ballein, to throw. We cannot here give a detailed account of the
history of the subject. Fortunately, however, an excellent one is to

[1] W. E. Leuchtenberg, *The Perils of Prosperity, 1914–1932* (Chicago and London,
1958), p. 181.

be found in an appendix to a definitive work on mathematical ballistics.[2]

Newton was considerably concerned with the motion of projectiles acted upon by gravity and retarded by air or water. Indeed, the
apocryphal story of the apple is perhaps an overly-simplistic
statement of his interest in the problem, but it at least serves to
show that the motion of a projectile was for him a paramount
example. His desire to show the universality of his law of gravita-
tion made the motion of a projectile a quite natural object for him
to want to study. The projectile is one of the few objects in fact
that the natural philosopher can experiment upon and thereby
test the relevance of his hypotheses. For this reason, and perhaps
others, we find an astonishingly large number of quite distinguished men working on the problem throughout the last 300 years
—including Galileo dropping balls from the Leaning Tower.
Much of Newton's work on the subject assumed that the retardation
of the air varied as the second power of the speed; this was due to
his belief that this was the actual law both in air and water. In the
Scholium to Section I, Book II of his *Principia* it is stated that "In
mediums void of all tenacity, the resistances made to bodies are as
the square of the velocity. . . . Let us, therefore, see what motions
arise from this law of resistance." Both Moulton and von Karman
have remarked that Newton's law is a reasonable approximation
for very high velocities. However, it remained for Euler, whom we
recall as playing a key role in lunar theory, to formulate the equations of motion for rigid bodies in general and projectiles in particular. He also developed two approximative techniques for solving
these equations.

Many other great figures worked on the various problems of
ballistics, and we note the familiar names of Clairaut, D'Alembert,
Lagrange, and Laplace, who was Examiner of the Royal Artillery
in 1784.

One of the central problems in ballistics is how to determine
the drag function, the retardation of the air, as a function of the
velocity. Various physicists and mathematicians have worked on
this ever since Newton. In the middle of the nineteenth century an
accurate method was perfected by Francis Bashforth in England.
Using his ideas, various ballisticians determined realistic drag
data, and in the twenty years from 1880 to 1900 a commission working in Gâvre, France, put together these results into what is known

[2] E. J. McShane, J. L. Kelley, and F. V. Reno, *Exterior Ballistics* (Denver, 1953),
pp. 742ff.

as the *Gâvre function*. This function formed the principal drag
function used during World War I for virtually all shells, even
though it was probably a poor approximation for many types.

In the era just before World War I a number of other approximations of doubtful validity were made to reduce the amount of
numerical work that needed to be done. Typical of the approximations made was the assumption that the density of the air was
constant with altitude. Others were also made so that the equations
of motion would assume a particularly simple form that could be
solved with little work. Moulton described the situation at the
start of the war thus: ". . . a hasty examination of the classical ballistic methods showed, not only that they were wholly inadequate
for current demands, but also they were not well suited to the solution of the problem, even under earlier conditions. They contained
defects of reasoning, some quite erroneous conclusions, and the
results were arrived at by singularly awkward methods." [3]

Another man, active in ballistics both in World War I and II and
a longtime professor of mathematics at Brown University, Albert A.
Bennett, described what went on as "wrenching the equations into
a form that could be solved by very simple means." The interested
reader may find the details of the so-called Siacci approximation
method and variants on it in an excellent little book on ballistics by
Bliss.[4]

The Siacci method had only very limited accuracy in real-life
situations. In the early part of World War I the German Navy designed and built a very powerful gun. When it was first fired it
was found that its maximum range was about twice as large as predicted by the Siacci method. This was so since much of the trajectory went through air only about half as dense as that at ground
level. This discovery led the German ballisticians to design the
long-range gun fired against Paris — Big Bertha, as it was sometimes
called. This gun was nine inches in caliber and fired at ranges over
ninety miles using a propellant designed by the distinguished
chemist Fritz Haber (1861–1924), who won the Nobel Prize for
his synthesis of ammonia. He worked for the Kaiser's government
and made several fundamental contributions: he increased the
supply of ammonia which was then converted into nitric acid for
high explosives, and he was chief of the chemical warfare service
of the German Army. It was he who directed the chlorine attack at
Ypres in 1915.

[3] F. R. Moulton, *New Methods in Exterior Ballistics* (Chicago, 1926), p. 1.
[4] G. A. Bliss, *Mathematics for Exterior Ballistics* (New York, 1944), pp. 27–41.

The United States Army assembled two first-class groups of men to work on ballistic problems during World War I. One group was in the Ballistics Branch of the Office of the Chief of Ordnance in Washington and was headed by Forest Ray Moulton (1872–1952) from April 1918. The other group, headed by Oswald Veblen, was first at Sandy Hook, and then in 1918 it was moved to the newly-established Aberdeen Proving Ground in Maryland. It is greatly to the credit of the relevant officials of the United States Army that they early became cognizant of the problem and brought in such eminent men to head up the work. Both these men were scientists of the first magnitude; they had been colleagues and friends at the University of Chicago, where both had been graduate students and then faculty members, the one in astronomy and the other in mathematics. Moreover both turned out to be great administrators and managers of people.

Moulton belonged to a remarkable family consisting of himself and four brothers: Harold, who was for many years a professor of economics at the University before becoming president of the Brookings Institution of Washington, D.C., during its period of great growth; Elton, a mathematician who had a distinguished career at Northwestern University, first as a professor and then as dean of the Graduate School; and Earl and Vern who became well-known executives in business. Among Moulton's most remembered intellectual accomplishments is the Planetesimal or Spiral-Nebula Hypothesis of the solar system, which he and the geologist T. C. Chamberlin enunciated.

From our point of view his great accomplishment was to play a signal role in putting ballistics on sound, scientific footings. He gives the following account:

> The subject was taken up anew by the author as a scientific problem requiring close co-ordination of adequate theory and well-conducted experiments. . . .
>
> The theory was developed as rapidly as numerous exacting administrative duties would permit. Before the end of June, 1918, the essence of this volume, except the last chapter, had been worked out and issued in blue print form. Some additions . . . were made by the author and others, particularly Professor Bliss. . . .
>
> The practical experiments required to parallel the theory were (a) proving-ground experiments and (b) wind-tunnel experiments. The proving-ground ballistic experiments were fortunately under

the direction of Professor (then Major) Oswald Veblen, who not only carried out the instructions of the Washington Office with vigor and success, but also exhibited a high order of initiative, insight into physical problems, and imagination, combined with unusual ability to get things accomplished under difficult circumstances. The results obtained under his direction were notable, and only the termination of the war prevented their being greatly extended. . . .

Proving-ground experiments give only the integrated effects of the factors that contribute to the final results, but theory must deal with differential effects. Consequently the high importance of wind-tunnel experiments was evident. Unfortunately the power required to give velocities above that of sound . . . is very great. It was learned, however, that the General Electric Company was planning to test at Lynn, Massachusetts, in the early autumn of 1918, some huge blowers for use in blast furnaces. . . . At a conference in Washington, in July, 1918, Dr. Moss promised the full cooperation of the General Electric Company. Immediately general plans for the experiment were worked out in conjunction with Professor (then Captain) A. A. Bennett; an appropriation of as much of $100,000.00 as might be required was secured to prosecute the experiments; upon application to Director Stratton, of the Bureau of Standards, Dr. L. J. Briggs was assigned to work on the experiment; . . . These experiments proved most difficult, as was expected, and have continued over several years.[5]

Notice Stratton appearing again in our tale. Also Lyman Briggs (1874–1963) is worth saying a word about. He was a physicist working in the Department of Agriculture when Stratton had him transferred to the Bureau of Standards, where he stayed. He became the Director in 1933 and Director Emeritus in 1945. His research career revolved closely around studies of objects in high velocity air streams and liquids under high pressures.

Let us return to Moulton. One of his great specific contributions was the general introduction of finite differences into ballistic computation. In England this was done by the great Karl Pearson (1857–1936), who pioneered so brilliantly in the applications of statistics, particularly to biology. For many years he was Galton Professor of Eugenics and Director of the Galton Laboratory of Eugenics at University College, London. But in the United States

[5] Moulton, *op. cit.*, pp. 2–4.

it was Moulton who brought to ballistics the numerical techniques of astronomy. He developed techniques that represented very real advances over previously known methods. What he did was to take a procedure of John C. Adams, whom we mentioned earlier as co-discoverer of Neptune, and make an important modification in it that now goes by the name of the Adams-Moulton method. This is a numerical procedure for solving systems of total differential equations with as much precision as is desired. Later we will discuss numerical methods in a little more detail and make clearer what this is all about. His work on this subject appears, among other places, in his book on *New Methods in Exterior Ballistics.*

Moulton left the University of Chicago in 1926 to become the financial director of the Utilities Power and Light Corporation of Chicago and, in 1932, the Director of Concessions of the Chicago World's Fair. Then in 1937 he became the permanent secretary of the American Association for the Advancement of Science, where he played a key role in acquiring and building up *Science* and *Scientific Monthly* and in making the AAAS the great organization it is today. In a tribute to him in 1963, one of the officials of the National Academy of Sciences-National Research Council referred to him as "the deGaulle of American science."

The other leader in the story of ballistics in the United States in World War I was Oswald Veblen (1880–1960). He was the grandson of a Norwegian family who migrated from Norway to Wisconsin in 1847 and there and in Minnesota raised a family of twelve children, one of whom was Thorstein Veblen (1857–1929), the sociologist and economist. Oswald Veblen was born in Iowa and took his Ph.D. at the University of Chicago in 1903. He stayed on there as an associate in the mathematics department for several years before receiving a call from President Woodrow Wilson and Dean Henry Fine of Princeton University to become one of the new preceptors there.

After many years as a professor at Princeton, Veblen and Albert Einstein were appointed the first professors at the newly-founded Institute for Advanced Study, where he remained until his retirement. The Institute was incorporated on May 20, 1930, and the two men were appointed two and a half years later. Veblen was also a trustee and later honorary trustee of the Institute until his death. His total contribution both to the Institute and the University as well as to the American mathematical scene was enormous. It was he who played an absolutely vital role in creating one of the greatest mathematics departments of the world at the university, and who was largely responsible for the selection of the matchless mathematics faculty at the Institute upon its creation: James Alexander, Albert Einstein, Marston Morse, John von Neumann, and Hermann Weyl. Not only did he create this unique department, he also persuaded the founders, Mr. Louis Bamberger and his sister Mrs. Felix Fuld, as well as the first director, Abraham Flexner, to concentrate the Institute's energies on post-doctoral men. Mr. Bamberger and Mrs. Fuld wrote: "The primary purpose is the pursuit of advanced learning and exploration in fields of pure science and high scholarship. . . ." In this connection it is germane to quote what Robert Oppenheimer wrote as Director of the Institute in 1955: "The immediate effects of the Institute's work are in knowledge and in men. The new knowledge and the ideas find their way into the worldwide communities of science and scholarship, and the men take their part throughout the world in study, in teaching, in writing and in discovering new truth. History teaches — and even the brief history of the Institute confirms — that new knowledge leads to new power and new wisdom, and alters the destiny and heightens the dignity of man." [6]

Veblen from the earliest times understood these things and saw very clearly the growing need for further mathematical training of young recipients of the doctor's degree, particularly before they were immersed in the daily struggle for survival as instructors with heavy teaching loads. Many years before in his early days at Princeton he had helped to protect Henry Norris Russell, one of the great astronomers and astrophysicists, from the university authorities when Russell wanted to devote some portion of his energies to research. At that time the authorities viewed a faculty member's role as being exclusively an instructor of undergraduates. Again somewhat later, in 1924, Veblen persuaded the National Research Council to establish post-doctoral fellowships in mathematics; for many years the selection committee consisted of himself, Gilbert Bliss of Chicago, and George Birkhoff of Harvard; these three men all received their Ph.D. degrees at the University of Chicago where they first became friends, and later each was to be a preceptor at Princeton and still later a leader of the American mathematical community. When President Harper brought Michelson to Chicago he also brought a young American mathematician, Eliakim H. Moore (1862–1932), who was to be the man who really started modern mathematics in the United States. Among others, Birkhoff, Bliss, and Veblen were his students; in fact, his students

[6] *The Institute for Advanced Study, Publications of Members, 1930–1954*, foreword by Robert Oppenheimer (Princeton, 1955).

formed a Who's Who of American mathematics for a whole generation.

While Veblen understood with absolute clarity the needs of young American mathematicians, he was no chauvinist or xenophobe. At various times he was an exchange professor at Oxford and on several occasions lectured in Germany. One of his colleagues, Deane Montgomery, in a touching appreciation of Veblen's life, quotes from Veblen's obituary to Dean Fine:

> Dean Fine was one of the group of men who carried American mathematics forward from a state of approximate nullity to one verging on parity with the European nations. It already requires an effort of the imagination to realize the difficulties with which the men of his generation had to contend, the lack of encouragement, the lack of guidance, the lack of knowledge both of the problems and of the contemporary state of science, the overwhelming urge of environment in all other directions than the scientific one. But by comparing the present average state of affairs in this country with what can be seen in the most advanced parts of the world, and extrapolating backwards, we may reconstruct a picture which will help us to appreciate their qualities and achievements.[7]

These words might with justice be said of Veblen also.

When Mr. Bamberger initially urged establishing the Institute in his home in the Oranges in New Jersey, it was Veblen who succeeded in persuading him to establish it in Princeton in the neighborhood of one of the great universities, since Veblen knew it could not flourish in a vacuum. His entire life was devoted to establishing the highest academic standards. "In this he would consider no compromises. Even though he became virtually blind in his later years he always retained a youthful interest in mathematics and science. . . . He was often amused by the comments of younger but aging men to the effect that the great period for this or that was gone forever. He did not believe it. Possibly part of his youthful attitude came from his interest in youth; he was firmly convinced that a great part of the mathematical lifeblood of the Institute was in the flow of young mathematicians through it."[8]

It is of some interest to note that Mrs. Veblen's brother was the Nobel Laureate in Physics, Owen Richardson, and her brother-in-

[7] D. Montgomery, "Oswald Veblen," *Bull. Amer. Math. Soc.*, vol. 69 (1963), pp. 26–36.
[8] *Ibid.*

law was another Nobel Laureate, Clinton Davisson, who did so much not only for physics by his experiments on electrons but also for establishing the Bell Telephone Laboratories as a pre-eminent industrial research laboratory. He as well as Arthur H. Compton were students of Richardson at Princeton.

It was Veblen who was responsible for bringing both Eugene P. Wigner (1902–   ), Nobel Laureate in Physics, and John von Neumann (1903–1957) to Princeton in January of 1930 with joint appointments in mathematics and physics. The physicist Ehrenfest recommended to Veblen that the two men be brought over together. This Veblen did with tremendous consequences to our story.

However, our aim at this time is not so much to describe Veblen's total career as that part of it devoted to the United States Government. This spans the two world wars. We have already summarized his role in the first of these, in which he was the organizer and creator of a group of first-class mathematicians and physicists including James W. Alexander, Gilbert A. Bliss, Thomas Gronwall (a Swedish scientist who came to Princeton after Birkhoff left), as well as the renowned Norbert Wiener and Harvey Lemon. Lemon was professor of physics at Chicago for many years and upon his retirement in 1950 became the Director of the Museum of Science and Industry. He, as well as Veblen, was to be influential at Aberdeen during both wars. (He is a brother-in-law of Birkhoff.) Lemon was the head of rocket development at Aberdeen during World War II. We postpone until later Veblen's roles in World War II and in subsequent times.

Moulton also created a first-class group in Washington, containing among others A. A. Bennett, whom we have already mentioned, and Dunham Jackson, who became one of America's important mathematicians and made great contributions all during his life to the theory of approximations, an interest stemming from his war work.

Several important things resulted from Moulton's pioneering work that are relevant to our history. Somehow he persuaded the Army and the Navy to send a number of especially promising young officers to the University of Chicago to work for the Ph.D. in mathematics with a specialty in ballistics. This is probably the genesis of the idea that not every officer has to be an all-purpose individual, as capable of commanding troops in the field as of understanding the requirements for a supersonic wind tunnel. Of course to this day only very few officers succeed in attaining high position by pursuing single-mindedly a technical career—Admiral Rickover being perhaps the prime example.

The success of the Moulton and Veblen operations resulted in ensuring the continuation of a decent Technical Branch to the Office of the Chief of Ordnance as well as a comparable Ballistic Research Laboratory at Aberdeen Proving Ground through the lean years between the wars. In particular, it enabled a man who had done hand calculating during the first war, Samuel Feltman, to rise to become what amounted to the permanent under-chief of the ballistics work in Washington until his untimely death in 1947 in a Labor Day automobile accident. Feltman acted for many years as the stabilizing and guiding force for the numerous Army officers—some technically trained, most not—who had tours of duty as chief of the Technical Branch. Through his vision the Ordnance Department had high tensile strength steel for manufacturing guns capable of firing at supersonic velocities during World War II. This obvious sounding desideratum, or indeed necessity, was anything but fifty years ago. It was not at all clear to most so-called experts that there would come a time when we would need to fire against rapidly moving targets. If anyone doubts this, he need only to be reminded that in 1914 the entire British Army had only 80 motor vehicles, and the United States was not much better off.

Another man who started in the same office and who was to be the scientific leader of the Aberdeen operation was Robert H. Kent (1886–1961). When the United States entered World War I, Kent was teaching electrical engineering at the University of Pennsylvania. He volunteered and became a captain in the Ordnance Department of the Army and was in Moulton's office from 1917 to 1919 where he became a close friend of Alexander. Then he was a civilian employee in the Washington office from 1919 to 1922; finally, he went to the Ballistic Research Laboratory, where he remained until his retirement in 1956. He was awarded the Medal of Merit by President Truman for his great accomplishments as Associate Director of the Ballistic Research Laboratory and the Potts Medal of the Franklin Institute.

The Moulton and Veblen tradition was clear enough in the minds of Feltman, Kent, and others and its value so understood that at the onset of World War II steps were immediately taken to bring an absolutely pre-eminent mathematician with a deep understanding of its applications to fill Moulton's old role in Washington. The choice made was impeccable. H. Marston Morse (1892–   )

professor of mathematics at the Institute for Advanced Study from 1935 till his retirement in 1962, was the man. His contributions were manifold and profound, as has been everything he has ever done. This is not the place to discuss them since they impinge so slightly on our main story, but we would be totally remiss not to mention them since they were so signal that President Truman honored Morse in 1945 by conferring on him the highest civilian award in the United States, the Medal of Merit, for his wartime accomplishments and in 1965 President Johnson presented him with the National Medal of Science for his total contribution to mathematics. Morse was most ably assisted in his work in Washington by William Transue, professor of Mathematics at the State University of New York.

It is of some interest to note that Morse was the friend and colleague of Birkhoff for many years at Harvard. Although Birkhoff had been a student of Moore at Chicago, he was really the intellectual disciple of Henri Poincaré (1854–1912), one of the great mathematicians of all time, and the one to whom Michelson appealed during his controversy with the English applied mathematicians over the Gibbs phenomenon. Morse in his beautiful appreciation of Birkhoff said: "F. R. Moulton's study of the work of Poincaré had something to do with Birkhoff's own intense reading of Poincaré. Poincaré was Birkhoff's true teacher." [9] To say just a word about Poincaré, whom E. T. Bell calls "The Last Universalist," we may quote Bell: "His record is nearly five hundred papers on *new* mathematics, many of them extensive memoirs, and more than thirty books covering practically all branches of mathematical physics, theoretical physics, and theoretical astronomy as they existed in his day. This leaves out of account his classics on the philosophy of science and his popular essays." [10] It should be noted that all this work was done between 1878 and 1912.

The great thing perhaps that Poincaré passed on to Birkhoff and to Morse was the notion of applying to astronomy the techniques of what was then called Analysis Situs and is now called Topology. Veblen also was to be heavily influenced by this work and himself wrote a memoir on Analysis Situs.

It is also of interest to read what Morse said about the mathematical leaders of the period: "The sturdy individualism of Dickson, E. H. Moore, and Birkhoff were representative of American

[9] Marston Morse, "George David Birkhoff and His Mathematical Work," *Bull. Amer. Math. Soc.*, vol. 52 (1946), pp. 357–391.

[10] Bell, *Men of Mathematics*, p. 538.

mathematics 'coming of age.' The work of these great Americans sometimes lacked external sophistication, but it more than made up for this in penetration, power, and originality, and justified Birkhoff's appreciation of his countrymen." [11]

The author must also mention a colleague of Morse's in Feltman's office, a mathematician of very great ability, Griffith C. Evans (1887–   ). His contribution can be measured in part by noting that he also was the recipient of the Presidential Certificate of Merit in 1948. He had little to do with the development of computers but much with the development of impeccable, intellectual standards in the Army. In the spring of 1971 the University of California, Berkeley, honored Evans by naming its new mathematics building after him.

It is worth noting at this time how American science – at least mathematics, which really started with Moore at the University of Chicago – went through a period of seeking its own identity, as indicated in Morse's remark, by embarking on a kind of chauvinistic campaign just as the United States was busily admitting millions of poor emigrants. Between 1820 and 1860 about five million came over; between 1860 and 1890 another thirteen and one-half million – in both waves the people came primarily from northern and western Europe; and between 1900 and 1930 almost nineteen million – mainly from southern and eastern Europe.[12] These people in the last wave were fleeing economic, political, social, and religious tyrannies in their home lands. Large numbers had the highest aspirations for their children, and to our great good fortune we had the educational facilities to accommodate these young people.

Then, with the Nazi tyranny, another wave came over – not only poor, uneducated workers but also famous scientists and scholars as well as younger ones. It was the duty of our scientific community to make room for these men, who have done superbly for America and in the process helped mould it. This transition, this opening of doors to Europeans, was of course not always effected without some resistance by excellent but short-sighted people. In any case, it was a movement which played a vital role in ending the possibility of a socially elite caste of university professors and allowed the democratic spirit of the United States to operate.

[11] Morse, *op. cit.*
[12] S. P. Hays, *The Response to Industrialism, 1885–1914* (Chicago, 1957), p. 95.

What we wish now to do is to set the stage for the works of Van-nevar Bush and his colleagues at MIT; of Howard Aiken at Harvard and his colleagues at IBM; and of George Stibitz and his associates at the Bell Telephone Laboratories. To do this we need to gain a little understanding of the degree of mathematical sophistication of the electrical engineer in the 1920s.

Ever since early days there has been a fairly close linkage between various branches of engineering and the cognate sciences. In many cases of course the links were formed by natural philosophers or scientists who incidentally made important inventions. Thus, in 1830 Joseph Henry (1797–1878) set up a telegraph cable a mile long. He also invented in 1829 the prototype of the electromagnetic motor which we all use in many ways today, as well as the electric relay which was fundamental both in telegraphy and then in telephony.

Henry was not an engineer but rather one of the greatest American physicists. He was conducting fundamental experiments on electricity at the same time as Faraday. In 1832 he became professor of Natural Philosophy at what was then known as the College of New Jersey and subsequently became Princeton University. In 1846 he left there to become the founding secretary of the Smithsonian Institution, a founder of the American Association for the Advancement of Science, a charter member of the National Academy of Sciences and for many years its president. He was at the Smithsonian from 1846 till 1878, during which time he pioneered in weather forecasting by receiving current meteorological data from around the United States by telegraph. This work resulted in the founding of the U.S. Weather Bureau. We shall see later how the forecasting problem was to become a major one for von Neumann and resulted in a revolution in our understanding of weather.

Both Henry and Faraday (1791–1867), whose experiments were to influence Maxwell profoundly, were essentially self-taught men, the one in the United States and the other in England. One was apprenticed to a watchmaker and the other to a bookbinder. In

Henry's case an encounter with a book on natural science, and in Faraday's attendance on some popular lectures by Sir Humphrey Davy (whom we mentioned earlier in connection with Babbage), led each to pursue careers in science with the happiest results. We have already briefly outlined Joseph Henry's career. Faraday became Davy's protégé and then his successor at the Royal Institution, where he worked for many years doing his beautiful experiments. E. N. daC. Andrade has described Faraday as "possibly the greatest experimental genius the world has known." He was also a wonderful expositor and initiated the series of lectures for children known by the charming Victorian title of Christmas Courses of Lectures Adapted to a Juvenile Auditory; he gave nineteen of these himself and they even today make first-rate reading. Finally, he also started the famous Friday evening discourses at the Royal Institution, one of which we will mention later.

Another example of this cross-linking of physics and electrical engineering may be found in Kelvin, whose work on the first trans-Atlantic cable was essential to its success. We have already cited his work on building computers. He also had an interest in the development of electrical power systems. In testimony in 1879 before a Select Committee of Parliament he suggested the use of what is described as the "Falls of Niagara" as a source of energy to drive engines which would generate power that could then be conducted hundreds of miles to where it was needed.

It would be impossible not to mention the discovery in 1845 by the German physicist Gustav Robert Kirchhoff (1824–1887) of the two laws that enable any student to express the behavior of an electrical network. These laws are implicit in Maxwell's equations.

Another remarkable instance is of another sort. In 1873 Maxwell published his world-famous *Treatise on Electricity and Magnetism* in which he predicted the existence of electromagnetic waves. Already by 1886 Heinrich Hertz (1857–1894) had showed their reality in a series of experiments that revolutionized both physics and electrical engineering. This is the work that led Guglielmo Marconi (1874–1937) to the invention in 1897 of wireless telegraphy. In fact as early as 1895 he had sent signals a distance of over a kilometer through the air.

Since it is not our purpose to attempt a history of electrical engineering, we will not go into further details on the developments of the subject. All we wished to suggest is that the symbiotic inter-relationships between various electrical engineers and physicists as well as the eagerness of the physicists of the period to make inventions resulted in electrical engineering assuming a considerable degree of mathematical sophistication at a fairly early age. It is noteworthy that this interest by scientists in engineering problems is one of the reasons why there was so much engineering development in, for example, the electronics field during World War II. In that period very many of the best physicists and engineers joined together at places such as the Radiation Laboratory, Massachusetts Institute of Technology, to invent new devices.[1]

By the time World War I ended there were a number of great centers of electrical engineering in the United States, all in need of computational assistance. In particular, we may single out Bell Telephone Laboratories, General Electric Co., and Massachusetts Institute of Technology, to mention the leaders. It should be emphasized that their need arose from the fact that the engineer could write down in mathematical form a description of the phenomenon he was discussing, but he had no comparable mathematical apparatus for analyzing this description. In other words the mere expression of a physical situation in mathematical terms does not *per se* lead to any deepening of his understanding. It only does if he can then use the machinery of mathematics to penetrate into the equations; this usually happens when these equations have already been studied or are special cases of known ones.

The power of mathematics in applications usually lies in revealing similarities or even identities between previously unknown material and well-established material. Thus, one of Kelvin's triumphs was to show that the diffusion of electric current through a cable could be reduced to Fourier's equation for the diffusion of heat:[2] this then made knowledge about heat immediately transferable to electric cables. Another instance of this sort of thing is the use of Kelvin's so-called cable equation appearing in the beautiful work of A. L. Hodgkin and A. F. Huxley on the conduction of nerve impulses along nerve fibers.[3] For this work they shared the Nobel award in 1963 in Medicine and Physiology.

Of course, such similarities are not always obvious or the known factors always well established. Then the power of mathematics to illuminate the situation diminishes. One of two things then happens: either there is a mathematical apparatus that enables the

---

[1] The reader who wishes to pursue the history of electrical engineering in greater depth may consult various histories such as the one by P. Dunsheath, *A History of Electrical Engineering* (New York, 1962).

[2] *Proc. Roy. Soc.*, 1855.

[3] Hodgkin, *The Conduction of the Nerve Impulses*, The Sherrington Lectures, VIII (Liverpool, 1964).

specifics to be handled or the mathematician is forced to admit defeat. In this latter case the computer is the tool of last resort.

In principle the computer can also be used as a heuristic tool to study a large class of related situations in the hopes of finding any regularities that may exist. This has not been done much as yet by mathematicians, but hopefully it will emerge in the near future as a tool for the mathematician to use in exploring novel, uncharted areas. As an illustration of this a rather extensive series of calculations was undertaken in the period just after World War II by Robert D. Richtmyer and Adele K. Goldstine for the Los Alamos Scientific Laboratory. When Enrico Fermi (1901–1954), the great physicist, looked over the numerical results, he perceived that the solutions depended hardly at all on a parameter that appeared in the equations. He therefore set this to zero and thereby so simplified the equations that he could write down a mathematical formula to express their solution. This led to a most significant increase in understanding of the phenomenon in question with an attendant ability to predict with minimal calculation what would occur. Unfortunately examples of this sort are very rare.

During the last years of the nineteenth and the early years of the twentieth century physicists and engineers the world over had been working at all levels of fundamentalness on electrical questions. They had succeeded in making the mathematical formulation of problems in circuit theory a routine part of the engineering curriculum, and excellent papers and texts of a mathematical nature were produced in the 1920s by such leaders in the field as Vannevar Bush of MIT, A. E. Kennelly of Harvard and MIT, J. R. Carson, G. A. Campbell, O. J. Zobel of the Bell Laboratories, and C. P. Steinmetz of General Electric Co. Of course this list is in no sense complete, but some of the names will perhaps evoke some memories. In his 1929 work on circuit analysis,[4] Bush showed how to make the tools, originally fabricated by a remarkable and singular Englishman Oliver Heaviside (1850–1927), accessible to the undergraduate study of electrical engineering. In this he was aided by Norbert Wiener (1894–1964), who wrote an elegant appendix to the book entitled "Fourier Analysis and Asymptotic Series." In this Wiener laid the mathematical groundwork for Bush's applications.

While it is not really germane to our story, it is impossible not to say a few words about Heaviside and Wiener. The former was pretty much a self-educated man and the nephew of Sir Charles

[4] *Operational Circuit Analysis* (New York, 1929).

Wheatstone. He became one of the most important figures in the field of electricity by developing a very elegant mathematical apparatus for handling equations and by his masterful use of this apparatus. He was a disciple of Maxwell and had an important effect on the scientific community. His mathematical analysis of what is called electromagnetic induction was remarkable and was so recognized by men such as Rayleigh.

Wiener is too well known to say much about here except to mention two things: first, he was a civilian and then an enlisted man in Veblen's group at Aberdeen during 1918–19; and, second, much of his finest mathematical work lay at the very fundaments of electrical engineering. Probably his presence at MIT had a great deal to do with the high degree of mathematical sophistication its engineering faculty has had. Bush, in the preface to the 1929 text cited above, says of his relationship to Wiener: "I did not know an engineer and a mathematician could have such good times together. I only wish that I could get the real vital grasp of the basic logic of mathematics that he has of the basic principles of physics."

These things having been said, we are in a position to enquire into computer advances and developments of the twentieth century. The first set of these took place at MIT and are due to Vannevar Bush, whose career we will sketch a little later.

As we mentioned earlier both Kelvin and then Michelson were much concerned with developing a device that could perform the operation of forming the integral of the product of two functions. In symbols,

$$\int_a^b f_1(x)f_2(x)dx.$$

This integral is the area under the curve $y = g(x) = f_1(x)f_2(x)$ between the lines $x = a$ and $x = b$ as shown in Fig. 9.

One reason for their interest in this integral is that when f_2 is specialized to be a sinusoidal function or pure tone of a given frequency the resulting integral is a Fourier coefficient; that is, it measures the amplitude of the overtone of the given frequency. These were basic to their harmonic analyzers.

Early in 1927, Bush, Gage, and Stewart of the Electrical Engineering Department at MIT announced the design, development, and construction of a device that not only could evaluate such integrals but also when several were interconnected could solve a variety of useful "problems in connection with electrical circuits,

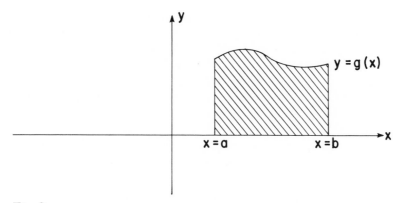

Fig. 9

continuous beams, etc., and certain problems involving integral equations." [5]

In fact this machine, like the others Bush designed, was an analog one; its two fundamental operations were to form continuously, as a variable x changes,

$$(1) \qquad\qquad F(x) = f_1(x)f_2(x),$$

$$(2) \qquad\qquad G(x) = \int_a^x f_1(x)f_2(x)dx.$$

Bush and his colleagues realized, as had Kelvin before them, that by what they call "back-coupling" equations such as

$$g(x) = \int_a^x f(x)g(x)dx$$

may be solved for g once f is known. This last equation is equivalent to a very simple differential equation and represents a first step on the way to solving the numerical problems of electrical engineering.

The basic device for integrating used by Bush to carry out operation (2) above is a variant on the watt-hour meter found in virtually all our houses to enable the power company to measure how much power we have consumed. This device is exactly what was needed to effect the second operation since it does evaluate the integral of a product, the two factors being current and voltage as a function of time. Bush and his colleagues saw that the meter with some modifications would still operate even if the current and voltage

[5] V. Bush, F. D. Gage, and H. R. Stewart, "A Continuous Integraph," *Journal of the Franklin Institute*, vol. 203 (1927), pp. 63–84.

were replaced by arbitrary functions of time that could be introduced as voltages which were proportioned to the heights of curves.

The device used to perform operation (1) above – multiplication
of two functions – was much less sophisticated. It was a fairly
simple mechanical linkage which permitted the multiplication of
two lengths and depended on an elementary geometric theorem.

We need not spend more time on this machine since it forms
only the first step in the direction Bush and his associates were
going, which clearly was to produce machines that could solve the
calculational problems of electrical engineering. He said in 1931:

> The status of physics and engineering at the present time is
> peculiarly favorable to a development of this sort. Electrical
> engineering, for example, having dealt with substantially linear
> networks throughout the greater part of its history, is now rapidly
> introducing into these methods elements the non-linearity of
> which is their salient feature, and is baffled by the mathematics
> thus presented and requiring solution. Mathematical physicists
> are continually being hampered by the complexity rather than
> the profundity of the equations they employ; and here also even
> a numerical solution or two would often be a relief.
>
> Not any one machine, nor even any one program of develop-
> ment can meet these needs. It was a long hard road from the
> adding machines of Pascal to the perforated card accounting
> machines of the present day. There must be much of labor and
> many struggles before the full ideal of Leibnitz can be consum-
> mated.[6]

MIT was not the only place where analog computers were being
built. There were several such projects. Typical of this activity
was the construction of several harmonic analyzers and synthe-
sizers; one was built in 1916 at what was then called the Case School
of Applied Science and another at the Riverbank Laboratories,
Geneva, Illinois.[7] Recall that Case was the school where Michelson
did some of his greatest work on the velocity of light.

In general Bush and his associates had a three-pronged program
in mind: the development of a so-called network analyzer which
would enable the engineer to solve "complicated simultaneous

---

[6] V. Bush, "The Differential Analyzer, A New Machine for Solving Differential
Equations," *Journal of the Franklin Institute*, vol. 212 (1931), pp. 447–488.

[7] The interested reader may find further references in F. W. Kranz, "A Mechanical
Synthesizer and Analyzer," *Journal of the Franklin Institute*, vol. 204 (1927),
pp. 245–262.

algebraic equations as they occur for example in the treatment of
modern power networks, by means of alternating current measurements made on an electrical replica of a power system; [8] the development of an optical means for evaluating a general class of integrals following up an ingenious suggestion of Wiener; and the
development of a device for solving ordinary differential equations.
This program was carried out in the period from 1927 through 1942
and was quite successful. Perhaps its only defect was that it committed — at least intellectually — the engineers at MIT to the analog
point of view to the exclusion of the digital one for many years.
This was of course subconscious but nonetheless present in the
minds of Bush's colleagues. It required the efforts of a subsequent
generation centered around Jay W. Forrester to break with this
tradition and push MIT into the modern era of the electronic digital
computer.

Let us return to the past. We have already briefly mentioned
Bush's first assay into the computing field with the device called a
continuous integraph and shown how this instrument could solve
certain first-order differential equations. However, the differential
equation most often met in physics and electrical engineering is
not of the first-order but rather of the second. The underlying
reason for this can perhaps be explained by a cursory discussion
of Newton's famous laws of motion for mechanical systems. These
laws typically express the acceleration of a particle in terms of the
applied forces, which depend on position and velocity. Now it is
well known that velocity is the rate of change of position and
acceleration the rate of change of velocity. Thus acceleration is a
second rate of change of position and this gives rise to equations
which are of the form

$$\frac{d^2x}{dt^2} = f\left(x, \frac{dx}{dt}\right);$$

these are said to be of the second order. Why Nature behaves in
this manner is a fascinating topic.[9]

Be that as it may, the continuous integraph was inadequate as a
tool for the engineer. Therefore Bush and Harold Hazen announced
in November of 1927 a revised instrument that could solve a single

[8] Bush, *op. cit.*, p. 448.

[9] The interested reader may wish to read an elegant and profound discussion of
the topic by Eugene Wigner entitled "The Unreasonable Effectiveness of Mathematics in the Natural Sciences," *Communications on Pure and Applied Mathematics*, vol. XIII (1960), pp. 1–14.

second-order differential equation.[10] Hazen was destined later to become head of the Electrical Engineering Department and then Dean of the Graduate School at the Massachusetts Institute of Technology until his retirement in 1967. He also headed the so-called Division 7 of the National Defense Research Committee, the civilian agency of the U.S. Government for the research and development of new weapons in World War II. Its history was very briefly this: In June 1940 the National Defense Research Committee was formed, and one year later as a result of an executive reorganization the Office of Scientific Research and Development was created as an agency of the Office of Emergency Management for the purpose of "coordinating the efforts of scientists and technical men in connection with many phases of the war effort, but it was also given the definite charge of pursuing aggressively the work that had already been started by NDRC; and for this purpose NDRC was incorporated in its organization." [11] Bush chaired NDRC from its inception, and then from 1941 to 1946 he was Director of OSRD and James Conant (1893–   ), president of Harvard, became chairman of NDRC. Conant was assisted by Karl T. Compton, president of the Massachusetts Institute of Technology. There were also various governmental members.

The OSRD, in addition to Bush as director, had an advisory council made up of Conant; Newton Richards, who headed the Committee on Medical Research and was also the vice president for medical affairs of the University of Pennsylvania; and Jerome C. Hunsaker, who headed the National Advisory Committee for Aeronautics and also headed the aeronautical engineering department at MIT. It also had governmental membership.

The machine that Bush and Hazen now produced was an interesting evolutionary step in the chain leading to the 1930 machine which could really handle the calculational problems of the electrical engineer in the 1930–1940 decade. Remember, the first Bush machine could only solve a first-order equation and its basic component was a modified Elihu Thomson watt-hour meter. This is a device that has two inputs both of which are electrical and an output that is a mechanical rotation. It was at that time technologically impossible, or extremely difficult, to convert this rotation into an electrical signal — more on this later, it plays a role in Bush's 1942 analyzer. Bush and Hazen therefore introduced a second integrat-

[10] V. Bush and H. L. Hazen, "Integraph Solution of Differential Equations," *Journal of the Franklin Institute*, vol. 204 (1927), pp. 575–615.

[11] Bush, *Endless Horizons* (Washington, 1946), pp. 110–117.

ing component: the integrator of James Thomson, which we have discussed earlier. They did this so that the output of the Elihu Thomson meter could serve as the input to a James Thomson integrator, which requires a mechanical rotation for this purpose.

Bush and Hazen give two interesting examples in their paper showing the power of their new instrument for handling difficult problems, one in electrical engineering, the other in mechanics. In both cases the essential character of their technique of feeding the output of the meter into the Kelvin integrator is displayed. They were able to obtain results that were valid "within 2 or 3 percent, and except when unfavorable gear ratios are used accuracy within 1 percent." The basic difficulty with the machine was perhaps its extreme inelegance on the one hand and on the other its inability to solve intercoupled systems of second-order equations. The inelegance arises from the need to have two different types of integrating devices, one for each integration; this evidently fixes the scope of the machine in a rather artificial and arbitrary way. It is therefore not surprising for us to see Bush busily engaged in a further development. Indeed by 1931 he was able to publish a paper describing a much more elegant machine which was named a differential analyzer.[12] It was totally mechanical in character, the watt-hour meter having been dropped. All devices were actuated by mechanical rotations and produced rotations for outputs. Interconnections were achieved by rotating shafts and gears. In 1942 MIT people were once again to find an electrical solution to part of the problem.

The new machine was in many respects a realization of what Kelvin would have built if he could have solved the problem of producing an amplifier for the mechanical rotations of the output shafts of his brother's integrators. As we mentioned earlier an integrator is a device made up of a disk turned by an electric motor upon which sits a wheel in a plane at right angles to the disk (Fig. 10). Motion of the disk causes the wheel to turn, but unless the wheel is pressed very hard against the disk the rotating wheel is not able to turn a heavy shaft.

The essential thing lacking to Kelvin and seized upon by Bush was a quite remarkable invention by C. W. Niemann, an engineer with the Bethlehem Steel Co.[13] This torque amplifier of Niemann's

---

[12] V. Bush, "The Differential Analyzer, A New Machine for Solving Differential Equations," *Journal of the Franklin Institute*, vol. 212 (1931), pp. 447–488. The machine was actually put into use in 1930.

[13] C. W. Niemann, "Bethlehem Torque Amplifier," *American Machinist*, vol. 66 (1927), pp. 895–897.

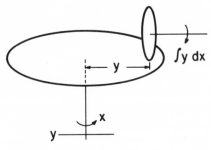

Fig. 10

was a beautiful solution to the problem that had plagued both
Kelvin and Bush. It has three shafts in it, which we may term the
drive, the control, and the work shafts. Of these, the first is driven
by an outside source such as an electric motor which simply turns
at a uniform rate; the second is, in our case, the output shaft attached to the little wheel sitting on the disk; and the third is the
heavy shaft or bus which must be turned. The whole point in
Niemann's device, of course, is by some ingenious trick to use the
power in the drive shaft to turn the work shaft under the guidance
and direction of the control shaft. In Niemann's words: "The work
shaft is directly coupled to the work to be done, as for instance the
elevating or training gears of a gun, a ship's rudder, or the steering
wheel of an automobile. The control shaft can be freely revolved
in either direction, with only a small amount of effort; the work
shaft maintains at all times its angular synchronism with the control shaft, and in addition exerts a heavy torque to overcome outside
resistance."

It is beside the point for us to try to describe the mechanical details beyond saying that it works somewhat analogously to a ship's
capstan. In such devices a motor rotates a shaft and when a sailor
desires to lift a heavy weight he passes a loop of rope attached to
the weight around the rotating shaft and takes hold of the free end.
By a slight force he tightens up his end and causes the rotating shaft
to lift the load; the more he tightens up the more power is exerted
by the motor. In our analogy the control is the sailor's end of the
rope. Niemann devised a very elegant variant on this principle to
make his amplifier.

Using Niemann's amplifier Bush now built a machine using
exclusively Thomson's integrators—in this machine there were
six of them. He also built in a multiplier, three so-called input

tables by means of which human operators could feed in arbitrary functions, an output table which produced graphical displays of the results of a calculation, as well as devices for adding the motions of shafts, and gear boxes by means of which constant factors could be introduced. All interconnections were made by means of long shafts whose rotations were proportioned to certain variables. The integrators and other devices could be interconnected by varying the manner of tying together the shafts. The whole instrument resembled nothing so much as a child's huge Erector set model. This analogy is really not so bad. Douglas Hartree, an English physicist whom we shall meet again, actually constructed such a device "so far as possible from standard Meccano parts, a model to illustrate the principles of the machine originally done more for amusement than for serious purpose." He further reports that it "was successful beyond our expectations. It gave an accuracy of the order of 2% and could be, and was, used for serious quantitative work for which the accuracy was adequate." [14] This description was given at one of the Friday evening discourses at the Royal Institution initiated by Faraday.

After the success of this machine Hartree, who was then a professor of physics at the University of Manchester, persuaded a public-spirited Midland businessman, Sir Robert McDougall, to give enough money to the university to have Metropolitan-Vickers Electrical Co., Ltd. build a large machine based on Bush's instrument and with Bush as consultant. This was followed by other copies in England, including one at the Mathematical Laboratory of the University of Cambridge built under the leadership of Sir John E. Lennard-Jones, professor of theoretical chemistry, by a number of young men who were to become leaders in the British digital computing field in the 1940s. One of them, F. C. Williams, was in 1939 to carry out a suggestion of the Nobel Laureate in Physics, P. M. S. Blackett, and build an automatic curve follower to replace the human operators at the input tables of the differential analyzer. This device was different in principle from a very

[14] D. R. Hartree, "The Bush Differential Analyzer and Its Applications," *Nature*, vol. 146 (1940), pp. 319–323; and Hartree, *Calculating Instruments and Machines* (Urbana, 1949). In both works, especially the latter, there are good bibliographies. The reader may also wish to consult Bush, "Instrumental Analysis," *Bull. Amer. Math. Soc.*, vol. 42 (1936), pp. 649–669, or C. Shannon, "Mathematical Theory of the Differential Analyzer," *Journal of Mathematics and Physics*, vol. 20 (1941), pp. 337–354.

interesting one designed and built in 1936 for the MIT instrument by Hazen, J. J. Jaeger, and G. S. Brown.[15]

We have singled out Brown, Hartree, Wilkes, and Williams because of their subsequent roles. It was under Brown's direction as head of the Servo Mechanisms Laboratory at MIT that Forrester was to make his great contributions to the field of electronic digital computers. Brown was later to be Dean of Engineering and to continue to play a key role in directing MIT's policies on computing. Hartree we will meet on a number of occasions as the man who started Great Britain off in the field; Wilkes was to be the man who, under Hartree's general impetus, built the first stored program computer modelled on an American design; and Williams was to be the developer of a type of storage device that was to make a whole generation of electronic computers possible.

Once again, the reader must forgive me for jumping ahead of myself. There are natural sequences of events, and it is sometimes difficult to drop a subject once it is started. But let us go back. Shortly after the inauguration of the Bush analyzer in 1930 both the Ballistic Research Laboratory at Aberdeen and the Moore School of Electrical Engineering of the University of Pennsylvania realized the fundamental significance of this device for their tasks: in the former case to solve ballistic differential equations, and in the latter, equations of electrical engineering. Both groups therefore independently approached Bush to see about getting copies of his machine. He was very willing and helpful, and in 1933 a design program was undertaken jointly by the Ordnance Department of the U.S. Army, the Moore School of the University of Pennsylvania, and the Massachusetts Institute of Technology. As a result of this program one design was agreed upon for both Aberdeen and the Moore School. The one for the Ballistic Research Laboratory was built under the supervision of Bush and his associates commercially, while the one for the Moore School was constructed at the university under a Civil Works Administration grant. This was one of the agencies created by President Roosevelt to provide a stimulus to the U.S. economy during the Great Depression of the 1930's by providing employment for skilled workers. The two machines were completed in 1935 and were to form the nexus between the Ord-

[15] J. E. Lennard-Jones, M. V. Wilkes, and J. B. Bratt, *Proc. Cambridge Phil. Soc.*, vol. 35 (1939), p. 485; P. M. S. Blackett and F. C. Williams, *ibid.*, p. 494; H. L. Hazen, J. J. Jaeger, and G. S. Brown, *Review of Scientific Instruments*, vol. 7 (1936), p. 353.

nance Department and the University of Pennsylvania that was to be so significant, not only to both of them but to the world, just a decade later.

Although it is not particularly relevant, it should be stated that differential analyzers were apparently built in Germany and Russia in the late 1930s, and the Moore School collaborated with the General Electric Co. to design one for the Russians. Thus, by 1940 Hartree could say: "There are now, so far as I know, altogether seven or eight full-size machines in operation in the world. There are also, in Great Britain, several smaller and less accurate model versions of the machine. . . ."[16]

We must also mention Sir Harrie S. W. Massey (1908–), who collaborated with Sir Nevill F. Mott (1905–) on the theory of atomic collisions in a well-known book on this subject in 1933. In 1938, following the lead of Hartree, Massey and his associates in the Physics Department at Queen's University, Belfast, built a small differential analyzer for the solution of a variety of physics problems.[17] In the paper they state that they will mainly use the machine for the solution of the so-called Schrödinger wave equation for a spherically symmetrical system. This is a simple differential equation of second order which appears in quantum mechanics.

In Bush's master program for computing, one phase was to be a so-called network analyzer. Such a machine is one that can be used to determine the characteristic behavior of a power system under a variety of conditions. In 1930, Hazen and his colleagues wrote: "The advantages of experimental methods over lengthy and usually impracticable mathematical calculations for the determination of these quantities have long been recognized." (It is interesting to remark in passing that in 1968 a group at IBM was to make a very extensive analysis for various power companies and the Edison Institute of the great black-out using precisely the mathematical calculations that were then possible as a result of the electronic computer.) Several network analyzers were built in miniature form before the one we now wish to mention was designed. After these preliminary designs the groups at the General Electric Co., in Schenectady, N.Y., and at MIT who had worked on such things

[16] Hartree, "The Bush Differential Analyzer and Its Applications," *Nature*, vol. 146 (1940).

[17] H. S. W. Massey, J. Wylie, R. A. Buckingham, and R. Sullivan, "A Small Scale Differential Analyser—Its Construction and Operation," *Proc. Royal Irish Acad.*, vol. XLV (1938), pp. 1–21.

collaborated to develop and construct the MIT Network Analyzer.[18]

The third area of computer development at MIT was to be in the development of what was called the Cinema Integraph. It too was aimed at evaluating by analog means integrals of the form

$$F(x) = \int_a^b f(x \pm y)g(y)dy$$

where g and f are known functions and x is a parameter. The basic idea for the machine was, as mentioned before, due to Wiener. It was first executed by T. S. Gray but "its speed of operation and accuracy, while good, were not all that could be desired." [19] The new machine was therefore designed and built using the same basic idea—which was an optical one that we will not pursue further since the interested reader can find the details in Hazen and Brown and since the utility of devices of this sort, while important at the time, soon became lessened with the advent of electronic, digital computers.

Finally in 1942 Bush and S. H. Caldwell were to build still another differential analyzer. In this one they attacked the problem of speeding up the time needed to *program* the machine for a new problem. In the earlier analyzers it was a major—often one- or two-day—chore to make all the physical interconnections required to describe uniquely and exactly a new problem. By means of pre-prepared punched paper tapes they were able to effect a major speed-up. "For the largest problem so far put on the machine, the entire assembly was done in 3 to 5 minutes—less than the time required for a single solution." [20] To make this possible the shafts,

[18] H. L. Hazen, O. R. Schurig, and M. F. Gardner, "The MIT Network Analyzer, Design and Application to Power System Problems," *Quarterly Transactions of the Amer. Inst. of Electrical Engineers,* vol. 49 (1930), pp. 1102–1114. It is noteworthy that Schurig was to be at the Ballistic Research Laboratory during World War II. Gardner was for many years a very distinguished member of the Electrical Engineering Department at MIT. From 1930 on he taught the course originally given by Bush on Operational Circuit Analysis and brought it up to date with the introduction of the so-called Laplace transform. This replaced the so-called Cauchy-Heaviside and Fourier transforms as a prime mathematical tool of the electrical engineer.

[19] H. L. Hazen and G. S. Brown, "The Cinema Integraph, A Machine for Evaluating a Parametric Product Integral, with an Appendix by W. R. Hedeman, Jr.," *Journal of the Franklin Institute,* vol. 230 (1940), pp. 19–44 and 183–205. The instrument described in this paper "was developed in part by G. S. Brown in partial fulfilment of the requirements for a thesis. The appendix contains numerical examples taken from a thesis by W. R. Hedeman, Jr."

[20] V. Bush and S. H. Caldwell, "A New Type of Differential Analyzer," *Journal of the Franklin Institute,* vol. 240 (1945), pp. 255–326.

which in earlier analyzers remembered and transmitted information between units, were replaced by electrical connections.

This machine worked successfully but was soon to be displaced by electronic digital computers. It should however be remarked that a substantial industry has grown up which makes very ingenious and rapid electrical analog machines that fill a role in the computing needs of the world. We do not have a place in our account for these machines, which are in many ways the logical successors to the 1942 MIT machine.

Before closing our discussion of analog computers once and for all, we need to mention the mathematical accomplishments and attempts of Douglas Hartree, whom we have already met. He was the physicist, first at Manchester and then at Cambridge, who was to lead Great Britain into the electronic era by his enthusiasm and senior position as Plummer Professor of Mathematical Physics at Cambridge and as the English pioneer in the use of the differential analyzer. To describe Hartree perhaps we can quote one of the important figures in quantum mechanics, Egil A. Hylleraas, a Norwegian physicist (1898–1965), who described him in these words: "This was V. Fock, the second father of the famous Hartree-Fock method, whose range is far beyond the two-electron problem. The former one, Douglas R. Hartree, I shall never forget, as being one of the kindest persons I ever met, and whose premature death I sincerely regretted. The words he used of his own father, the other Hartree, whose name appears on some joint publications, that he was the most wonderful artist in numerical calculations he ever knew, may well be turned toward himself." [21]

One of Hartree's great accomplishments was to attempt to solve so-called partial differential equations with the help of the differential analyzer. He was probably the first man to do serious calculations of this sort and thereby to set the stage for what was to happen with the advent of the electronic computer. In order to make a little clearer what he did, we need to say a few words about total and partial differential equations.

---

[21] E. A. Hylleraas, "Reminiscences from Early Quantum Mechanics of Two-Electron Atoms," *Review of Modern Physics*, vol. 35 (1963), pp. 421–431. This paper was one of a series of articles read at an International Symposium on Atomic and Molecular Quantum Mechanics in honor of Professor Hylleraas and arranged by the University of Florida at Sanibel Island, 14–19 January 1963.

The electrical engineers and the ballisticians were almost without exception primarily concerned with descriptions of these sorts: given an electrical network of whatever complexity, describe the current and voltage as functions of time at one or several places in the circuit; given a fuse-shell-gun combination, describe as a function of time the position and velocity of the projectile when the gun is fired at various elevations and initial velocities. These descriptions can be achieved by the numerical treatment of systems of so-called *total* or *ordinary* differential equations, and as we mentioned earlier there are techniques for doing this that are well known and go back to John Couch Adams, Euler, Moulton, and others.

There is however another type of physical situation whose mathematical treatment is much more complex. First, an illustration: Suppose we have a rod one end of which is in an ice bath and the other in a fire. We now ask how the temperature of the rod varies as a function both of each point on the rod and of the time; i.e., at each instant of time and each place on the rod we desire to know the temperature. In the previous example there were one or more dependent variables — current and voltage for circuits, position and velocity for projectiles — but only one independent variable, time. In the heat conductive case, there is one dependent variable, temperature, but several independent ones, position along the rod and temperature. The mathematical formulation for problems of the latter type leads to what are called *partial* differential equations. The heat conduction case is one first given analysis in Fourier's brilliant treatise, *Théorie analytique de la chaleur,* mentioned above (p. 46).

There are however many other examples. One is the study of the pressure, density, and temperature of the air at each point on and in the vicinity of a wing of an airplane. Such questions also give rise to partial differential equations. Still another, and our last, is provided by quantum mechanics. Here even simple situations can lead to rather complicated mathematical formulations. This is due to the fact that a particle can no longer be said to be at a fixed point but only that, with certain probabilities, it is at certain points.

We may conclude with the understanding that the ballisticians, engineers, and physicists who made use of differential analyzers for the solutions of systems of total differential equations were attacking only a part of the total spectrum of important and even urgent problems. It was to be Hartree's role to try to use the differential analyzer for handling the other part. He made this attempt with a group of students and colleagues at the University of Manchester before and during the war. The results of this ambitious but not entirely successful venture are interestingly summarized by Hartree in his lecture series at the University of Illinois in 1949.[22]

As we shall see later, Illinois was one of the leaders in the then just emerging electronic digital field; this leadership was provided by Abraham H. Taub, a mathematician about whom more later; Ralph Meagher, an engineer whom we will meet again; and Louis Ridenour, a physicist who was both Dean of the Graduate School at Illinois and prominent in the scientific councils of the U.S. Air Force until his untimely death.

There are broadly speaking three types of partial differential equations, called parabolic, elliptic, and hyperbolic. Of these, the heat conduction problem typifies the former type; the electrical potential on a surface, the second; and the aerodynamical problem mentioned before, the third. All three have markedly different mathematical behaviors since they have such diverse physical ones. Hartree had some success with the first type, recognized his inability to deal with the second by means of a differential analyzer, and mentions that "a few attempts have been made to apply it to equations of 'hyperbolic' type, with only partial success." [23]

Hartree's difficulty in part lay in his apparent lack of awareness of a totally new mathematical phenomenon arising in the study of partial differential equations from a numerical point of view. This phenomenon was first discovered and understood by Richard Courant, Kurt Friedrichs, and Hans Lewy in a now classic paper written in 1928.[24] In a later section we shall attempt a brief discussion of what is now known as numerical analysis, the study that concerns itself with numerical phenomena. Now however we may close this very short look at the work on partial differential equations done by Hartree and his colleagues at Manchester and by Lennard-Jones and his at Cambridge. While these results were not too important in an absolute sense, they were in another: they represented the first concerted attack on extremely complex physical and engineering phenomena by computational rather than by experimental means. The men who did this deserve to be mentioned

---

[22] Hartree, *Calculating Instruments and Machines*, chap. 3.

[23] *Calculating Instruments*, pp. 33–34.

[24] Courant, Friedrichs, Lewy, "Über die partiellen Differenzen-gleichungen der Physik," *Math. Ann.*, vol. 100 (1928–29), pp. 32–74.

here and remembered by numerical analysts for their pioneering spirit.

We may note, in closing this section, that it is in a curious sense interesting that so few British, or indeed any other nationality, physicists or applied mathematicians worked on computers in the period between Kelvin and Hartree. This span of about 50 years was a particularly brilliant one for British physics. Sir Joseph J. Thomson (1856–1940) of the Cavendish Laboratory and Trinity College was one of the world's great physicists. He was one of the men who led physics into the study of sub-atomic phenomena by his beautiful experiments. In his earlier days he also wrote a book on electricity and magnetism that is often called the third volume of Maxwell. He became the Cavendish Professor in 1894 and stepped down in 1919, when he was succeeded by one of his many brilliant students, Sir Ernest Rutherford (1871–1937), who was himself a giant among giants. There were many other great names, but none worked on computers.

Apparently the need to specialize more had already by the beginning of this century begun to limit the scope of even the most brilliant men. The age of universalists was disappearing. Coupled to this delimiting of areas of work came another phenomenon which perhaps turned the great physicists with few exceptions away from computer developments. This was the entire subject of modern physics starting with J. J. Thomson's work and including that of Max Planck, Albert Einstein, Niels Bohr, Arnold Sommerfeld, as well as Paul A. M. Dirac, Erwin Schrödinger, Max Born, Werner Heisenberg, and Pascual Jordan. The directions followed by these men were away from the highly quantitative ones of Kelvin; they were trying to understand and to explain in qualitative terms the nature of the microcosmos. This led them generally away from the quantitative and hence from computers. An exception was the calculation of Egil Hylleraas, who carried out a great calculation by hand and thereby showed that the new quantum mechanics led to numerical results in accord with experiment.[25] This calculation of the so-called ground state of helium yielded a result that was slightly lower than the experimental one and subsequent calculations using modern electronic machines have given refinements to his result. However, in general few detailed numerical calculations were carried out before Hartree.

[25] E. A. Hylleraas, *Zeits. f. Physik*, vol. 65 (1930), p. 209.

As early as 1928 Hartree had proposed what is sometimes called the Hartree self-consistent field method, or the Hartree-Fock method, and had done extensive calculations.[26] Another approach, known as the Thomas-Fermi method, to the same problem was numerically undertaken on the differential analyzer by Bush and Caldwell.[27]

A word more about Egil Hylleraas: his work was of the greatest importance to the newly developing quantum or wave mechanics. He says of it:

> As is well known, wave mechanics at once reproduced all correct results obtainable from Bohr's theory, and the use of its much more convenient perturbation theory added considerably more, however, not always in the strict numerical sense. Now, particularly by Max Born, it was agreed that the simplest crucial test of the correctness of wave mechanics in general was to be found in its application to the helium atom — in particular to the ground state.
>
> As is well known from Sommerfeld's exposition of the matter in his *Atombau und Spektrallinien,* the Bohr theory, applying a definitely inconsistent *ad hoc* model of the atom with its two electrons in strictly opposite positions with respect to the nucleus, led to a numerical value of about 28 eV for the ionization energy of the first electron. On the other hand, a simple perturbation treatment of the Schrödinger equation . . . led to a much lower value of about 20.3 eV. The true value of 24.46 eV, as known from spectroscopic measurement, was about in between. Hence there was a broad gap of about 4 eV to be filled up.[28]

---

[26] D. R. Hartree, "The Wave Mechanics of an Atom with a Non-Coulomb Field," *Proc. Camb. Phil. Soc.,* vol. 24 (1928), part I, pp. 89–110; part II, pp. 111–132; part III, pp. 426–437; see also J. C. Slater, *Physical Review,* vol. 35 (1930), and V. Fock, "Näherungsmethode zur Lösung des quantenmechanischen Mehrkörper problem," *Zeits. f. Physik,* vol. 61 (1930), pp. 126–148.

[27] V. Bush & S. H. Caldwell, *Physics Review,* vol. 38 (1931), p. 1898; L. H. Thomas *Proc. Camb. Phil. Soc.,* vol. 23 (1926), p. 542; E. Fermi, "Eine statistische Methode zur Bestimmung einiger Eigenschaften der Atoms und ihr Anwendung auf die Theorie der periodischen Systems der Elemente," *Zeits. f. Physik,* vol. 48 (1928), pp. 73–79, and "Statistische Berechnung der Rydbergkorrektionen der s-Terme," *ibid.,* vol. 49 (1928), pp. 550–554. The interested reader may wish to consult L. I. Schiff, *Quantum Mechanics* (New York, 1949), p. 273, or E. C. Kemble; *The Fundamentals of Quantum Mechanics* (New York, 1937), p. 777.

[28] Hylleraas, "Reminiscences," p. 425. The symbol eV stands for electron volt, a standard unit used to measure the amount of energy gained by an electron in going from a given potential to another that is one volt higher.

Hylleraas goes on to say:

A systematic attack on the ground-state problem of the helium
atom had been planned by Max Born . . . since Born himself had
no preference for numerical work.

When Professor Born first suggested to me that — as he said — I
was the right one to go on with the helium problem, I felt of
course greatly flattered. . . .

One thing which I noticed fairly soon was that solutions must
exist which depend on only three coordinates, instead of the full
number of six. . . . When confronted with this really useful simplification Born asked: "What does that mean? Let us consult
Wigner!"

Eugene Wigner was already at that time a central person
among the young Göttingen pioneers. He was suspected to be
familiar with some kind of black magic, called group theory.
This was several years before the culmination of the so-called
"Gruppenpest," when every paper on wave mechanics in order
to be taken seriously had to start by stating the "group character"
of the subject. . . .

In connection with these mathematical aspects of the theory, it
might be just to mention the valuable support for the whole
Göttingen school provided by the two famous Göttingen mathematicians, Richard Courant and David Hilbert, and occasionally
also Hermann Weyl, a visitor from Zurich. The excellent book by
Courant and Hilbert, *Methoden der mathematischen Physik,*
may be known to most physicists working in the field of quantum
mechanics and, in those early days, it was of course more badly
needed than it ever has been since.

Courant was an excellent lecturer, playing on his audience like
an instrument. Hilbert was quite different. As a professor emeritus
and "Geheimrat," his lectures were given only accidentally and
voluntarily, and out of pure interest for the new developments
in physics. He never was in a hurry; on the contrary, he rather
seemed to like to taste repeatedly his own sentences. He was
extremely popular, and it was a real pleasure to listen to his
mild voice and look into his white-bearded gentle face. To him,
the inventor of Hilbert space, the pathways leading from matrix
to wave mechanics, and *vice versa,* were of course no secret, and
this he expressed in the funny way of shaking his head, saying,
"Die Nobel-preise liegen ja auf der Strasse." [29]

[29] *Ibid.,* pp. 425–426.

This gentle Norwegian goes on to describe his calculation in these words:

> In this room was installed a 10 x 10 automatic electric desk computer, an excellent Mercedes Euclid, but strong and big as a modern electronic computer and, hence, with the faculty of giving out not only acoustic waves, but even respectable shock waves. Now we all of us know that it does not work very well just to do one's job. In order to gain fame, it may be as important to make some noise about what you are doing, and in this respect the Mercedes Euclid helped me quite excellently. . . .
>
> The end result of my calculations was a ground-state energy of the helium atom . . . of 24.35 eV which was greatly admired and thought of almost as a proof of the validity of wave mechanics, also, in the strict numerical sense. The truth about it, however, was in fact that its deviation from the experimental value by an amount of one-tenth of an electron volt was . . . a quite substantial quantity and might as well have been taken to be a disproof. . . .
>
> The discrepancy continued to bother me for a long time. . . .
>
> This change of coordinates had, to my astonishment and to my great satisfaction as well, almost the effect of a miracle. . . . The troublesome discrepancy already told of, disappeared entirely on the electronic volt scale, although still considerable in spectroscopic units. . . . The rest became a matter of tedious and accurate calculations as improved from time to time by many authors, also myself, and, in particular, Chandrasekhar and Herzberg and their various co-workers. . . . In the most recent time we have to point to the indeed wonderful calculations as performed by Kinoshita and Pekeris.[30]

[30] *Ibid.*, pp. 426–427. Kinoshita and Pekeris, using modern computers, made several refinements of Hylleraas's work.

Chapter 11　　　　　　　　　　Adaptation to

　　　　　　　　　　　　　　　　　Scientific Needs

We have now finished with our main discussion of analog computers and wish to turn back to digital ones. In doing this we have purposely omitted mention of a variety of analog machines of great ingenuity such as Mallock's machine for solving systems of linear equations. The interested reader is again referred to the works of Murray and d'Ocagne.[1]

It is perhaps worth remarking that a Spaniard, Torres Quevedo (1852–1936) proposed in 1893 an electromechanical solution to the ideas of Babbage. His machine may be viewed as an extremely tentative step in the chain of development of digital devices. It is described in a note in a popular French journal.[2]

It remained however for Howard Aiken of Harvard University and Clair D. Lake of IBM, on the one hand, and George R. Stibitz and his associates, Ernest G. Andrews and Samuel B. Williams of the Bell Telephone Laboratories, on the other, to carry out Babbage's program of a century before. Before discussing these developments, we need first to return to our earlier tale of astronomy in general and celestial mechanics in particular.

We saw that as far back as Leibniz the astronomers had a substantial interest in and need for accurate tables. Moreover, we saw how this need runs as a motif through Babbage's work; in fact, it continues right through modern times. The analog computers had little appeal for the astronomer, as we recall. His need was for error-free tables with an accuracy that transcended that attainable by machines such as the differential analyzer. What has he been doing all the while?

We mentioned that the Scheutz machine was sold in 1856 to an observatory in Albany, New York, and that it was there until 1924. There is some confusion over its use in Albany. Archibald states: "There it remained unused until sold in 1924"[3] whereas d'Ocagne

---

[1] F. J. Murray, *The Theory of Mathematical Machines*, and M. d'Ocagne, *Le Calcul simplifié*.

[2] *La Nature*, June 1914, p. 56.

[3] R. C. Archibald, "P. G. Scheutz, Publicist, Author, Scientific Mechanician . . . ," p. 240 (and see above, pp. 15–16).

says: "où elle a été utilisée pour le calcul de tables de logarithmes, de sinus et de logarithmes-sinus." [4] Archibald takes very sharp issue with d'Ocagne and also with H. H. Aiken, who apparently got his information from d'Ocagne's book on this matter. From the evidence there is every reason to believe that Archibald was correct. However, from our point of view it does not matter, since the Scheutz machine did in fact, somewhere, calculate various mathematical tables and print them mechanically. We have also mentioned Wiberg's improvement on that machine.

It was not until 1928 that the Hollerith tabulating machines were to find extensive use in the construction of astronomical tables, and the name associated with this development was Leslie John Comrie (1893–1950), a New Zealander educated at Cambridge in mathematics and astronomy. He taught briefly at Swarthmore College and Northwestern in the 1920s before becoming first Deputy Superintendent of H. M. Nautical Almanac Office of the Royal Naval College in 1926. Later, from 1930 through 1936, he was the Superintendent. He was in his day a superb calculator knowing absolutely all there was on the use of the Brunsviga, the Hollerith, and the National machines for astronomical computation. By 1928 we find Comrie using the Hollerith tabulator to make tables of the moon's position.[5] This work of Comrie's is most important. It marks a transition point: punch card equipment originally intended for statistical and business purposes is now found to be valuable, even essential, for advanced scientific purposes as well. This is the next giant step forward.

Perhaps we should pick up the very brief account of lunar theory we gave earlier [6] and bring it up to date. As we mentioned there Charles-Eugene Delaunay published a theory of the moon's position of "great mathematical elegance and carried out to a very high degree of approximation." [7] It was this same great astronomer who presented "avec éloge, à l'Académie des Sciences de Paris" the machine that Wiberg had developed as an improvement on the Scheutz machine. This machine was examined with great care by a commission of the Academy consisting of Claude Mathieu, a famous professor of astronomy at the Collège de France; Michel

[4] d'Ocagne, op. cit., p. 76.

[5] L. J. Comrie, "On the Construction of Tables by Interpolation," Royal Astronomical Society, Monthly Notices, vol. 88 (April 1928), pp. 506–623, and "The Application of the Hollerith Tabulating Machine to Brown's Tables of the Moon," ibid., vol. 92 (May 1932), pp. 694–707.

[6] Above, pp. 27ff.

[7] Moulton, Celestial Mechanics, p. 364.

Chasles, a great mathematician and a founder of the Faculty of Science at the Sorbonne, as well as a Fellow of the Royal Society and president of the French Academy; and Delaunay. Their report is very detailed and was published by the Academy of Sciences.[8] The device was used to produce tables of logarithms and of interest. The former are very extensive and a copy of them, one of the two known to be extant, is in the library of Brown University as well as a copy of Scheutz' tables "dedicated to Charles Babbage by his sincere admirers, George and Edward Scheutz." [9]

It was not until 1878 that the next step forward was taken when George W. Hill (1838–1914), the superb American astronomer and mathematician, published his lunar theory "based on new conceptions, and developed by new mathematical methods." He was for very many years — 1859 until 1892 — an assistant in the office of the *American Ephemeris and Nautical Almanac* and for a short while a lecturer at Columbia University. Cajori says: "Following a suggestion of Euler, Hill takes the earth as finite, the sun of infinite mass at an infinite distance, the moon infinitesimal and at a finite distance. The differential equations which express the motion of the moon under the limitations adapted are fairly simple and practically useful." [10] This work of Hill's was to have the greatest importance both for lunar theory and for our history. About ten years after Hill undertook his investigations, Sir George H. Darwin (1845–1912), a son of the naturalist and Plumian Professor of Astronomy and Experimental Philosophy at Cambridge, encouraged Ernest W. Brown (1866–1938) to study Hill's work. Brown, who was a fellow of Christ's College, Cambridge, from 1889 to 1895, did so with memorable results. Brown left Cambridge and came to the United States; he taught for a few years at Haverford before going to Yale University, where he stayed from 1907 until his retirement in 1932.

Moulton in describing Brown's work says: "As it now stands the work of Brown is numerically the most perfect Lunar Theory in existence, and from this point of view leaves little to be desired." [11] And so it was to remain for many years. In 1931 Wallace J. Eckert

[8] Académie des Sciences de Paris, *Comptes Rendus*, vol. 56 (1863), pp. 330–339.

[9] M. Wiberg, *Tables de Logarithmes Calculées et Imprimées au moyen de la Machine à Calculer* (Stockholm, 1876). According to Archibald, these tables were exhibited at the Philadelphia Exhibition of 1876. Also see George and Edward Scheutz, *Specimens of Tables, Calculated, Stereomoulded and Printed by Machinery* (London, 1856).

[10] F. Cajori, *A History of Mathematics* (New York, 1919), p. 450.

[11] Moulton, *Celestial Mechanics*, p. 365.

(1902–1971), Brown's disciple and perhaps the greatest modern figure in numerical astronomy, received his Ph.D. at Yale under the direction of Brown and began what was to be an ever-continuing interest in the calculational aspects of celestial mechanics. This interest was to be closely coupled to the continuing development of computers, as we shall see soon.

Eckert first went to Columbia in 1926 as an assistant in astronomy; after receiving his doctor's degree he became an assistant professor and began to develop a computing laboratory at the university. The genesis of this Computing Bureau is of some interest, since it marked the first step in the movement of IBM out of the punch card machine business into the modern field of electronic computers. In 1929 Thomas J. Watson, the chief executive officer of IBM, was persuaded by Benjamin D. Wood, who headed the university's Bureau of Collegiate Educational Research, to establish the Columbia University Statistical Bureau. It was fitted up by Watson with a set of the standard machines of the period. Eckert says: "The computing laboratory which is now operated by the Bureau was developed by the writer, first in the Columbia University Statistical Bureau and later in the Department of Astronomy of Columbia University. The directors of these institutions, Dr. Ben Wood and Dr. Jan Schilt, have given every encouragement." [12]

By 1930 the success of this laboratory was such that Watson authorized the construction of a special tabulator known under a variety of names, one of which was "Difference Tabulator." This machine, installed in 1931, was to be the modern version of the Babbage-Scheutz-Wiberg instrument. "No two jobs were alike in this interesting workshop. . . . Among the clients of the Bureau were, besides Columbia and [the Carnegie Foundation], Yale, Pittsburgh, Chicago, Ohio State, Harvard, California, and Princeton." [13]

In 1933 Eckert succeeded in persuading Watson of the need to enlarge the laboratory both from a machine and organization point of view. The result was to be the Thomas J. Watson Astronomical Computing Bureau. This Bureau was, as Eckert wrote, "a scientific non-profit-making institution which was organized for the use of astronomers. . . . [It] is a joint enterprise of the American Astronomical Society. The International Business Machines Corporation, and the Department of Astronomy of Columbia University.

[12] W. J. Eckert, *Punched Card Methods in Scientific Computation* (New York, 1940), p. iii.

[13] J. F. Brennan, *The IBM Watson Laboratory at Columbia University: A History* (New York, 1970).

Its operation is in charge of a Board of Managers, the majority of whom are appointed by the American Astronomical Society." [14]

The Board of Managers of the Bureau when it was founded in 1937 consisted of Eckert of Columbia, as director, E. W. Brown of Yale, T. H. Brown of Harvard, Henry Norris Russell of Princeton, and C. H. Tomkinson of IBM. (This latter gentleman was in fact the one who conceived of the idea of this new Bureau.) These gentlemen had an Advisory Council consisting of other notables in astronomy including T. E. Sterne of Harvard, who was at Aberdeen during World War II, where his knowledge of punch card techniques was to have a decisive influence, as we shall see. An American, he took his Ph.D. at Trinity College in Cambridge and then came back to the United States where he soon became a member of the staff and later a professor at Harvard University. Russell we will again see at Aberdeen where he was a member of the Scientific Advisory Committee. His contributions there were of great scientific value.

The requirements laid down by Eckert for emendations of the standard IBM machines to make them more useful for scientific work forced the company to develop an attitude of flexibility toward scientific users of machines which was to have great consequences for the electronic computer developments both at the University of Pennsylvania and at the Institute for Advanced Study, as well as for the company itself. Indeed, it may have been absolutely essential to its later success.

In Eckert's book we see that already by 1940 he had progressed quite far in his understanding of the value of digital computers for science and had shaped up his life's program. Regarding computers he said: "The Electric Punched Card Method, known also as the Hollerith, or Electric Accounting Machine Method, is now generally known to the business and statistical worlds. Its use in physical science has, however, been confined largely to astronomy. The purpose of this book is to show the possibilities of the method in scientific computation."

We see by the beginning of World War II a fair competence in the United States and Britain for calculation by the use of punch card machines. When in 1940 Eckert left Columbia to become director of the Nautical Almanac, he found no modern apparatus there at all for computation. He was instrumental in automating and modernizing the antiquated operation of that office. The first

[14] Eckert, *op. cit.*, pp. iii and 129f.

result of that effort was the production of an Air Almanac as an aid to navigators of airplanes.

We shall say more about Eckert and his accomplishments later. At this point we need to go back and discuss some other points before going forward. We have hopefully seen that the continuing need of astronomers for tables—particularly lunar ones—has provided a continuing impetus for the development of digital computers, and that this has gone on since at least Leibniz' time. Moreover, this need has found expression, and has had an impact on IBM with numerous consequences, a few of which we discuss later in the appropriate place.

By 1937 there were two other men in the United States who had developed interests in electromechanical digital computers: Howard H. Aiken, then a graduate student in physics at Harvard University, and George R. Stibitz, at that time a mathematician at the Bell Telephone Laboratories. In an unpublished memorandum apparently written in the fall of 1937, Aiken outlined his aims for a computer. He indicated four points of difference between punched card accounting machinery and calculating machinery as required in the sciences. The requirements are these: ability to handle both positive and negative numbers; to utilize various mathematical functions; to be fully automatic in operation—no need for human intervention; to carry out a calculation in the natural sequence of mathematical events.

Aiken goes on to say: "Fundamentally, these four features are all that are required to convert existing punched-card calculating machines such as those manufactured by the International Business Machines Company into machines specially adapted to scientific purposes. Because of the greater complexity of scientific problems as compared to accounting problems, the number of arithmetical elements involved would have to be greatly increased." [15]

Apparently this proposal of Aiken's came to the attention of Professor T. H. Brown at Harvard, whom we mentioned earlier as a member of the Board of Managers of the Watson Computing Bureau at Columbia. He had Aiken visit with Eckert and his associates, and this catalysed a reaction which led to an agreement for a collaborative development between Aiken and a group of IBM engineers under Clair D. Lake. The project started in 1939 and by 1944 was completed. On August 7 of that year "Mr. Thomas J. Watson, on behalf of the International Business Machines Corpora-

---

[15] H. H. Aiken, "Proposed Automatic Calculating Machine," edited and prefaced by A. G. Oettinger and T. C. Bartee, *IEEE Spectrum*, Aug. 1964, pp. 62–69.

tion, presented Harvard University with the IBM Automatic Sequence Controlled Calculator." [16] In a 1946 issue of *Nature*, L. J. Comrie headed his review of the book just cited as "Babbage's Dream Come True." He says:

The black mark earned by the government of the day more than a hundred years ago for its failure to see Charles Babbage's difference engine to a successful conclusion has still to be wiped out. It is not too much to say that it cost Britain the leading place in the art of mechanical computing. Babbage then conceived and worked on his "analytical engine," designed to store numbers and operate on them according to a sequence of processes conveyed to the machine by cards similar to those used in the Jacquard loom. This, however, was never completed.

The machine now described, "The Automatic Sequence Controlled Calculator," is a realisation of Babbage's project in principle, although its physical form has the benefit of twentieth century engineering and mass-production methods. Prof. Howard H. Aiken (also Commander, U.S.N.R.) of Harvard University inspired the International Business Machines Corporation (I.B.M.) to collaborate with him in constructing a new machine, largely composed of standard Hollerith counters, but with a super-imposed and specially designed tape sequence control for directing the operations of the machine. The foremost I.B.M. engineers were assigned to the task; many of their new inventions were incorporated as basic units. When the machine was completed, Thomas J. Watson, on behalf of the Corporation, presented it to Harvard University — yet another token of the interest I.B.M. has shown in science. Would that this example were followed by their opposite members in Great Britain! One notes with astonishment, however, the significant omission of "I.B.M." in the title and in Prof. Aiken's preface, although President Conant's foreword carefully refers always to the I.B.M. Automatic Sequence Controlled Calculator. [17]

This machine, which was electromechanical in nature, contained 72 counters for storing numbers, each of which was made up of 23 digits plus a sign. In addition there were 60 registers controlled by manually set switches in which constants could be stored. The machine took about six seconds to do a multiplication and about

[16] *A Manual of Operation for the Automatic Sequence Controlled Calculator* (Cambridge, Mass., 1946), foreword by James Bryant Conant.
[17] L. J. Comrie, *Nature*, vol. 158 (1946), pp. 567–568.

twelve for a division. There were also three units for calculating logarithms, exponentials, and sines (or cosines).

The machine was controlled by means of a paper tape which contained the instructions or orders for the machine arranged in serial order. In each instruction there were three parts: one which stated where the data to be operated upon were to be found; another which stated where the result was to be stored; and the third which stated what operation was to be performed. Comrie says regarding the machine's speed:

> Prof. Aiken estimates that the calculator is nearly a hundred times as fast as a well-equipped manual computer; running twenty-four hours a day, as it does, it may do six month's work in a day. Perhaps his examples, chosen for their simplicity, do not do the machine justice, because they could be done almost as quickly, and certainly more economically, with a Brunsviga and a National.
>
> The question naturally arises: Does the calculator open up new fields in numerical and mathematical analysis — especially in such pressing problems as the solution of ordinary and partial differential equations, and the solution of large numbers of simultaneous linear equations? It is disappointing to have to record that the only output of the machine of which we are informed consists of tables of Bessel functions, which are not difficult (to the number of figures required in real life) by existing methods and equipment. If the machine is to justify its existence, it must be used to explore fields in which the numerical labour has so far been prohibitive.[18]

Aiken describes his collaboration with the IBM engineers as follows:

> Our first contact . . . was with Mr. J. W. Bryce. Mr. Bryce for more than thirty years has been an inventor of calculating machine parts, and when I first met him he had to his credit over four hundred inventions — something more than one a month. They involved counters, multiplying and dividing apparatus, and all of the other machines and parts which I have not the time to mention, which have become components of the Automatic Sequence Controlled Calculator. . . .
>
> With this vast experience in the field of calculating machinery,

[18] *Ibid.*, p. 568.

our suggestion for a scientific machine was quickly taken and
quickly developed. Mr. Bryce at once recognized the possibilities. He at once fostered and encouraged this project, and the
multiplying and dividing unit included in the machine is designed by him.

On Mr. Bryce's recommendation, the construction and design
of the machine were placed in the hands of Mr. C. D. Lake, at
Endicott, and Mr. Lake called into the job Mr. Frank E. Hamilton
and Mr. Benjamin M. Durfee, two of his associates.

The early days of the job consisted largely of conversations —
conversations in which I set forth requirements of the machine
for scientific purposes, and in which the other gentlemen set
forth the properties of the various machines which they had developed, which they had invented, and based on these conversations the work proceeded until the final form of the machine came
into being.[19]

[19] *Manual of Operation,* Foreword (quoting Aiken).

Chapter 12

Renascence and

Triumph of Digital Means of

Computation

Concurrent with Aiken's collaboration with Lake and his associates, George R. Stibitz of the Bell Telephone Laboratories, with the help of two most able engineers, Samuel B. Williams and later Ernest G. Andrews, also of those laboratories, was busily engaged on a similar task. Both groups started in 1937 and Stibitz's team produced its first device in 1940; this "partially automatic computer" was shown to the American Mathematical Society at a fall meeting at Dartmouth College, where appropriately so many important things have been done in the computer field—including the appointment of John Kemeny, the inspiration for much modern work in the computing field, to the presidency of the college.

During the period of the war Stibitz was a technical aide in the NDRC where he was to play a major role in designing various relay-operated digital computers for the Armed Forces. These machines were also built by the Bell Telephone Laboratories. The culmination of this work was to be a general purpose, relay-operated, digital computer. To describe this machine takes us ahead of our story, but it is not unreasonable to do so. In his description of the machine F. L. Alt, who was one of the mathematicians at Aberdeen and the one who was in charge of the Bell machine there, says:

Before entering upon a description of this machine, a brief enumeration of its "ancestors" may be in order. The basic idea underlying all machines in this family, that using telephone switching equipment for computing purposes, was conceived several years before the outbreak of the last war by George R. Stibitz, then on the staff of the Bell Telephone Laboratories. The Company decided to test the idea on a small scale by building, for its own use, a machine capable of performing addition, subtraction, multiplication, and division of complex numbers. Under the impact of the war-time demand for large-scale computing, the Company next developed the "Relay Interpolator," a machine consisting mainly of about 500 telephone relays, together with some teletype equipment used for transmitting numbers into and out of the machine and for directing the operations. The next machine to be built was the "Ballistic Computer," containing about 1300 relays, more elaborate and more complex than the Relay Interpolator but still, like all the earlier machines, a special-purpose device, designed to carry out only a few special kinds of computations. Finally in 1944 the Bell Telephone Laboratories undertook to develop the all-purpose computing system with which this article is concerned. This system differs from its predecessors in size — it contains over 9000 relays and about 50 pieces of teletype apparatus, covers a floor space of about 1000 square feet, and weighs about 10 tons — as well as in flexibility, generality of application, reliability, and ability to operate automatically without requiring the presence of human operators.[1]

This machine stored and used numbers containing seven decimal digits; the Harvard-IBM machine used 23, and the standard IBM punch card units used 8. Its speed of operation was not great. Alt states that "the calculator will perform an addition in 0.3 seconds, a multiplication in 1 second, a division or square root in perhaps five seconds on the average (the exact computing time depends on the digits occurring in the operation). However, the routine control requires about two seconds to read the order for any of these operations. Thus the calculator remains idle part of the time, and the reading of orders becomes the controlling element in determining the time requirements of a problem."[2]

This machine stored numbers by means of electromechanical devices known as relays. A relay is in essence nothing other than a two-position switch which is operated by a control signal. A relay may be so arranged that in one position it *closes* one set of circuits and *opens* another set while in its other position it opens the first set and closes the second. It is possible by combining relays to perform all the standard logical functions and this is what is indeed done. There is no particular need for our going into much detail on how this is done at this time. The interested reader may consult, among others, Stibitz and Larrivee for some interesting examples.[3]

There were several important developments resulting from the approaches of Comrie, Eckert, Aiken, and Stibitz. The first and fore-

[1] F. L. Alt, "A Bell Telephone Laboratories' Computing Machine," parts I and II, *Math. Tables and Other Aids to Computation*, vol. 3 (1948), pp. 1–13 and 69–84. The interested reader may also wish to read G. R. Stibitz and J. A. Larrivee, *Mathematics and Computers* (New York, 1957).

[2] Alt, *op. cit.*, p. 79.

[3] *Mathematics and Computers*, chap. 6.

most is the renascence of interest in the digital means for scientific computation; this had lain dormant for perhaps a century under the impact of the highly useful analog machines we have discussed. It was now to emerge again and, with the linking of electronics to digital machines, to become almost totally dominant. The next most important development was the fact that Watson, and through him the IBM Corporation, had learned or sensed the fundamental importance of scientific calculation to our society and hence to IBM; in the author's opinion, this was to be one of the fundamental reasons for the phenomenal success of that company in the future. Another — or perhaps the other — reason was the conviction of Thomas J. Watson, Jr. that the electronic approach was *the* correct one for IBM to follow.

The machines at Harvard and the Bell Telephone Laboratories came too late in time, however; they were too close in time to the so-called ENIAC, the first electronic computer, and therefore they failed in their purpose in a real sense. The publication, for example, of the *Manual of Operation* of the IBM Automatic Sequence Controlled Calculator occurred in the year the ENIAC was placed in operation and the Bell Relay Calculator tried unsuccessfully to compete with the ENIAC from the date of its installation in 1946.

We will discuss the implications of this in detail later, but suffice it to say at this point that the speed factor for multiplying between the Harvard machine and the ENIAC was roughly 500 to 1 in favor of ENIAC, and the factor was even worse for the Bell machine. In spite of various arguments about the importance of reliability as against speed, the factor of 500 was to be determining and doomed the electromechanical approach. (Up to this point we have mentioned multiplying rates for computers and used these as an index for measuring overall speed. In a later section we will justify this *figure of merit*.)

It is somewhat ironical that the culmination of Babbage's dream should come just too late, and that in a sense he and his ideas never reached the fulfillment that he always yearned for. It seems to the author clear that part, at least, of Babbage's hostility toward the world grew out of what he regarded as its rejection of his intellectual achievements. Thus even after a century his ghost has never been laid to rest.

Neither Comrie nor Wallace Eckert was to become wedded to technology; rather, they were primarily users. Comrie never went beyond his earlier adaptation of commercial equipment to scientific purposes, and in his later life he headed a small service com-

puting company which lacked the resources to acquire major equipment. However, he never seemed to the author to be embittered
by this and always found fascination in his numerical experiments
and tasks. Eckert, as we shall see, went with IBM in 1945 and
founded a major new organization there that served to focus that
company's attention just at the apposite moment on the technological problem of the period. He himself was to make a unique astronomical achievement by his calculations and his refinements of the
mathematical underpinnings of lunar theory. We discuss this again
later.

Stibitz retired from Bell Telephone Laboratories to academic life
and interested himself in a variety of problems ranging from electronic music to biomedical models. He does not seem to have
designed any major digital machines since he left Bell. The Laboratories, however, built an improved version of his machine for
internal use.[4] Moreover, Bell went on to become a major user and
developer of electronic digital techniques for computation — both
commercial and scientific — and for communication. In fact, it has
been often said that without these techniques the U.S. telephone
system would not have grown to its present size and state of perfection, since there are not enough girls in the entire country to
handle the calls involved by old-fashioned means. However that
may be, Andrews, Stibitz, and Williams must have played a key
role in making the Bell Laboratories aware of the potentialities of
the digital approach; Williams even designed some electronic,
digital circuits.

The machine at Harvard became known as Mark I, since Aiken
formed a Computation Laboratory there which produced a number
of other machines for the U.S. Navy and Air Force known as Marks
II, III, and IV. Mark II was designed and built by the Harvard
group for the Dahlgren Proving Ground of the U.S. Navy, Bureau
of Ordnance. Its design was started in November of 1944, and the
machine was installed at Dahlgren shortly thereafter. It is another
electromechanical device using relays. It handled 10 decimal digit
numbers and could store about 100 of them internally.[5] It had a
multiplying speed comparable to other electromechanical devices
and was therefore doomed to extinction.

The symposium on large-scale digital machinery took place in
January 1947 and is noteworthy in that the various papers presented

[4] M. V. Wilkes, *Automatic Digital Computers* (London, 1956), p. 134.
[5] Proceedings of a Symposium on Large-Scale Digital Calculating Machinery,"
The Annals of the Computation Laboratory, vol. xvi (Cambridge, Mass., 1948),
pp. 69–79.

show the diversity existent at that time between the advocates of the electromechanical and the electronic approaches. This period is perhaps a watershed one.

In December of 1949 another symposium was held at Harvard to celebrate the near-completion of the so-called Mark III calculator for Dahlgren; it was also known as ADEC—Aiken Dahlgren Electronic Calculator. It was a decimal machine with 16 digits per number and with a multiplying rate of about 80 such operations per second. Thus already by 1949 Aiken had shifted over from electromechanical devices to electronic ones. This machine was completed in March 1950 and was then removed to Dahlgren where operation on a production basis was started about a year later.[6] Finally, Aiken was to design and build the Mark IV for the Air Force. It was started in 1950 and completed in 1952 and was housed at Harvard. Prof. Alwin Walther (1898–1967) of the Technical Institute of Darmstadt was much influenced by this design when he built his machine. In fact, Walther's machine bears marks both of Aiken and Stibitz; it was begun in 1951 and completed in 1959.[7]

The fundamental importance of Howard Aiken's work lies in the building up of a laboratory at Harvard where young men could be given academic training in circuit and component design for electronic digital computers. This was to be the training ground for a large number of first-rate men active in the field today. The list of these men is very long and their personal accomplishments exceptional. It has been said of this laboratory's work: "The Harvard Computation Laboratory under the leadership of Professor Aiken introduced many new components and methods of design: for example, the use of a logical algebra (or algebra of logic) in the design of switching circuits. Fortunately the work at Harvard has been well documented in the volumes of the Computation Laboratory published by the Harvard University Press."[8]

In this connection it is also worth remarking that Claude E. Shannon, the founder of what is often called Information Theory, in his master's thesis showed in a masterful way how the analysis of complicated circuits for switching could be effected by the use of Boolean algebra.[9] This surely must be one of the most important

---

[6] Staff of the Computation Laboratory, *Description of a Magnetic Drum Calculator* (Cambridge, Mass., 1952).

[7] W. de Beauclair, *Rechnen mit Maschinen, Eine Bildgeschichte der Rechentechnik* (Braunschweig, 1968), p. 149.

[8] C. V. L. Smith, *Electronic Digital Computers* (New York, 1959), p. viii.

[9] C. E. Shannon, "A Symbolic Analysis of Relay and Switching Circuits," *Trans. AIEE*, vol. 57 (1938), p. 713.

master's theses ever written. It was done under the direction of
Bush, Caldwell, and F. L. Hitchcock, all of whom were on the
faculty at MIT, and the former two of whom we have already
mentioned. The paper was a landmark in that it helped to change
digital circuit design from an art to a science.

It is not altogether surprising that the machines at the Harvard
Computation Laboratory were not themselves to be progenitors of
future lines. In the early stages of a vital new field very many novel
ideas will be developed and built into devices. The acid test of
the discriminating scientific marketplace then acts to weed out the
unfit, and this is the technological analogue of the Darwinian Survival of the Fittest or Natural Selection. When a new and basic
idea first appears, we may expect a great proliferation of changes
on the theme and this will go on until a new steady state is reached.
This occurs when the various new changes have competed with
each other and the unfit have perished and the best ideas in the best
have come together into a few types. By 1955 a survey was to show
88 domestic electronic digital computing systems; by 1957, 103;
by 1961, 222; and by 1964, 334.[10]

This discussion has carried us forward in time beyond where we
wish to be, but it suffices to complete the discussion of the ambitious electromechanical machines at Harvard and the Bell Telephone Laboratories. It is time now to move into the second part of
our story, in which we shall review the exciting developments that
took place during World War II.

[10] M. H. Weik, *A Survey of Domestic Electronic Digital Computing Systems*,
Ballistic Research Laboratories Report No. 971 (Aberdeen Proving Ground, Decem-
ber 1955); *A Second Survey of Domestic Electronic Digital Computing Systems*,
B.R.L. Report 1010 (June 1957); *A Third Survey of Domestic Electronic Digital
Computing Systems*, B.R.L. Report 1115 (March 1961); and *A Fourth Survey of
Domestic Electronic Digital Computing Systems*, B.R.L. Report No. 1227 (January
1964).

1. Reconstruction of machine designed and built in 1623 by Wilhelm Schickard of Tübingen. Writing to the astronomer Johann Kepler, Schickard proudly proclaimed that it "immediately computes the given numbers automatically, adds, subtracts, multiplies, and divides. Surely you will beam when you see how [it] accumulates left carries of tens and hundreds by itself or while subtracting takes something away from them." (PHOT: IBM)

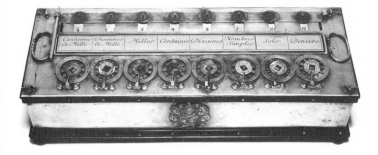

2. Calculating machine built by Blaise Pascal in 1642, when he was only twenty years old. His calculator could perform addition and subtraction but not the non-linear operations of multiplication and division. (PHOT: IBM)

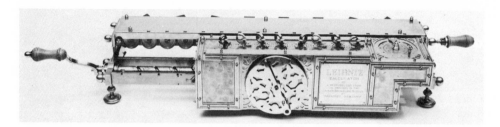

3. Replica of a calculator invented by the philosopher Leibniz in 1673. A device known as the Leibniz wheel enabled the calculator to perform the operation of multiplication automatically by repeated additions. (PHOT: IBM)

4. Difference Engine of Charles Babbage, "the irascible genius." The photograph is of a replica of the first of two machines built by Babbage, neither of which was ever entirely completed. In building his tabulator Babbage sought to automate "the intolerable labor and fatiguing monotony" of calculating tables and to replace fallible people by infallible machines. The Difference Engine was very specialized in its function, but his Analytical Engine was in concept a general purpose computer and incorporated a number of the principles that characterize the modern computer. (PHOT: IBM)

5. Automated loom of Joseph Marie Jacquard. Holes punched in the cards according to the desired pattern allowed hooks to come up and pull down threads of the warp, so that when the shuttle passed through it went over some threads and under others. The device was an immediate success, and by 1812 there were more than 11,000 Jacquard looms in use in France. (PHOT: IBM)

6. Difference engine built in 1853 by Pehr Georg Scheutz of Stockholm. It was exhibited in London in 1854 and subsequently in Paris. In 1856 it was purchased for the Dudley Observatory in Albany, N.Y. An exact duplicate of the machine was used by the Register-General's office in London to produce the life expectancy tables used for many years by English insurance companies. The photograph is of a replica made for IBM. (PHOT: IBM)

7. Lord Kelvin's tide predictor. This machine was based on Kelvin's earlier harmonic analyzer, the object of which, he said, was "to substitute brass for brain." (PHOT: Reproduced by courtesy of the National Museum of Science and Industry, London)

8. Hollerith tabulating equipment used in the Eleventh Census of the United States in 1890. Counting, sensing, punching, and sorting units are shown. Cards containing 288 locations at which holes could be punched ran under a set of contact brushes which completed an electrical circuit wherever holes appeared. The 1890 census handled the records of 63,000,000 people and 150,000 minor civil divisions. The Hollerith system proved highly successful, and the superintendent of the census was able to announce the total population just a month after all returns arrived in Washington. (PHOT: IBM)

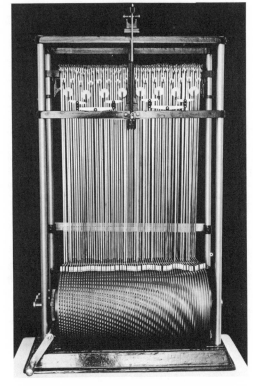

9. Replica of the Michelson-Stratton harmonic analyzer. Developed in 1898 by Albert A. Michelson and Samuel W. Stratton, the machine was capable of handling Fourier series of 80 terms. Fed the Fourier coefficients of a trigonometric series, it could produce a graph of the sum function and also perform the reverse process. (PHOT: IBM)

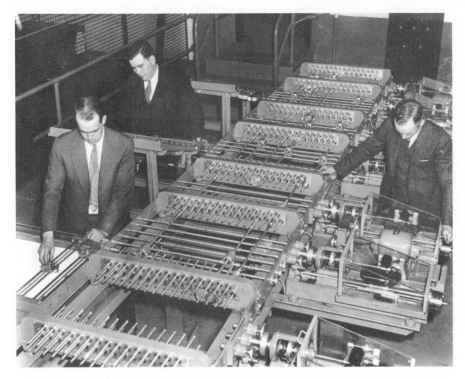

10. C. J. Weygandt, C. C. Chambers, and J. G. Brainerd operating the differential analyzer at the Moore School of Engineering, 1935. This machine was based on the analyzer developed by Vannevar Bush and his colleagues at MIT. (PHOT: The Smithsonian Institution)

11. Harvard-IBM Mark I, the Automatic Sequence Controlled Calculator. Completed in 1944, this electromechanical digital computer contained 72 counters for storing numbers, each of which was made of 23 digits plus a sign. In addition, there were 60 registers controlled by manually set switches in which constants could be stored. The machine took about six seconds to do a multiplication and about twelve for a division. There were also three units for calculating logarithms, exponentials, and sines (or cosines). Developed by Howard H. Aiken of Harvard and a group of IBM engineers under Clair D. Lake, the Mark I was a major advance from relatively simple punch-card accounting machinery to the complex machinery required for scientific use. (PHOT: IBM)

12. View of part of the ENIAC, the first electronic digital computer, operational in December 1945. This historic computer contained 18,000 vacuum tubes, 70,000 resistors, 10,000 capacitors, and 6,000 switches. It was 100 feet long, 10 feet high, and 3 feet deep. In 1946 it was moved to the Ballistic Research Laboratory in Aberdeen, Md., and in 1955 it became part of a permanent exhibit at the Smithsonian Institution. (PHOT: The Smithsonian Institution)

13. Corporal Irwin Goldstein setting function switches on the ENIAC. (PHOT: The Smithsonian Institution)

14. John von Neumann (left) and J. Robert Oppenheimer at the dedication of the Institute for Advanced Study computer, 1952. Known variously as the IAS, Princeton, or von Neumann machine, this computer helped to realize the ideal of a fundamental tool for heuristic investigations in areas of pure mathematics where the burden of algebraic computation is prohibitive by normal human methods. (PHOT: Alan W. Richards)

Part Two

Wartime Developments:
ENIAC and EDVAC

Chapter 1   Electronic Efforts prior

                             to the ENIAC

As we saw in the last pages of Part One, the electromechanical digital devices were doomed to extinction by the advent of electronics. It is curious how delicate are the timing and balance of what we now refer to as research and development. There seems to be an optimal time for discovery, as well as an optimal period for the perfection of the idea. Thus, if one is late in starting or dilatory in carrying out the task, one may, as John Couch Adams almost did in connection with the discovery of Neptune, find oneself "scooped." If, on the other hand, one undertakes the task of finding or inventing something too soon before this optimal point, one needs to expend too much energy and therefore frequently fails. An outstanding example of this is probably Babbage. The industrialization of England had not yet proceeded far enough to warrant the work he undertook, and he broke himself in the process. Another example is a physicist named John V. Atanasoff, who was on the faculty at what was then called Iowa State College as an associate professor of mathematics and physics from 1926 to 1945.

During the latter part of the 1930s Atanasoff started some of his graduate students on ways of solving linear differential and integral equations. This led him to look very carefully into Bush's work on the differential analyzer. (Indeed, he believes he coined the expression "analog computer.") This examination apparently convinced him in 1937 that the correct mode of computation was by electronic digital means. Writing in 1966, Atanasoff said he had then realized the high speeds that could be attained electronically and how this could be used to permit serial processing of data; he also claimed to have hit upon numbers base 2 as the one to use.[1] He indicated that this use of base 2 had been previously suggested for mechanical devices by a French author but did not name him. In any case he and a colleague, Clifford Berry, built an electronic machine to solve systems of simultaneous linear equations. In 1940 the Direc-

[1] Letter, Atanasoff to Miss D. V. Henry, Engineering Publications Office, Iowa State University, 10 June 1966.

tor of the Agricultural Experimental Station of the College was
writing to its Committee on Patents regarding "a patentable development, a new 'computing machine for the solution of linear
algebraic equations.'" [2] The 15 January 1941 issue of the *Des
Moines Tribune* showed a photograph of Berry holding a piece of
the machine. The accompanying article said that the entire machine
was to contain over 300 vacuum tubes when completed in about
a year.

In the communication to the Engineering Publications Office,
which seems never to have been published, Atanasoff displays a
keen understanding of the fundamental importance of large linear
systems of equations. Indeed, von Neumann and I chose this topic
for the first modern paper on numerical analysis ever written
precisely because we viewed the topic as being absolutely basic
to numerical mathematics. [3] In his account Atanasoff stated: "In this
way the solution of large systems of linear algebraic equations constitutes an impcrtant part of mathematical applications. . . . The
solution of general systems of linear equations with a number of
unknowns greater than ten is not often attempted. But this is precisely what is needed to make approximate methods more effective
in the solution of practical problems."

Interestingly enough, Atanasoff realized very early on the importance of the well-known method of Gaussian elimination for solving
systems instead of that of determinants. The latter is important in
pure mathematics as a means of displaying in a very elegant closed
form the solutions of a system. (This dichotomy between what is
elegant or beautiful in pure mathematics and in numerical mathematics is fundamental, and we shall see a number of instances of
this later.) Much of Atanasoff's early work shows considerable
sophistication, although this is occasionally combined with a certain naïveté, as for example when he discusses how to convert from
one number base to another. He failed to realize apparently how
trivial a calculation it is to effect this conversion.

He did appreciate very explicitly the need to store quantities in
his device and chose for this purpose large numbers of little capacitors — in those days they were called condensers — which are
devices for storing electrical charges. The earliest versions of
these devices were known as Leyden jars, and they are the standard
means for storing electrical charges in a circuit. He also knew that

[2] Letter, R. E. Buchanan to Committee on Patents, 21 November 1940.
[3] J. von Neumann and H. H. Goldstine, "Numerical Inverting of Matrices of High
Order," *Bull. Amer. Math. Soc.*, vol. 53 (1947), pp. 1021–1099.

the charges in these devices gradually leak away, and so he felt it desirable "to arrange the mechanism so that the machine would jog its memory at short intervals." This in somewhat primitive form is what was done by von Neumann and the author in their general-purpose machine at the Institute for Advanced Study a decade later.

Atanasoff envisaged his machine as performing addition and subtraction for its basic operations and compounding multiplication and division out of these linear operations by successive additions and subtractions in a manner reminiscent of Pascal. To speed up the execution of these operations he also planned to build into his machine the operations of shifting numbers one place to the right or left—this corresponds to multiplying or dividing by a factor 2. Furthermore, he planned to solve systems of up to 30 simultaneous equations by the usual Gaussian scheme which may be described as follows. Take the first equation of a system such as

$$ax + by + cz = d,$$

$$ex + fy + gz = h,$$

$$ix + jy + kz = l,$$

solve it for x and then eliminate it from the other equations; this results in a system of two equations in the two unknowns y and z; this process is then repeated until a fixed value for z is found. When this value is substituted into the equation giving y in terms of z, y is found; similarly x is also. This procedure is generalizable to any number of equations.

Atanasoff contemplated storing the coefficients of an equation in capacitors located on the periphery of a cylinder rotating at one revolution per second and using punch cards for the storage of intermediate results. He apparently had a prototype of his machine working "early in 1940." This machine was, it should be emphasized, probably the first use of vacuum tubes to do digital computation and was a special-purpose machine; i.e. one designed to do a specific task, just as Babbage's Difference Engine was.

This machine never saw the light of day as a serious tool for computation since it was somewhat premature in its engineering conception and limited in its logical one. Nonetheless it must be viewed as a great pioneering effort. Perhaps its chief importance was to influence the thinking of another physicist who was much interested in the computational process, John W. Mauchly. During the period of Atanasoff's work on his linear equation solver,

Mauchly was at Ursinus College, a small school in the environs of Philadelphia. Somehow he became aware of Atanasoff's project and visited him for a week in 1941. During the visit the two men apparently went into Atanasoff's ideas in considerable detail. The discussion greatly influenced Mauchly and through him the entire history of electronic computers. Just about at the time Atanasoff confided his ideas to Mauchly the latter left Ursinus to take a post at the Moore School of Electrical Engineering at the University of Pennsylvania. Atanasoff also apparently had ideas for a more general-purpose electronic digital computer and discussed his nascent ideas with Mauchly on this occasion.[4]

[4] Atanasoff recently (June 1971) testified regarding his early work in considerable detail during a case that is *sub judice*. While it is hardly germane for our purpose to recount his testimony, the interested reader may wish to look it up in Honeywell vs. Sperry Rand, U.S. District Court, District of Minnesota, Fourth Division. In this same case many of the other figures in the early history of the electronic computer either have or will testify. The transcript of the testimony when it completed as well as the opinion of the court in this case will therefore form a most valuable archive of material for the period.

Chapter 2

The Ballistic Research
Laboratory

Before following up on this story, we need to digress long enough
to bring things into their proper temporal sequence. To do this it
is convenient first to introduce a number of people whose careers
are tied in closely with our story.

In 1910 Hermann H. Zornig graduated from Iowa State College
and was commissioned in the regular Army of the United States
that summer. He subsequently did graduate work at the Massachusetts Institute of Technology and the Technische Hochschule in
Charlottenberg, Germany, where he had the unique opportunity
to study under one of the greatest figures in ballistics, Carl J. Cranz
(1858–1945). On completion of his studies he did important work
in metallurgy at Watertown Arsenal and in explosives and propel-
lants at Picatinny Arsenal and served from 1927 to 1930 as assistant
military attaché in Berlin. Then in 1935 he had the vision to create
out of diverse operations at Aberdeen the Research Division of the
Proving Ground, which in 1938 was renamed the Ballistic Research
Laboratory. He served as its director from 1935 to 1941.

Before going to Aberdeen, Zornig headed the Ammunition Division of the Technical Staff of the Office of the Chief of Ordnance.
While there he encouraged a young officer, Leslie E. Simon, trained
at West Point and MIT, to apply statistical techniques to quality
control of ammunition. Simon's work in this field was outstanding
and received recognition from a number of organizations, and in
1949 he received the Shewhart Medal for Quality Control. Simon
became assistant director to Zornig and succeeded him as director
of the Ballistic Research Laboratory holding this position until
1949. He rose to the rank of major general, an indication of the
significance of the role of science and technology in the military.

The third member of the trio of regular army officers, all of whom
obtained their master's degrees at MIT, was Paul N. Gillon. He
became executive officer of the Ballistic Research Laboratory in
1939, and then when Simon became director, assistant director. He
held this position until he went to the Research and Materials
Division of the Office of the Chief of Ordnance, first as its deputy

chief and then as chief. Later he was Director of Research at Watertown Arsenal and then he founded and headed the Office of Ordnance Research, which was located on the campus of Duke University. This office subsequently became the Office of Army Research
(Durham), the office charged with administering Army research
contracts with universities.

At the founding of the Ballistic Research Laboratory Robert H.
Kent, whom we met earlier, and L. S. Dederick were appointed as
associate directors and were the chief civilians in the laboratory.
Two of Col. Zornig's greatest contributions to the laboratory were
the choice of Oswald Veblen, who served as the chief scientist
throughout World War II, and the establishment of a Scientific
Advisory Committee of leading American scientists. The original
members of the committee were: Hugh L. Dryden (d. 1965), one
of the great aerodynamicists of our time and a leading figure in the
creation of the National Aeronautics and Space Agency; Albert W.
Hull (d. 1966), "the world's most prolific inventor of electron
tubes"; [1] Bernard Lewis, a physical chemist who was then at the
U.S. Bureau of Mines and a great authority on combustion and
explosion; Henry N. Russell (d. 1957), whom we mentioned earlier
and who was one of the world's greatest astronomers; Isador I.
Rabi, Nobel Laureate in Physics and one of the elder statesmen of
science in the United States; Harold C. Urey, Nobel Laureate in
Chemistry and another of the scientific elder statesmen; Theodore
von Karman (d. 1963), one of the great aerodynamicists; and John
von Neumann (d. 1957), one of the great mathematical universalists. There were a few changes in this group during the war when
various equally distinguished scientists, including George Kistia-
kowsky, came onto the committee; but basically it was a group of
scientists of the highest stature who helped ensure the integrity
and quality of the work undertaken at the laboratory.

Before going forward with the events leading up to the first electronic digital computer, let us look for a moment at the financial
side of research and development in the U.S. Army in the period
between the two World Wars.

One of Veblen's last tasks in World War I was to make a number
of recommendations to guide his successors in carrying on bal-
listical research and development. This blueprint was followed
with the results we have read. He had postulated a level of financial
support which was not followed because of severe budgetary cuts

[1] National Academy of Sciences, *Biographical Memoirs*, vol. XLI, pp. 215–233.

during the post-war recession of the 1920s. By fiscal 1923 Ordnance was allocated only about $6,000,000 for all purposes, and this allocation declined still further during the next five years. By fiscal 1928 it came back up to $6,000,000 and stayed at about this level until 1937. At that time it leaped up to $17,000,000 as a result of the European situation. It then went rapidly to $177,000,000 in 1940.[2] In the light of these figures we can measure the importance Ordnance in that era placed on research and development work. In the period from 1922 to 1937 about $1,000,000 per year was allocated for this purpose, which is percentage-wise quite adequate, the average per annum total appropriation for Ordnance in that period being about $11,000,000. But in the period from 1937 through 1940 the R&D budget never reached the $2,000,000 mark.[3] In spite of these small budgets it is amazing that such high caliber people did such excellent work—the division contained only about thirty people in 1935. One must give enormous credit to Zornig and to Kent for this.

Let us recall that in 1935 the Research Division acquired its copy of the Bush differential analyzer. Interestingly, the prime man at Aberdeen in this acquisition was Major James L. Guion, who headed ballistic computation there and who was one of Moulton's students.[4] From the beginning Paul Gillon, then a lieutenant, was to display a particular interest in computational work, and it is highly to his credit that on the eve of World War II he realized the great importance of digital computation to the construction of firing and bombing tables. He was also one of the few leaders of the Laboratory to realize the fundamental importance of these tables to the using services (we shall discuss later what these tables were for). He therefore persuaded Simon and Zornig to approach IBM to obtain a set of punch card machines for the laboratory's use.

Zornig, Simon, and Gillon approached John C. McPherson of IBM, who had a considerable interest in the use of punch card machines for scientific calculation. McPherson had made a number of calls on Kent and Dederick in 1940–41, and they had developed a scientific rapport, so that it was quite natural for the officials at Aberdeen to approach him. He worked out an arrangement which resulted in bringing into the Ballistic Research Laboratory a set of standard punch card machines to do ballistic calculations. Also, in 1944, two special multiplying machines were built by IBM and

[2] Ballistic Research Laboratory, "Ballisticians in War and Peace," unpublished manuscript kindly made available by Dr. John H. Giese, Chief of the Applied Mathematics Division of the Ballistic Research Laboratories.
[3] Ibid.   [4] M. H. Weik, "The ENIAC Story," Ordnance (1961), pp. 3–7.

installed at Aberdeen. Parenthetically, it is interesting to note that as a result of the success of this installation about fifteen other government agencies and contractors set up similar ones. Notable among these was the Los Alamos Scientific Laboratory. This marks then a renascence of interest in the digital approach, which had proved so important in the days of Bliss, Gronwall, Moulton, and Veblen, but which had been temporarily superseded by the acquisition of the differential analyzer in 1935. This idea of doing calculation digitally was to culminate in the ENIAC, the first electronic digital computer, and eventually in the gigantic computer industry of today.

However, the day of the digital machine had only partially arrived. More or less at this time Gillon had contracted with Harold Pender (d. 1959), the first dean (1923–1949) of the Moore School of Electrical Engineering of the University of Pennsylvania, to take over the differential analyzer there for the duration of the emergency. He also had started at that school a program to train in ballistic computation women who had science degrees. This was undertaken as part of an extensive program instituted by the U.S. Government to train people for technical positions in various aspects of the war effort. Thus, for example, both John W. Mauchly and Arthur W. Burks, whom we will soon meet, took electrical engineering training at the Moore School and both did so well they received calls to the faculty of the School. The government called its program ESMWT, Engineering, Science, Management, War Training. In general this effort was very successful and enabled substantial numbers of men and women to qualify for technical posts in areas where there were great shortages. For his part, Gillon's foresight in anticipating the computational load that was to fall on the laboratory was to be decisive for the United States, and he greatly deserved the Legion of Merit Medal he received for his total contribution to computation.

The scientific staff of the Ballistic Research Laboratory was reorganized on a wartime basis in 1941–42 with the highest standards of scientific excellence. With Veblen's help a considerable number of top-flight scientists were brought in from the university world to enlarge the laboratory's cadre: *in mathematics*, A. A. Bennett, was brought back from Brown University; H. B. Curry, Pennsylvania State; H. H. Goldstine, Michigan; J. W. Green, Rochester; F. John, the Courant Institute; J. L. Kelley, Notre Dame; D. H. Lehmer, Berkeley; H. Lewy, Berkeley; E. J. Mc-Shane, University of Virginia; C. J. Morrey, Berkeley; A. P. Morse,

Berkeley; E. Pitcher, Lehigh University; I. J. Schoenberg, Pennsylvania; L. Zippin, Queens; *in physics*, L. A. Delasasso and G. Breit, Wisconsin; T. H. Johnson, Bartol Research Foundation; H. B. Lemon, Chicago; R. G. Sachs, Purdue; L. H. Thomas, Ohio State; J. P. Vinti, Worcester; *in astrophysics and astronomy*, L. E. Cunningham, Harvard; E. P. Hubble, Mt. Wilson; D. Reuyl, Virginia; M. Schwarzchild, Columbia; T. E. Sterne, Harvard; R. S. Zug, Carleton; *in physical chemistry*, J. H. Frazer, Buffalo; J. E. Mayer, Columbia. There were many others who came later but who are not included, both because it would unduly lengthen the lists and because the author's memory has failed him.

These lists, incomplete as they certainly are, still suffice to show the extremely high standards Veblen set in choosing his scientific staff. It is also of course true that the laboratory in 1941–42 already had many very excellent men on its staff who had been recruited earlier and who have not been mentioned. Writing to his wife in 1942, the great astronomer Hubble said: "I am more and more impressed with the place. It is not the home of genius, but it knows the answers to many problems and how to get the answers for others. And best of all, it fights for high standards — when a conclusion is reached and a statement formulated, you treat it with respect or get into trouble." [5]

A description of my own experience may perhaps serve as a typical example of the manner in which the staff was recruited. I obtained my Ph.D. at the University of Chicago in 1936 and then for a number of years was a research assistant in the mathematics department there. During the years from 1936 through 1939 and each summer thereafter, I worked with Gilbert A. Bliss (d. 1951), the chairman of the department and a mathematician with wide-ranging interests that grew out of his profound understanding of what is called the calculus of variations. This is the subject which enables one to answer questions such as this: among all simply closed curves of a given length to find the one enclosing the largest area; or, among all curves joining two points in a vertical plane to find the curve down which a bead would descend from the upper to the lower point in the least time. Both of these are classical problems, and the latter one is particularly famous since "the systematic development of the theory of the calculus of variations really began" with this problem. It was originally formulated by Galileo in 1630, was solved by John Bernoulli (1667–1748) in 1697 in an ingenious but highly special fashion, and then solved by Ber-

---

[5] National Academy of Sciences, *Biographical Memoirs*, vol. XLI, p. 182.

noulli's brother James (1654–1705) in the same year by means which "were sufficiently powerful to be effective for a large variety of maximum and minimum questions." [6]

During World War I Bliss had made a great contribution to the mathematical theory of ballistics by introducing perturbational or variational techniques into the subject and made possible the study of the effects of small changes in various parameters on the flight of a projectile. Typical of such changes are variations in the density or temperature of the air, in the weight of the projectile, in the temperature of the propellant change, in the velocity, wind, etc. At various times he taught graduate courses in subjects such as ballistics, quantum mechanics, etc. in which he had an interest. Fortunately for me Bliss allowed me to teach his courses for several years; and thus I had the unique opportunity for a young man of that period to teach, and therefore to learn, a considerable spectrum of mathematical subjects, both pure and applied. During my association with Bliss we principally worked on writing his magnum opus on the calculus of variations.[7] This was, incidentally, also the subject of Morse's work for many years. Bliss was more or less the master of the classical theory, and Morse, using the techniques of topology, the modern one.

It was while I was Bliss's assistant that I became acquainted with Edward McShane, Marston Morse, and Oswald Veblen. These acquaintanceships were later to develop into what I hope were very real friendships. In fact, at one point Morse invited me to come to the Institute for Advanced Study as his assistant. I preferred however to leave Chicago to go to the University of Michigan to start a teaching career. This was a period in the economic history of the United States in which positions in universities were scarce and considerable numbers of young men competed vigorously for them, many very good ones being forced to leave the academic world for careers in business. I was fortunate enough to have taught a course in exterior ballistics and to have worked with Bliss on his book. Thus, when in July of 1942 I was called into the service, Bliss wrote Veblen suggesting the desirability of having me transferred to the Ballistic Research Laboratory. Veblen agreed, and on 7 August 1942 I reported for duty to the laboratory. I was then a first lieutenant and was assigned at once to Gillon who had charge of all ballistic computations.

[6] Bliss, *Calculus of Variations*, Carus Mathematical Monographs (Chicago, 1925), pp. 174–175.

[7] *Lectures on the Calculus of Variations* (Chicago, 1945).

On September 1 of that year Gillon and I went to inspect the small activity at the Moore School. We found things in a not very good state. This stemmed from three causes: first, this project, like many other new ones, suffered from growing pains; second, the faculty chosen by the University to teach mathematics consisted of several quite elderly professors emeriti who were no longer up to the strain of teaching day-long courses; and third, the cadre of trained people sent up from Aberdeen to run the differential analyzer and to do the other things needed to prepare firing and bombing tables was in need of leadership. On or before 30 September 1942, I was placed in charge of the entire operation in Philadelphia and soon thereafter my wife, Adele K. Goldstine (d. 1964), who plays a key role in the story, and I moved to Philadelphia. At this point Gillon was transferred to the Research and Materials Division of the Office, Chief of Ordnance. This office, of which Gillon was deputy chief, was the one to which the Ballistic Research Laboratory reported technically, and its staff included, among others, Feltman, Morse, and Transue.

Dean Pender had assigned the task of liaison with Ordnance to John Grist Brainerd, who was then a professor in the Moore School and who was later to be Director—the title of the head of the School was changed from Dean for an organizational reason. (Brainerd just retired in the spring of 1970 from this post and is now University Professor of Engineering.) Brainerd was perhaps the best qualified member of the faculty for this purpose. He combined a considerable interest in computation—particularly in using the differential analyzer [8]—with substantial ability as a leader of men and a manager of affairs. He did an excellent job of handling this assignment, which was soon to occupy him full time. At all times it was a distinct pleasure for me to deal with this honest, kindly, and well-meaning gentleman. He undoubtedly deserves the credit for being the university's key man in the manifold relationships that were to be developed between Aberdeen and the university. While we did not always agree upon all issues in the first instance, his sense of fair

---

[8] J. G. Brainerd, "Note on Modulation," *Proc. IRE*, vol. 28 (1940), pp. 136–139; Brainerd and C. N. Weygandt, "Solutions of Mathieu's Equation—I," *Phil. Mag.*, vol. 30 (1940), p. 458; Brainerd, "Stability of Oscillations in Systems Obeying Mathieu's Equation," *Jour. Franklin Institute*, vol. 233 (1942), pp. 135–142; Brainerd and H. W. Emmons, "Temperature Effects in a Laminar Compressible-Fluid Boundary Layer along a Flat Plate," *Jour. App. Mech.*, vol. 8 (1941), pp. 105–110; "Effect of Variable Viscosity on Boundary Layers with a Discussion of Drag Measurements," *Jour. App. Mech.*, vol. 9 (1942), pp. 1–6.

play was such that he was always open to reason, and the rapport
between us was always most cordial.

In the early stages of the university-government relationship,
Prof. Carl C. Chambers, who was to be the second dean of the
Moore School, was involved both with the training courses and the
differential analyzer. He had however so many other responsibilities and governmental projects that he stepped out of this
relationship. He was later to retire as dean and become the vice-
president of the university for engineering affairs, a post he holds
today.

Another key figure was Brainerd's computational collaborator,
Prof. Cornelius J. Weygandt, Jr., who was in charge of the differential analyzer and who was to perform yeoman's service in keeping this machine operating and in devising ways to improve its
performance. In both these capacities he was invaluable to the
government. An account of the period says of him and his work:

> Fortunately, at this time there was a very talented group at the
> Moore School under the direction of Professor Brainerd and as a
> result of Lieutenant Gillon's discussions . . . , assistant Professor
> Weygand [sic] undertook to develop an electronic torque ampli-
> fier to replace the mechanical torque amplifiers on the Bush
> differential analyzers. This work was eminently successful. . . .
>
> In addition, photoelectric followers were developed by the
> Moore School group for both the input and output tables of the
> analyzer. As a result . . . the productive capacity of these ana-
> lyzers at both the Moore School and at Aberdeen was enhanced
> by at least an order of magnitude.[9]

The training of women to become computers or calculators for
the Ballistic Research Laboratory was, as indicated above, the first
problem. I persuaded Brainerd to terminate the Moore School's
arrangements with the elderly instructors, and a new scheme was
worked out whereby Adele K. Goldstine, my wife; Mildred Kramer,
the wife of Samuel Noah Kramer, distinguished scholar of Sumer
at the university and Clark Research Professor of Assyriology there;
and Mary Mauchly (d. 1946), wife of John W. Mauchly, formed the
teaching staff for this program. Somewhat later these arrangements
were changed, but the program was by then so well founded that
it did not falter. After a while the reservoir of talent in the Baltimore
and Philadelphia areas dried up, and Mrs. Goldstine made recruit-

---

[9] Weik, "The ENIAC Story."

ing trips to colleges all over the Northeast. Finally, however, the
WACs — Women's Army Corps — were formed, and after much
travail a number of them were made available to the Ballistic
Research Laboratory. They were trained in Philadelphia and went
to work at Aberdeen. As a result of somewhat herculean efforts the
Ballistic Research Laboratory had an adequate staff of trained
people to prepare the firing and bombing tables then so urgently
needed by the U.S. Army and Air Force (total personnel was about
200, of whom half were in Philadelphia). The center of gravity for
firing-table work shifted from Aberdeen to Philadelphia as a result
of the training program. Early in the war days I was fortunate to
secure the services of John V. Holberton as my civilian assistant.
Holberton was excellent in handling the myriad personnel problems that arose; the computer trainees liked and trusted him and
deservedly so.

As I have said, one of the main functions of the Ballistic Research
Laboratory was the production of firing and bombing tables and
related gun control data. It is worth saying a few words about such
tables so that the reader will have some conception of what was
being undertaken. The automation of this process was to be the
raison d'être for the first electronic digital computer.

Basically, a gunner is in possession of one fundamental piece of
information and a number of ancillary ones: the former is the
location of a target and hence a distance (range) and an angle from,
let us say, north. His gun however is in aiming respects like a
surveyor's instrument in that it can be rotating both in a horizontal
and in a vertical plane through predetermined angles. Thus he
needs to convert his range into an angle in the vertical plane
through the gun. This conversion is done by the firing table, the
main functions of which is to tell the gunner at what angle to elevate
his piece to reach a certain distance. This is necessary since, of
course, one does not point a gun directly at a target. Instead, one
fires up into the air, and the projectile arches up and then down in
an orbit somewhat parabolic in shape. The horizontal angle, the
azimuth, is directly measured and is determined purely geometri-
cally.

Secondarily, the gunner possesses certain other data that play a
minor but not necessarily insignificant role: he is furnished with
information as to head or tail winds, cross winds, air density and
temperature — all as a function of altitude — the weights of his shells
and propellant charges and their temperatures and possibly a few

other facts. The tables enable him to factor these considerations into a final angle of elevation and angle of deflection.

These tables sometimes existed as small booklets that fitted into a gunner's pocket or as automata—small highly specialized and usually analog computers—attached to guns, which accepted as input radar locations plus the secondary information and without human intervention positioned the guns. These devices were introduced during World War II because of the speed requirements imposed as a result of airplane targets. However, the same material needed to be calculated in any case since the *director,* as the fire control apparatus was called, had built into it the ballistic tables and merely looked up entries with great speed and used this information to control electrical motors. Bombing tables and bombsights were entirely analogous.

Thus, to make a long story short, there was need to calculate these sorts of tables for every combination of gun, shell, and fuse. Let us now try to make some rough estimates of how much work it takes to make a table, since this will lead us into a somewhat deeper understanding of the calculational process in general.

A digital computer is a device which performs the elementary operations of arithmetic: addition, subtraction, multiplication, and division. A fair way to estimate the time required to do a given calculation is to determine how many of each of these operations is performed and to know how long it takes to do each. It turns out that in many scientific calculations there are about as many additions and subtractions as multiplications and divisions; but usually computers require considerably longer to do one of the nonlinear operations, multiplication and division, than the linear ones, addition and subtraction. In the case of the first electronic computer, the time required for a multiplication was 14 times that for an addition. Thus, it is not a bad criterion for comparing two computers to ask how they compare in multiplying speed. This is of course a very crude way to measure so highly complex a device as a computer, but it serves quite well for a first orientation in a difficult area. To estimate the total calculation time it is not unreasonable to estimate the total time spent on multiplications and then to multiply this by a factor of about 3. In the case of a typical scientific calculation the factor was 2.57.[10] Thus, if a calculation takes one hour to do the multiplying involved, the total calculation time—all

[10] H. H. Goldstine and J. von Neumann, "Blast Wave Calculation," *Comm. Pure and Applied Mathematics,* vol. III (1955), pp. 327–354.

operations including transferring of data and the other "bookkeep-ing" sorts of things—will be around three hours.

With this understood, let us ask about the multiplication rates for a few machines.[11] A human being multiplying two 10-digit numbers by hand without any mechanical aids takes about 5 minutes. The traditional desk calculator of the 1940s spent about 10–15 seconds on this task, and thus such a machine provides a speed-up of between 20 and 30 times. This of course is the value of such a device. However, in both cases the human must then record the results by hand on paper, whereas all the automatic machines do this without human intervention. This probably slows the hu-man down by a significant factor—let us say 2. Thus, when we come to compare total calculation times we must assume the human will require perhaps 6 times the multiplying time instead of 3 as in the cases of automatic machines.

The Harvard-IBM machine we mentioned earlier took about 3 seconds per 10-digit multiplication and was therefore only around 3 to 5 times faster than a human with a desk calculator on the basis of our, perhaps overly crude, figure of merit. The Mark II, the second machine built at Harvard, had two multipliers and could attain a multiplying speed of about 0.4 seconds—a speed-up over the human with desk calculator of between 25 and 40 times. The Bell Telephone computers had a rate of about 1 second and were thus about 10 to 15 times faster than the human *cum* desk machine. F. L. Alt said of these computers: "In comparison with hand com-puting, the speed of this machine is about five times that of a human computer with a good (fully automatic) desk machine. Since the machine can usually handle two problems simultaneously, its output, hour for hour, is equivalent to that of ten human compu-ters." [12] As we shall see, the ENIAC was to have a speed perhaps a thousand times greater than that of the Bell Laboratories' machine.

Now let us see what these numbers mean in terms of a ballistic trajectory, which is the basic element needed to make a firing or bombing table. Such a calculation is a way of determining via Newton's Laws of Motion the shape of the path taken by a shell,

[11] For a more detailed analysis the interested reader may consult H. H. Goldstine and J. von Neumann, "On the Principles of Large-Scale Computing Machines." This appears in von Neumann's Collected Works (New York, 1963), vol. v, pp. 1–32. (See p. 216, n. 10, for additional comment on this paper.)

[12] Alt, "A Bell Telephone Laboratories' Computing Machine." It is fair to state that the machine could run on a 24-hour basis, and this gives it another factor of per-haps 3—the doubt is because it sometimes got into difficulties when running un-attended and it stopped.

as a function of time, as it moves from the gun to some point, say on the ground. Of course, as the angle of elevation or muzzle velocity is changed this path will change. A typical trajectory or path required the order of 750 multiplications, and took a differential analyzer anywhere from 10 to 20 minutes to calculate it with a precision of about 5 parts per 10,000, which was about the utmost in accuracy attainable by such a machine. This precision corresponds to about four decimal digits and therefore means that the analyzer did the equivalent of 750 multiplications in 10 to 20 minutes. Thus it had an effective rate of 0.8 to 1.6 seconds per 4-digit numbers or 2 to 4 seconds for 10-digit ones, under the assumption that multiplication times are proportional to the number of digits. This indicates that the analyzer was roughly in the speed range of the relay machines such as the Bell Telephone computing machine.

The human *cum* desk calculator (10 seconds per multiplication) would then spend about 2 hours on the multiplying; and, with our estimate of a factor 6, about 12 hours doing an individual trajectory. This was about right, perhaps a little low. The Harvard-IBM machine (3 seconds) required about 2 hours; the Bell machine (1 second), about 2/3 hour; and the Mark II (0.4 seconds) about 1/4 hour. The differential analyzer took, as we have said, about 10–20 minutes.

These numbers, however crude, serve to show to some extent what was achievable by electromechanical devices; this was a good deal, as we see by noticing that the most advanced device was about 50 times faster than a human with desk machine. None of these was sufficient for Aberdeen's needs since a typical firing table required perhaps 2,000–4,000 trajectories—assume 3,000. Thus, for example, the differential analyzer required perhaps 750 hours—30 days—to do the trajectory calculations for a table; and the reader may estimate the time requirements for the digital machines easily.

These estimates reveal a situation that was unsupportable both because the volume of work was too large and, perhaps more importantly, because the work had to be done very promptly to avoid delays in putting weapons into the hands of the troops in the field. For these reasons Gillon and I were constantly on the lookout for better—quicker and more accurate—ways to expedite table calculation. Indeed, I did then and still do view my primary task in Philadelphia as being, first, one of finding new apparatus to expedite the production of firing and bombing tables, and, second, to make possible numerical explorations of exterior ballistics looking toward a more accurate mathematical description of the situation. The former task is of course a production type activity involving known mathematical procedures and is therefore highly routine although monumental in scope. In many ways it is akin to the astronomical table-making of Babbage, Scheutz, Wiberg, and others. The latter task, on the other hand, is a much more tentative one; it involves much trial and error and is therefore akin to the typical physicist's "quick and dirty" need.

We shall see later how the first requirement was met by a solution that did not do much for the second one, and how their lack resulted in another solution that started modern developments. Before discussing these things we need to say a few words about operational differences between analog and digital apparatus so that the reader will understand what is involved in subsequent pages.

We noted earlier that the nineteenth-century physicists had attained an ability to describe in mathematical form quite complex
machinery and were therefore able to invert this process. Given a
mathematical formula, they were, in principle at least, able to
invent a machine exactly describable by the formula. This is what
the analog computer is. What are the limitations on this idea?
There are fundamentally three aspects to the problem having to do
with the generality, the accuracy, and the speed of analog equipment.

While it is in principle possible to find a machine capable of
carrying out a given calculation, it is in practice sometimes very
hard. We saw that Kelvin could not realize practically his idea for
solving differential equations; and it was not done for over fifty
years, until Bush came along with his succession of devices. Thus
we see that the analog approach is ill-suited for wide varieties of
scientific calculations. Indeed we saw above (p. 90) that Bush
envisioned a number of quite different machines to try to span the
spectrum of needs of the mathematical engineer. This idea was not
wholly successful and lacked considerably in flexibility. In fact,
when Hartree tried to solve partial differential equations on a differential analyzer, he was forced to a mixed approach, analog in
one independent variable but digital in the other independent
variables.

Given a machine whose operation is describable by a mathematical formula, we quickly run up against two realities of engineering:
every machine is built out of parts having certain tolerances and
capable of running at certain speeds. These determine the accuracy
and speed with which the mathematical system can be solved. Not
only is there an upper limit on how exactly one can smooth physical
surfaces, cut gears, etc., there is also a degradation that sets in with
age when the parts become less accurate, so that the precision of
the total machine's response diminishes. Moreover, it is often true
that the faster a mechanical device is driven the less accurately it
depicts the mathematical situation, with the result that often accu-

racy and speed are linked together in a disastrous, or at least unhappy, way. So it was with the differential analyzer. If it was run
slowly, its accuracy was considerably better than when it was run
fast.

These restrictions on analog equipment are not shared by the
digital approach. Indeed what is this approach and what are its
limitations? To understand it we should first discuss a little the
nature of mathematical formulations of problems irrespective of
how they are to be solved. They normally involve varieties of mathematical operations, implicit definitions and limiting processes.
What are these things?

In addition to the elementary operations of arithmetic, there are
many much more complex mathematical operations such as, for example, square root, sine, logarithm, differentiation, integration.

Often quantities receive their expression in a mathematical
formulation implicitly. Thus, for example, to write

$$ax^2 + bx + c = 0$$

is a way of defining a quantity x; in fact, the equation is a statement
that for some quantity x the equality is valid. It is then one of the
quantities given by the formula, well known from our childhoods,

$$x = [-b \pm (b^2 - 4ac)^{1/2}]/2a.$$

Another example of an implicit definition is this: $y = y(x)$ is the
curve passing through the point $(0, 1)$ whose slope at each point is
equal to its height above the x-axis. Here we know a certain characteristic of a curve, and this fact contains in itself the detailed
shape of the curve. Indeed the curve is $y = \text{Exp}(x)$.

Finally, many mathematical operations or quantities are expressed as limits of sequences of more elementary ones. Thus, for
example, the well-known quantity π which we all met in geometry
classes may be expressed with the help of a sequence of regular
polygons of increasing numbers of sides all inscribed in a given
circle of unit radius. The lengths of the perimeters of these polygons are a set of numbers which steadily approach a given one,
which we call 2π. There are many other examples of definitions
made this way, but this should suffice to make the point.

To repeat: the mathematical formulation of a problem may result
in equations which involve operations that are transcendental and
definitions that are implicit as well as processes that are defined by
infinite sequences. This formulation has nothing to do with the
type of computing equipment, analog or digital, used to solve the

problem. Rather, it is in the nature of the problem being formulated. However, once it is decided to handle the problem by
numerical calculation on a particular kind of machine then all
transcendental operations must be replaced by elementary ones —
those operations the machine views as elementary, i.e. which the
computer can handle directly; all implicit definitions must be replaced by explicit ones involving finite and constructive procedures; and limiting processes must be truncated and replaced by
finite sequences of elementary operations.[1]

The operations that a given computer views as elementary are
those that are "built in." Thus, the Bush differential analyzer has as
its elementary operations differentiating, integrating, and solving
certain (essentially implicit) differential equations. The operation
of multiplication is not elementary for that machine but is compounded out of integration as

$$uv = \int u\,dv + \int v\,du.$$

Again, the human with a desk calculator usually does not have
square rooting as an elementary operation unless he happens to
have a table handy. He will have to treat this as "transcendental"
and form it by a set of elementary operations.

These things having been said, we may now be more precise
about what is meant by the digital approach. It is the realization
that a machine can be built to imitate the human method of calculating: to count and to build up the elementary operations — addition, subtraction, multiplication, division — by counting. Not only
can this be done but it may be shown that, in general, mathematical
formulations may be handled by means of these elementary operations. This is perhaps another way of paraphrasing Kronecker's
statement that "God created the integers, all else is the work of
man" or his other one that "all results of the profoundest mathematical investigation must ultimately be expressible in the simple
form of properties of integers."

It would carry us much too far afield to take up at this point the
delicate philosophical questions raised by mathematical logicians
on what are called computable numbers. But suffice it to say that
for our purposes numerical mathematics can be built up out of the
elementary processes of counting, and therefore that this approach
has a very real sense of universality or general purposeness about it.

---

[1] See J. von Neumann and H. H. Goldstine, "Numerical Inverting of Matrices of
High Order," *Bull. Amer. Math. Soc.*, vol. 53 (1947), pp. 1021–1099. In particular
Chapter I of that paper discusses the points we now are making.

Not only does the digital approach have this property, it has another which is also very attractive: to attain a given number of decimal places is always possible even though the cost may be great tedium. Note that this is not true of the analog approach where the accuracy of our knowledge of physical constants and the ability to build precisely are fundamental limitations. Thus, one can in principle effect a calculation digitally using 20 digits instead of 10 merely by increasing the quantity but not the quality of the labor, whereas to go from 4 to 8 digits on the differential analyzer is quite impossible.

Why then did Maxwell and Kelvin turn their backs on Babbage's work? The reason lies in speed considerations. Analog devices turned out in the ages of mechanical and electromechanical technologies to be decisively faster than digital ones. We recall (above, p. 11) that one of Babbage's machines was just about as fast as a tireless human, and we saw that a differential analyzer is about fifty times faster than a human.

Thus the digital approach would be a superb one if only it were very much faster. How this was achieved is essentially the theme of the rest of this book. We shall devote ourselves to the details of how this was effected in so short a time that a whole vast profession has grown up from nothing to one involving hundreds of thousands of people in a mere quarter century.

First, we should understand that either electromechanical relays or vacuum tubes can be used to build devices for remembering one of two possible states, i.e. a binary digit, and these can be compounded in a variety of ways to build a machine. We shall say more on this later. The fundamental difference between the two approaches—electromechanical and electronic—is entirely one of speed, although in the early days some felt it was reliability. A very fair appraisal of the situation is this:

A few years after the early work of Aiken and Stibitz on electromechanical components as computing elements, a group at the University of Pennsylvania began to study the applications of vacuum tubes to computation. This group, which included John Mauchly and J. P. Eckert of the University and Herman Goldstine, then of the Army Ordnance Department, was confronted with much more difficult problems than were the workers with the more conventional kinds of equipment. Whereas relays had been in commercial use for many years, and both the relays themselves and the circuits in which they would operate were well understood the use of electron tubes as relays was novel. Circuits for this application had to be developed, and the speeds that were sought were far above anything hitherto attempted in the computer field.[2]

Let us now accept the fact that one can build circuits either from relays or vacuum tubes to do computation, and let us ask how and why they differ in speed. After all, electricity flows pretty much with the velocity of light through wires. Why then is there a difference? To answer this we must understand or accept as fact that there are two time factors which enter into the total reaction time of a circuit involving a relay or vacuum tube. First, there is a time required to actuate the relay or tube and, second, a time to actuate the resistors, capacitors, and inductors making up the remainder of the circuit.

In the case of a relay it takes anywhere from 1 to 10 milliseconds —a millisecond is 1/1,000 second—to cause a relay physically to open or close. This time is due to the inertia of the mechanical parts, and it is in the nature of macroscopic things that most mechanical motions do take times in the millisecond, or longer, time scale. In the case of a vacuum tube—or of a transistor—all motion is at the microscopic level. The elements being moved are electrons, and they have masses of 9×10^{-28} grams as compared to relay contacts whose masses are probably about 1 gram. Thus there is essentially no inertia to be overcome, and a vacuum tube can be said to operate almost instantaneously for our purposes.

The time required to actuate the other parts of a circuit, e.g. to charge up the capacitors, is determined by the values of the resistances, capacitances, and inductances. It can be made very small. In the case of the ENIAC this time was on the order of 5 microseconds —a microsecond is 1/1,000,000 second. Thus the electronic technique is at least a thousand times faster than the electromechanical. In fact, at the present time (1971) electronic circuits can be made to operate in nanoseconds—billionths of a second. This is the reason why the first and most primitive electronic computer, the ENIAC, was several hundred times faster than the best of the relay machines.

What about reliability? Various people had argued *a priori* that the ENIAC could never operate because the tubes would fail so often that the "mean free path" between breakdowns would be measured in seconds. To avoid this the tubes with normal filament

[2] G. R. Stibitz and J. A. Larrivee, *Mathematics and Computers*, pp. 54–55.

voltages of 6.3 volts were operated at 5.7 volts and never turned off (turning on and off causes filament wires to change shape and sometimes as a result to create short circuits) and the plates and screens were operated at 25% of their rated power levels. Furthermore, all tubes were operated as on-off devices. This means they were driven to be either conducting by bringing their grids to a slightly positive level or non-conducting by forcing their grids much below their normal cut-off levels. In other words, no tube was ever used to represent a number by the magnitude of its output signal.[3]

These steps plus a number of others proved to be sufficient, and after the machine got into operation the failure rate was about 2 or 3 tubes per week. It was very reliable and very few hours were lost per week in tracing down the defective tubes. The greatest difficulty came about when on rare occasions two tubes would fail simultaneously. Then the observed symptoms were always highly anomalous, and von Neumann once jokingly described keeping the ENIAC operative as "fighting the Battle of the Bulge every day."

How reliable were the electromechanical machines of that era? Alt in his description of the Bell Laboratories' machine says on this score: "An extrapolation of the experience gained from the smaller relay-type machines of the Bell Telephone Laboratories leads to the expectation that the frequency of machine failures will eventually be at most one per day and possibly as low as one or two per week." We may therefore conclude not only that electronic devices provided an enormously faster solution to the computational problem but also that they were at least as reliable as electromechanical machines. These two facts spelled rapid doom for the electromechanical approach. The electronic machine allowed one to undertake calculations that were unthinkable with an electromechanical machine. Moreover, since both electronic and electromechanical systems had around two failures per week, the former was on a per operation basis enormously more reliable – i.e. because of the speed factor the probability that a given operation would be faulty was around a thousand times smaller for the electronic than for the electromechanical machine.

But why are these great speeds so important? Are they really necessary or have they been devised only to keep the new machine busy? Let us start the discussion by stating categorically that without the speed made possible only by electronics our modern com-

[3] See A. W. Burks, "Electronic Computing Circuits of the ENIAC," *Proc. IRE*, vol. 35 (1947), pp. 756–767.

puterized society would have been impossible: machines that can do as much as ten or twenty or thirty or even a hundred humans are very important but do not revolutionize modern society. They are extremely valuable and help greatly to ease the burden on humans, but they do not make possible an entirely new way of life.

Perhaps the best way to measure the need for speed is to examine one or two typical calculations and to count how much work is involved in carrying them out.

First let us take the ballistic trajectory problem we discussed a few pages ago. Here the position of a projectile—actually a point endowed with mass—in space is calculated as a function of time. There are three coordinates needed to locate the object at each of a sequence of times. It is not unreasonable to choose about 50 different times and to assume there are 15 multiplications to be done at each of these times. (This is how we arrived at our estimate of 750 multiplications per trajectory earlier.) The important thing to notice for this example, which is typical for the problems of Newtonian mechanics, is that there is one independent quantity, time. As soon as we move into more modern disciplines such as aero- or hydrodynamics, the situation becomes more complex. In problems of this sort—weather prediction is not atypical—there are more independent quantities. Usually what is wanted is a description of how some physical quantity such as pressure varies as a function not only of time but also of spatial position. Thus, one may quite reasonably ask about, let us say, the pressure at a set of points in the part of space under scrutiny at a given set of times.

First suppose there is only one spatial dimension, i.e. that our phenomenon proceeds spherically out from some source and spreads uniformly in all directions. Then we need examine just one spatial direction. For an actual case 100 spatial values and the order of 3,000 times were used; and around 50 multiplications were performed at each space-time point. Thus the amount of work involved is now around 5,000 multiplications for each different time; since there are 3,000 different times, the total amount of work has now gone up to 15 million multiplications. If this calculation were to be attempted on an electromechanical machine with a one-second multiplier, it would take the machine around a half year just to do the multiplying and therefore probably around two years to do the actual calculation.

Let us turn to the weather prediction calculation. Here it was, at least in early days, assumed that space is nearly two-dimensional; that is, for the first weather calculation—it was done on the ENIAC—

North America was divided up into 15 by 18 points, and each real
24-hour day was divided up into 8 time intervals. Without bother-
ing to estimate multiplications, we know it took just about 24 hours
to do a 24-hour forecast on the ENIAC.[4] Now unless a forecast for a
day can be performed in a time somewhat shorter than that day,
the forecast is of no real use to society since the whole point of
prediction is to notify people before the fact of what will happen.
Hence for calculations involving as independent variables either
one or two spatial and one temporal variable the need for speed is
essential.

We may now generalize and perhaps see the underlying principle. To do this let us suppose that at each point in space — be it
zero-, one-, two-, or three-dimensional — we do n multiplications,
that there are t time intervals, and $1, s, s^2$, or s^3 space points depending upon whether our space is 0, 1, 2, or 3 dimensional. Then the
total amount of calculating is proportional to nt, nts, nts^2, nts^3,
respectively. Thus for $n = 15$, $t = 50$, $s = 100$, which are unreason-
ably low estimates, we have: for zero spatial dimensions, about 750
multiplications; for one, about 75,000 — this is very low, as we saw
earlier; for two, about 75 million; and for three, about 750 million.
The amount of work is mounting exponentially, and nothing but
extreme speeds can make problems of these sorts attainable in any
reasonable sense.

Clearly efforts such as are typified by numerical weather prediction would be quite impossible without electronic computers.
This is then the whole point of the modern machines. It is not
simply that they expedite highly tedious, burdensome, and lengthy
calculations being done by humans or electromechanical machines. It is that they make possible what could never be done before! The electronic principle did much more than free men from
the loss of hours like slaves in the labor of calculating; it also
enabled them to conceive and execute what could not ever be done
by men alone. It is what made possible putting men on the moon —
and bringing others back safely from an abortive trip there.

[4] J. G. Charney, R. Fjörtoft, and J. von Neumann, "Numerical Integration of the
Barotropic Vorticity Equation," *Tellus*, vol. 2 (1950), pp. 237–254; or von Neumann,
Collected Works, vol. VI, pp. 413–430.

The reader should keep constantly in mind that at this time a truly definitive history of the electronic computer cannot be written since many things that happened in the early days (1943–1957) are still controversial and others are perhaps still classified. In setting down this account I have however felt that these dangers are less than the advantages that will accrue to later historians in having the detailed account of one of the few participants in this period who is still alive and in possession of a very complete set of the original documents. As I stated in the Preface, my effort here is to set out this account as objectively as possible, resting my various points on existing documents which I have given to the Library of Hampshire College in Amherst, Massachusetts. I hope that the result will in some measure furnish the present-day reader with a coherent and accurate account of the field and the future historian with material to put the formative years into perspective.

Sometime in the fall of 1942 I first became acquainted with John W. Mauchly, who displayed considerable interest in Aberdeen's computing problems. Mauchly in fact had a dual interest in computation. While still at Ursinus College he had not only worked with a harmonic analyzer but he had also thoroughly familiarized himself with Atanasoff's ideas. He also had thought a fair amount about the application of statistics to a diverse number of fields including weather prediction, which he frequently discussed with men such as the late Sam Wilks of Princeton University. His concern with these applications did not mature in the usual way and result in the production of papers on the subject, but it did suffice to keep him thinking about machines to handle the underlying mathematical tasks.

During 1941 Mauchly was so stimulated by his conversations with Atanasoff that he was sketching in his laboratory notebook various emendations to Atansoff's ideas. By August 1942 he had advanced in his thinking enough to write a brief memorandum summarizing his ideas; this was circulated among his colleagues and perhaps most importantly to a young graduate student, J. Presper

Eckert, Jr., who was undoubtedly the best electronic engineer in the Moore School. He immediately, as was his wont, immersed himself in the meager literature on counting circuits and rapidly became an expert in the field. This was to have inestimable import just a year later.

Mauchly and I had fairly frequent and mutually interesting conversations about computational matters during the fall of 1942. These talks served to emphasize to me Mauchly's point about the "great gain in the speed of the calculation . . . if the devices which are used employ electronic means for the performance of the calculation, because the speed of such devices can be made very much higher than that of any mechanical device." [1]

In March 1943 I indicated my very considerable interest in all this to Brainerd, who made available Mauchly's ideas and his own judgment that they were not unreasonable. I then conferred on the problem at some length with Gillon, and we agreed on the desirability of the Ordnance Department underwriting a development program at the Moore School looking toward the ultimate production of an electronic digital computer for the Ballistic Research Laboratory. Gillon in his positive and enthusiastic way pushed the matter forward with great celerity. Specifically on 2 April 1943 at my request Brainerd prepared a report for submission to Gillon and Simon which could and in fact did form the basis for a contractual relationship between the Moore School and the Ordnance Department. [2]

From that moment on things moved very fast indeed. On 9 April Brainerd, accompanied by Eckert and Mauchly, attended a conference at Aberdeen with the director of the laboratory, Col. Leslie E. Simon. This meeting had been preceded by others. One of these involved Simon, T. H. Johnson, Veblen, and myself at which Veblen after listening for a short while to my presentation and teetering on the back legs of his chair brought the chair down with a crash, arose, and said, "Simon, give Goldstine the money." He thereupon left the room and the meeting ended on this happy note. Thus the 9 April meeting was foreordained to be fairly successful, and it was. There was some concern voiced over the large number of tubes the

[1] Memo, Mauchly to Brainerd entitled "The Use of High-Speed Vacuum Tube Devices for Calculating," August 1942, included as Appendix A by Brainerd in a report to Aberdeen dated 2 April 1943. The dating of August 1942 is somewhat peculiar since Brainerd's acknowledgment of the document bears the date 1/12/43.

[2] J. G. Brainerd, "Report on an Electronic Diff. [sic] Analyzer, Submitted to the Ballistic Research Laboratory, Aberdeen Proving Ground, by the Moore School of Electrical Engineering, University of Pennsylvania, First Draft, April 2, 1943."

machine would contain—around 18,000. Johnson, who headed the electronics work at the Ballistic Research Laboratory, very properly expressed his apprehensions on this point, and this resulted in various meetings with NDRC and RCA people. The work began 31 May 1943, and a definitive contract was entered into on 5 June 1943. From this it must be clear that Gillon's evaluation was correct; he said: "I do know, however, that the proposal was brought to me in Washington on Friday, 8 April 1943, by Herman Goldstine and Grist Brainerd. I felt that the proposal represented an exceedingly important venture which should be backed by Ordnance to the hilt. With the excellent help of Sam Feltman I succeeded in swiftly obtaining the funds and authorization to proceed and Sam, Herman, and I got the contract through in record time." [3]

Just before the start of work there was a final meeting at the Moore School chaired by Gillon at which L. E. Cunningham, the astronomer, T. H. Johnson, and I from the Ballistic Research Laboratory, and Brainerd, Eckert, and Mauchly from the Moore School attended. At this meeting Gillon named the proposed machine the Electronic Numerical Integrator And Computer and gave it the acronym ENIAC. Much of substance was accomplished then. Cunningham and I reviewed with the group the operational needs of the laboratory as viewed mathematically; Brainerd and his colleagues discussed the then current state of the art and the proposed units for the machine and costs and completion dates. The last topic was the viewpoint of the Electronic Division of the NDRC. This group was largely under two sets of influences: one being the Stibitz view, which was digital but electromechanical, and the other the Caldwell one, which was analog but partially electronic. However it was pretty well decided to go ahead at this meeting in the face of the not negligible weight of the NDRC groups. This matter was still being discussed in October 1943, as we will see later. In any case a few days later Brainerd was writing to Pender to acquaint him formally with what had transpired and to obtain his "approval for continuing the discussions, the goal of which would be a definite proposal which might be submitted to the University's Executive Committee." [4] In the same memo he states that the development cost would run about $150,000.

It is of some interest to read today the views of experts in the design and development of computers in the 1940s. In the fall of

[3] Letter, Gillon to W. D. Dickinson, Jr., 19 November 1958. Dickinson was the head of administration at the Ballistic Research Laboratory.

[4] Memorandum, Brainerd to Pender, Proposed Army Project, 26 April 1943.

1943 we find two internal memoranda of the NDRC regarding electronic computation that very well summarize the world—or at least the U.S.—view on the subject:

> From the description given by Colonel Gillon, it would appear that the specifications of the equipment . . . substantially duplicates the specifications of the Rapid Arithmetical Machine program. . . .
>
> I feel that I should add one further comment. Apparently my earlier letter to you gave Colonel Gillon an incorrect impression of the extent of our program. He quotes my statement "We simply did not believe that we were in a position to build anything that was practical on the basis of our knowledge of the art." By that, I meant that there was not a sufficient body of adequate basic technique to permit the immediate construction of a well-engineered machine. It did not mean that we did nothing at all. On the contrary, we devoted about three years to research in basic circuits and components and were in the process of studying designs for the combination of such equipment into an overall machine when the work was suspended for the duration.
>
> As far back as 1939, we realized that we could build such a machine and possibly to make it work, we did not consider it practical. The reliability of electronic equipment required great improvement before it could be used with confidence for computation purposes. . . .
>
> Another thing which caused us to discard the idea of building a complete machine immediately was the very large number of small parts required. Colonel Gillon speaks of the machine containing thousands of tubes. In addition it contains thousands of resistors, condensers, and wiring terminals. One of the basic objectives of our "grass roots" program was to simplify the elements of the machine.[5]

And in the other memo:

> It was pointed out, I believe, in the memorandum sent to Colonel Gillon, that the speed of operation of a computing mechanism has, in itself, little significance. The essential problem is that of obtaining sufficient load-carrying capacity as economically as possible. Thus even though a particular machine may not be extemely fast, the load may be divided among several

[5] Memorandum, S. H. Caldwell to Harold L. Hazen, 23 October 1943.

such units, thus obtaining all the overall speed that may be required. . . .

I see no reason for supposing that the relay device RDAFB is less broad in scope than the ENIAC, since each is a numerical calculator and can presumably perform exactly the same operations provided switching and storage problems in connection with the ENIAC can be solved. I think the ideal equipment would probably be a combination of relay and electronic devices, but I am very sure that the development time for the electronic equipment will be four to six times as long as that for the relay equipment.[6]

These represent rather typically the views of the engineering and applied mathematical communities at this juncture in history. This is perhaps why the Moore School accomplishment was to be so great. The risks involved were incredible and were taken by Brainerd very coolly, without flinching. He wrote to his dean asking permission to proceed thus: "It has been pointed out . . . that this is a development project and that there is no certainty that the desired result can be achieved. It is, however, a reasonable chance." [7]

In my opinion the contributions made at the beginning were these: (1) Mauchly's technical foresight and courage in realizing that the time was ripe to introduce electronic techniques into digital computation. (2) Brainerd's calm acceptance of the risks for the University of Pennsylvania in the face of the contrary opinions of his peers at other famous engineering institutions. (3) My appreciation of the importance of Mauchly's raw ideas: realization that they constituted a decisive way to cut the Gordian knot into which the firing and bombing table work was tied; and ability to obtain stable financial backing for the enterprise. (4) Gillon's enthusiastic and decisive backing of the project at all times but especially in its nascency when it was so urgently needed. Later of course others, as we shall see, made fundamental contributions, and the roles of those just cited were to evolve.

To illustrate Gillon's accomplishment we may quote from a memorandum he wrote in rebuttal to the ones quoted above:

14. The fact that the ENIAC is electronic in character will enable it to operate at a speed much greater than that of the RDAFB. Present indications are that . . . in the particular case of

[6] Memorandum, G. R. Stibitz to W. Weaver, 6 November 1943.
[7] Memorandum, Brainerd to Pender, 26 April 1943.

the exterior ballistic equations, the time for obtaining a result would be of the order of one minute, whereas the time required for the RDAFB for the same operation would be comparable to that of the Bush Differential Analyzer and would be of the order of 45 minutes. Consequently the strongest reason for the development of the ENIAC is the anticipated speed of operation. . . .

16. To recapitulate the foregoing, it appears that the ENIAC if successfully developed will completely satisfy the original criteria. . . .[8]

To gain some rough measure of the magnitude of the risks, we should realize that the proposed machine turned out to contain over 17,000 tubes of 16 different types operating at a fundamental clock rate of 100,000 pulses per second. This latter point means that the machine was a synchronous one, receiving its heart-beat from a clock which issued a signal every 10 microseconds. Thus, once every 10 microseconds an error would occur if a single one of the 17,000 tubes operated incorrectly; this means that in a single second there were 1.7 billion $(= 1.7 \times 10^9)$ chances of a failure occurring and in a day $(= 100,000$ seconds) about 1.7×10^{14} chances. Put in other words, the contemplated machine had to operate with a probability of malfunction of about 1 part in 10^{14} in order for it to run for 12 hours without error. Man had never made an instrument capable of operating with this degree of fidelity or reliability, and this is why the undertaking was so risky a one and the accomplishment so great. Indeed to this day the computer represents man's most complex device. He has never before or since produced a device where the probability of failure has to be so low, unless it be the space capsules with all their attendant computers. A. W. Burks has provided some interesting statistics: "In addition to its 18,000 vacuum tubes the ENIAC contained about 70,000 resistors, 10,000 capacitors, and 6,000 switches. It was 100 feet long, 10 feet high, and 3 deep. In operation it consumed 140 kilowatts of power."[9]

Above all others the man who made it possible to achieve the almost incredible reliability needed for success was J. P. Eckert. He was the chief engineer and had Mauchly as his consultant. The roles of these two men are interesting to look back on and attempt to evaluate. Eckert fully understood at the start, as perhaps none of his colleagues did, that the overall success of the project was to depend entirely on a totally new concept of component reliability

[8] Letter, Gillon to Hazen, 7 October 1943.
[9] A. W. Burks, "Electronic Computing Circuits of the ENIAC."

and on utmost care in setting up criteria for everything from quality of insulation to types of tubes.

To typify what I mean let me quote from a set of engineering dicta issued by Eckert in 1943 to his engineers: "In order to obtain the desired tube life (which we hope will be at least 2,500 hours), the ratings in the tube manual should be modified as follows: . . . The plate voltage should be kept at not more than 50% of the rated maximum voltage. The plate current should be kept at not more than 25% of the rated maximum current."

In every other component, resistors, condensers, wiring boards, tube sockets, etc., Eckert set most exigent standards and insisted that there be no exceptions. He also developed basic counting circuits and saw to it that all his engineers used these in their designs.

Eckert's standards were the highest, his energies almost limitless, his ingenuity remarkable, and his intelligence extraordinary. From start to finish it was he who gave the project its integrity and ensured its success. This is of course not to say that the ENIAC development was a one-man show. It was most clearly not. But it was Eckert's omnipresence that drove everything forward at whatever cost to humans including himself.

The work on the project actually began, as said before, on 31 May 1943, and the definitive contract was dated 5 June 1943. It was stated in this document that the University "in cooperation with and under the direction of representatives of the Ballistic Research Laboratory . . . shall engage in research and experimental work in connection with the development of an electronic numerical integrator and computer. . . ." The University agreed to furnish copies of reports and in the event "the contract results in the fabrication and completion of any part or unit . . . [it] shall be delivered to the Government. . . ." [10]

Some confusion seems to exist in people's minds as to which agency of the government was responsible for the ENIAC. The entire contract was under the jurisdiction of the Ordnance Department of the War Department. However, since the actual allocation of responsibilities was complex, perhaps a few words should be said

---

[10] Fixed Price Development and Research Contract No. W-670-ORD-4926 between the United States of America and the Trustees of the University of Pennsylvania. There were to be in all twelve Supplements to this contract, the last being dated 16 November 1946. The total amount expended under the contract including its supplements was $486,804.22.

on the point. First, the actual agency which handled the contract was the Philadelphia Ordnance District, one of the units of the Ordnance Department set up to procure material for that department. Second, the technical responsibility was vested in the Ballistic Research Laboratory, Aberdeen Proving Ground, Maryland. Third, the overall responsibility for all phases including funding was in the Technical Division of the Office of the Chief of Ordnance. With this multifaceted set up it is a testimonial to Gillon and Simon that things went so smoothly. Both men knew, trusted, and admired each other and therefore did whatever was needful without question.

On 9 July of 1943 I was formally appointed as the representative of the Ballistic Research Laboratory. From then on until I left the Army I was to give almost my full time to the research and development work specified in this and another contract to be discussed later. As mentioned earlier, it was due in great measure to Brainerd's fine sense of fairness and his invariable poise that relations were so good between the university and Aberdeen and Washington. There were certainly many — seen retrospectively — minor crises but all but one were solved in such a manner that no lasting ill-will developed. The one which was not of this sort arose over patents and helped to fragment the Moore School staff, as we shall see in the proper place.

At the beginning, at least, Mauchly was to continue to play a key role in the ENIAC project. He alone of the staff of the Moore School knew a lot about the design of the standard electromechanical IBM machines of the period and was able to suggest to the engineers how to handle various design problems by analogy to methods used by IBM. Then as time went on his involvement decreased until it became mainly one of writing up of patent applications. Mauchly was at his absolute best during the early days because his was a quick and restless mind best suited to coping with problems of the moment. He was fundamentally a great sprinter, whereas Eckert not only excelled at sprinting but was also a superb marathon runner.

Eckert was very fortunate in having a first-class engineering group associated with him largely composed of men whom Brainerd had assembled over the war period. It consisted of the following persons: Arthur Burks, Joseph Chedaker, Chuan Chu, James Cummings, John Davis, Harry Gail, Adele Goldstine, Harry Huskey, Hyman James, Edward Knobeloch, Robert Michael, Frank Mural, Kite Sharpless, and Robert Shaw.

There has been considerable controversy over exactly who invented the ENIAC and no definitive account can be written on this at the present time. Posterity will have to make this judgment, if it can be made at all. What we can do here is to ignore these legalistic considerations, not because they are unimportant to some but rather because they are unresolvable now. Instead we can record roughly the principal intellectual contributions to the ENIAC made by these engineers.

In the first place, Eckert's contribution, taken over the duration of the project, exceeded all others. As chief engineer he was the mainspring of the entire mechanism. Mauchly's great contributions were the initial ideas together with his large knowledge of how in principle to implement many aspects of them.

Instead of my trying to summarize each person's contributions to what was, at least to me, a joint effort, let me just say that the senior engineers were Arthur Burks and Kite Sharpless, who somehow divided the overall systems responsibility with each other and with Eckert and Mauchly and who designed large pieces of the machine. Next in order were John Davis and Robert Shaw, who also made major contributions to pieces of the machine such as the accumulators and function tables. The others were also important to the project and not one of them could have been easily dispensed with. Their contributions were entirely noteworthy, and hopefully some day a definitive assessment can be made so that each person connected with the staff can receive his due credit. However each in his own way performed one of the labors of Hercules and should be remembered for it.

Chapter 5   The ENIAC as a

Mathematical Instrument

At this point we ought to pause and describe the ENIAC as a mathematical instrument so that the reader may better understand what it did and how. In doing this we should understand that the machine was unique and primitive; most of the underlying architectural and organizational ideas were abandoned after it was completed, and therefore we are in some sense dissecting a dinosaur.

A paper written at the time describes the machine generally in these terms:

> The machine is a large U-shaped assemblage of 40 panels . . . which together contain approximately 18,000 vacuum tubes and 1,500 relays. These panels are grouped to form 30 units . . . each of which performs one or more of the functions requisite to an automatic computing machine.
>
> The units concerned mainly with arithmetic operations are 20 accumulators (for addition and subtraction), a multiplier, and a combination divider and square rooter.
>
> Numbers are introduced into the ENIAC by means of a unit called the constant transmitter which operates in conjunction with an IBM card reader. The reader scans standard punched cards (which hold up to 80 digits and 16 signs) and causes data from them to be stored in relays located in the constant transmitter. The constant transmitter makes these numbers available . . . as they are required. Similarly results . . . may be punched on cards by the ENIAC printer unit operating in conjunction with an IBM card punch. Tables can be automatically printed from the cards by means of an IBM tabulator.
>
> The numerical memory requirements of the machine are met in several ways. Three function table units provide memory for tabular data. Each function table has associated with it a portable function matrix with switches on which can be set 12 digits and 2 signs for each of 104 values of an independent variable. . . . Numbers formed in the course of a computation and needed in

subsequent parts of a computation can be stored in accumulators. Should the quantity of numbers formed during a computation and needed at a later time exceed the accumulator storage capacity, these numbers can be punched on cards, and, later, can be reintroduced by means of the card reader and constant transmitter.[1]

The arithmetic units, the accumulators, were not unlike the electromechanical digital machines of that era in their mode of operation. These electromechanical devices contained counter wheels which could be turned one stage at a time by the reception of an electric signal. Electronic ring counters are analogous to these wheels. To explain their action let us first mention what is called a flip-flop or trigger circuit. This basic electronic memory device consists of a pair of vacuum tubes so interconnected that at each instant of time exactly one of the pair is conducting (i.e. current is flowing through the tube) and the other is non-conducting (no current is flowing). When one of the tubes is conducting we may say the flip-flop is in the "1" state and when the other is, the "0" state.

A counter in the ENIAC consisted of a linear array of flip-flops connected in such a fashion that at each instant (a) exactly one flip-flop of the array was in the "1" state, all the others being in the "0" state; (b) when a pulse was received by the counter this flip-flop was returned to the "0" state and its successor in the array was turned from the "0" to the "1" state; (c) there was a provision to reset the counter so that always a pre-assigned flip-flop, called the first stage, was set to the "1" state.

All the ENIAC counters were ring counters, which means simply that the first and last stages were so interconnected that whenever the counter was in its last stage and a pulse was received it cycled to its first stage, i.e. the successor of the last stage in the array was the first stage. This made a ring counter very like a counter wheel in principle of its action, but it was very much faster.

Since ENIAC was a decimal machine capable of handling numbers of 10 (decimal) digits each plus signs, each accumulator—the units that effected the addition and subtraction operations—contained 10 ring counters each of 10 stages and a 2-stage ring counter for the sign of the number. The stages of the 10-stage counters corresponded to the digits 0, 1, . . . , 9, respectively. The counters in

[1] H. H. and Adele Goldstine, "The Electronic Numerical Integrator and Computer (ENIAC)," *Mathematical Tables and Other Aids to Computation*, vol. II (1946), pp. 97–110.

each accumulator were interconnected by "carry" circuits so that when any counter advanced through stage 9 back to 0 a pulse was injected into the next counter to the left signaling the fact that a carry took place.[2]

Numbers were transmitted in the ENIAC in the form of trains of pulses. All ten decimal digits and the sign of a given number were transmitted in parallel over eleven conductors, a given digit $d(0 \leq d \leq 9)$ being represented by a train of d pulses sent over the appropriate conductor. On the sign conductor no pulses were sent for a positive number and 9 for a negative one.

When a number was received by an accumulator, it was added to the prior contents of that unit as indicated above. This leaves open the question of how subtraction was performed. This operation was performed as a type of addition through the representation of negative numbers by what are called "complements." This mysterious sounding concept is actually quite simple. As the accumulators were built, any integer between 1 and $10^{10} - 1$, inclusive, was uniquely represented by the 10 decade counters but 0 and 10^{10} were identical, the largest number that can be held in 10 decade counters being $9,999,999,999 = 10^{10} - 1$.

This being so the sum $(10^{10} - N) + N = 0$ as far as the decade counters were concerned. For convenience the quantity $(10^{10} - N)$ was referred to as the *complement* of N with respect to 10^{10}. Thus, the complement of the positive integer 1843 is $10^{10} - 1843 = 9,999,998,157$. Instead of transmitting the number -1843, the ENIAC transmitted the complement of 1843, i.e. $9,999,998,157$, on the 10 decade conductors and 9 pulses on the sign conductor to indicate it was a negative number.

We can now see how, by addition, subtractions were handled. Perhaps a few illustrations will make things clearer. Consider first the sum 8429 plus 1843

$$P\ 0\ 0\ 0\ 0\ 0\ 0\ 8\ 4\ 2\ 9$$
$$P\ 0\ 0\ 0\ 0\ 0\ 0\ 1\ 8\ 4\ 3$$
$$\overline{P\ 0\ 0\ 0\ 0\ 0\ 1\ 0\ 2\ 7\ 2};$$

next the difference 8429 minus 1843

$$P\ 0\ 0\ 0\ 0\ 0\ 0\ 8\ 4\ 2\ 9$$
$$M\ 9\ 9\ 9\ 9\ 9\ 9\ 8\ 1\ 5\ 7$$
$$\overline{P\ 0\ 0\ 0\ 0\ 0\ 0\ 6\ 5\ 8\ 6};$$

[2] Actually this is a slight oversimplification. For details see the article cited in the previous note.

and lastly the difference 1843 minus 8429

$$
\begin{array}{l}
\text{P 0 0 0 0 0 0 1 8 4 3} \\
\underline{\text{M 9 9 9 9 9 9 1 5 7 1}} \\
\text{M 9 9 9 9 9 9 3 4 1 4.}
\end{array}
$$

The symbols P and M, of course, symbolize plus and minus and stand for 0 and 9 pulses, respectively. This is why in the second example the answer contains a P: in the sign counter the sum of the 0 pulses and the 9 is 9 and the carry from the preceding stage is 10 which is treated exactly as 0 is. In the third example, however, there is no carry from that stage and hence the sign counter receives 9 pulses, which is M.

An addition or subtraction took 200 microseconds = 1/5,000 of a second. This was, not unnaturally, called an addition time. The multiplier unit was able to do a multiplication in at most 14 addition times or ~3 milliseconds = 3/1,000 of a second. To achieve this speed it had built into it an electronic device that remembered the familiar multiplication table of our childhoods. Division is, in the decimal system, a more complex business, and it took about 143 addition times or ~30 milliseconds = 3/100 of a second. Similarly for square root. The meaning of these speeds was well expressed by L. J. Comrie:

Tick . . . Tick . . . Tick . . . Tick.

Three seconds . . . a thousand multiplications finished. The clock ticks on . . . In an hour a million will be finished.

That is a picture of the present, not a dream of the future. It doubtless exceeds the bounds of the imagination of Jules Verne, Wells, or Tennyson, who wrote:

> For I dipt into the future, far as human
> eye could see,
> Saw the Vision of the world, and all the
> wonder that would be.[3]

It remains now to describe how one instructed the ENIAC to do a given task. This was a highly complex undertaking and was one of the reasons why it was to be a unique machine. This aspect of the machine was unsatisfactory, as the evolutionary process was to reveal. Most units contained individual program controls which were capable of noting that the unit was to perform a given opera-

[3] L. J. Comrie, "Calculating—Past, Present, and Future," *Future*, Overseas Issue (1947), pp. 61–68.

tion such as addition and also generally to signal when this was completed. Usually several program controls were needed to carry out completely an arithmetic operation. Thus, if the number stored in Accumulator 1 was to be added to that in Accumulator 2, a control on 1 had to be instructed to transmit its contents and a control on 2 to receive. Parenthetically, a channel between the two accumulators had also to be established so that the numbers could actually flow between them.

Let us program a trivial but perhaps illustrative example to show how this all took place. The example we choose is to form a table of squares. Thus at the start we assume Accumulator 1 contains n and Accumulator 2 contains n^2 and we show how to form $n + 1$ and $(n + 1)^2$. This is done in Fig. 11.

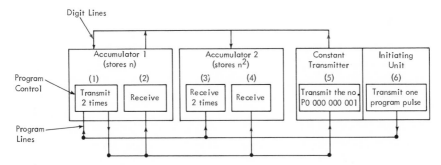

Fig. 11

At the start we may suppose that Accumulator 1 contains n and Accumulator 2 contains n^2 and that somehow — by pushing a button — the initiating unit transmits a pulse causing Accumulator 1 to transmit its contents twice and Accumulator 2 to receive it twice. Then, upon the execution of this operation, Accumulator 2 will contain $n^2 + 2n$ and Accumulator 1 will send out a signal which causes the constant transmitter to send out a 1 and both accumulators to receive it. Thus Accumulator 1 now contains $n + 1$ and Accumulator 2 contains $n^2 + 2n + 1 = (n + 1)^2$. At this point the program stops since no other program pulses are transmitted.

How could this be used in an iterative fashion to build up a table for $n = 0, 1, \ldots$? This was the function of a unit called the Master Programmer. It contained 10 program controls each of which could count program pulses and switch program connections. To illustrate its use let us suppose we desired to compute and print n, n^2 for $n = 0, 1, \ldots, 99$.

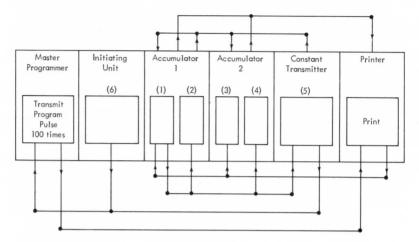

Fig. 12

In Fig. 12 the program controls (1), (2), . . . (6) are as before. Note the differences in the actions, however, caused by the presence of the Master Programmer. The accumulators are assumed to be cleared to zero at the start. Then the action of the Initiating Unit is to start the Master Programmer to count the printings and to order the first one to take place; this records $n = 0$, $n^2 = 0$. The result is to cause the accumulators to do as before; i.e. Accumulator 1 sends out 2×0 and Accumulator 2 receives this number causing the Constant Transmitter to send a 1 into each and to advance the Master Programmer one step. Now we print $n = 1$, $n^2 = 1$ and repeat our fundamental cycle as before. This goes on until 100 pulses have been emitted by the Master Programmer, at which time the last printing takes place and the process ceases. This yields a table of n vs. n^2 for $n = 0, 1, . . . , 99$, as desired.

This illustration was of course trivial, but it does serve to illustrate the operation of the system. We cannot afford to spend more time on it, but the interested reader may read more in considerable detail.[4]

At the inception of the project some of the engineers from the project and I had the good fortune to make several visits to the RCA Research Laboratories in Princeton, N.J., where we were shown a number of very interesting developments by Dr. V. K.

[4] See, e.g., the Comrie article cited in n. three above; A. W. Burks, "Electronic Computing Circuits of the ENIAC, *Proc. IRE*, vol. 35 (1947), pp. 756–767; A. K. Goldstine, *Report on the* ENIAC (*Electronic Numerical Integrator and Computer*), *Technical Report 1*, 2 vols., Philadelphia, 1 June 1946.

Zworykin and Dr. J. A. Rajchman. We will have more to say of these men later. At that time we saw a pilot model of an electronic digital fire control computer designed by RCA for Frankford Arsenal. Although this project never reached maturity, it contained at least one device that was adopted by the Moore School people and that was fundamental to the ENIAC.

This was the so-called function table, which was a way of storing a table of numbers in the form of an electrical network of resistances. It was able to operate at the normal speed of the ENIAC and made possible ENIAC's operation. At a later time, as we shall see, these tables were made part of a very ingenious remodeling of the ENIAC. It should be noted that these function tables were apparently invented independently by Rajchmann at RCA and Perry O. Crawford, Jr., then at MIT and now with IBM.

At the beginning there were of course any number of meetings and discussions, both about the ways to handle various functions electronically and about the mathematical characteristics and desiderata of the machine itself. In connection with the latter, Cunningham was to be of greatest value. It soon emerged that the machine would be much more useful than just a device for solving the differential equations of exterior ballistics. It gradually became clearer that the great advantage of the digital approach was that the ENIAC was going to be a truly general-purpose device. This meant that the machine would be inestimably more useful than had been originally considered—and indeed it was. It was to play a key role in a number of diverse fields and problems.

Gradually a first goal of producing a pair of accumulators was agreed upon as a test as to whether it made sense to proceed. At this juncture I was hospitalized with infectious hepatitis but in Brainerd's words: "Cunningham said you were running what amounted to an office by your bedside in the afternoons."[5]

After a few months of getting the "bugs" out of the two accumulators, Dean Pender was able to write me the following:

> The actual operation of two accumulator units of the ENIAC has been a source of satisfaction to all of us. Some of the details have been described to you over the telephone. I thought I would take this opportunity to say that the results of the tests have been gratifying not only in a general way but also in very specific manners. For example, the actual operation has indicated that the numerous different types of subsidiary units which go into

[5] Letter, Brainerd to Goldstine, 25 May 1944.

an accumulator have been properly designed. No major mistake has turned up. . . .

The present situation is that we have two accumulators together with the necessary auxiliary equipment, so that we have been able to do such things as solve the second order differential equation for a sine wave and that for a simple exponential, as well as go through the processes of addition, subtraction, automatic repeated addition, etc. These we have been able to do at the rated speed of the machine, which is so rapid that it is impossible to follow the results on indicating lights, or at a slower pace which enables the results to be followed. . . .

On the whole I believe there is reason for moderate optimism. This is, of course, a research and development project, and there is always the possibility that an unforeseen problem will develop which may range from serious to impossible. However, up to the present time we have met no insuperable difficulties, and see nothing in the future which should be more difficult than has already been overcome in the past. . . .[6]

This point, roughly one year after its inception, was the watershed of the project. In that time the project had gone from a highly speculative idea to an engineering status that ensured its success. Although an inordinately — virtually daily — large number of minor crises were still to be met and overcome, it was clear from mid-1944 on that the machine would be finished successfully.

The only part of the machine the Moore School felt it was incapable of developing — much less of building — was the unit needed for introducing and withdrawing data, as well as one for printing the results in human-readable form. Unless these problems could be reasonably solved there was grave danger that the utility of the ENIAC would be considerably lessened. It therefore seemed completely reasonable to think in terms of punch card equipment for this purpose. Accordingly, Gillon visited Thomas J. Watson on 28 February 1944, to engage his help. As indicated above (p. 112), Watson was always extremely sensitive and responsive to novel ideas that would extend the frontiers of digital computation. In this case he put the Moore School people in touch with Messrs. J. W. Bryce and A. H. Dickenson to work out a satisfactory solution. Gillon wrote:

I wish first to express to you my sincere personal appreciation of the many courtesies extended to me. . . .

[6] Letter, Pender to Goldstine, 3 July 1944.

As you know, I had a further discussion that afternoon with Mr. Bryce, and arranged for a conference. . . .

This conference brought together interested members of your staff, of this office, of the faculty of the Moore School of Electrical Engineering, and of the Ballistic Research Laboratory. . . .

From the survey so far made, it appears that a very satisfactory solution to our problem is possible with your assistance. I feel certain that the necessary details connected with the work can be swiftly arranged.

Although the undertaking from your viewpoint may appear to be a comparatively small one, its intrinsic importance relative to our overall ballistic computing problem is very great.[7]

The last paragraph quoted from the letter is amusing to read today. As it turned out, the undertaking from Mr. Watson's and IBM's viewpoint was to be crucial. In a certain sense this was Mr. Watson's tentative introduction to the world of electronic computing.

In order to provide a proper speed matching between the ENIAC's vacuum tube circuits and the IBM punches it was decided to build the Constant Transmitter, previously mentioned, out of relays of the sort being used so successfully by the Bell system. Accordingly, Gillon also approached Dr. Oliver E. Buckley, then president of the Bell Telephone Laboratories. I was able to write Gillon in May of 1944 that "The IBM and Western Electric people have been very cooperative. . . ."[8]

It is not without interest to record once again the need for the ENIAC as felt by Gillon and myself in 1944. In a draft of a letter to Dr. Buckley I wrote:

The Ballistic Research Laboratory is charged with preparation of all firing and bombing tables as well as all director and cam data for the Army Air Forces and Army Ground Forces, and consequently has one of the largest computing staffs in the country constantly engaged in the preparation of these firing and bombing data. This Laboratory also carries on basic research in ballistics, which of course requires in many cases carrying out of exceedingly large scale computing projects.

In addition to a staff of 176 computers in the Computing Branch, the Laboratory has a 10-integrator differential analyzer at Aberdeen and a 14-integrator one at Philadelphia, as well as a

[7] Letter, Gillon to T. J. Watson, 4 March 1944.
[8] Letter, Goldstine to Gillon, 26 May 1944.

large group of IBM machines. Even with the personnel and equipment now available, it takes about three months of work on a two shift basis to turn out the data needed to construct a director, gun sight, or firing table. . . .[9]

About six months later I compiled some other statistics to show the delays being experienced in producing the tables that were then so urgently needed. I summarized these in the following:

(4) The following summary for the week ending August 15, 1944 is representative of the quantity of this work, in the computing Branch alone, . . . :

	Completed	In Progress
Ground Gunfire Tables	10	16
Bombing Tables	1	28
Aircraft Fire Tables	3	16
Ballistic Tables		6
Miscellaneous Major Computations	1	8

The number of tables for which work has not been started because of lack of computational facilities far exceeds the number in progress. Requests for the preparation of new tables are being currently received at the rate of six per day.[10]

[9] Draft letter, Gen. G. M. Barnes to Dr. O. E. Buckley, around 1 February 1944.
[10] B.R.L. Memo, 16 August 1944. This was apparently a working memo to justify further efforts on Ordnance's part in pushing digital computing.

Chapter 6   John von Neumann and

the Computer

John Louis Neumann was born into a well-to-do family in Budapest on 28 December 1903, during the last socially brilliant days of that city under the Hapsburgs. His father Max, a banker, was a partner in one of the city's important private banks and was able to provide well for his children both intellectually and financially. He was ennobled in 1913 by the Emperor with the Hungarian title of *Margattai*, which young von Neumann later Germanized to *von*. His father and mother, Margaret, had three sons, John, Michael, and Nicholas, of whom John was the eldest.

While still very young, von Neumann showed tremendous intellectual and linguistic ability, and he once told the author that at six he and his father often joked with each other in classical Greek. He studied history as an avocation and became a first-rate historian. Later he was to concentrate on the culture of Byzantium, and he had a truly encyclopedic as well as profound knowledge of this as well as a number of other societies.

One of his most remarkable capabilities was his power of absolute recall. As far as I could tell, von Neumann was able on once reading a book or article to quote it back verbatim; moreover, he could do it years later without hesitation. He could also translate it at no diminution in speed from its original language into English. On one occasion I tested his ability by asking him to tell me how the *Tale of Two Cities* started. Whereupon, without any pause, he immediately began to recite the first chapter and continued until asked to stop after about ten or fifteen minutes. Another time, I watched him lecture on some material written in German about twenty years earlier. In this performance von Neumann even used exactly the same letters and symbols he had in the original. German was his natural language, and it seemed that he conceived his ideas in German and then translated them at lightning speed into English. Frequently I watched him writing and saw him ask occasionally what the English for some German word was.

His power of remembrance was a great help to his strong sense of humor because he was able to remember any story he wished. He

thereby built up an unparalleled storehouse of anecdotes, limer-
icks, and funny happenings. He loved to lighten up otherwise
informal but serious discussions with just the right stories. In this
he had a Lincoln-like quality. Whenever friends visited him they
always tried to bring along some new stories as a sort of present to
him.

He also greatly enjoyed people's company, and his house was
the scene of most wonderful parties and dinners. He had a keen
sense of responsibility to the temporary members of the Institute
for Advanced Study and felt an obligation to introduce them socially to their colleagues. Accordingly, at least once a week he and
his wife would entertain a houseful of people — ranging from young
men just out of graduate school to visiting older distinguished
scientists passing through Princeton. He delighted in entertaining
his guests with anecdotes or relevant quotations, usually humor-
ous, from history. In this he was superb. He had considerable ability at telling stories, particularly very long ones, so that his hearers
were almost breathlessly waiting for the denouement.

The Bela Kun communist uprising in Hungary in 1919 found the
von Neumann family fleeing to a place they owned in Venice.
There was apparently little doubt that his father feared for the
safety of his family at the hands of the communists. This experience
had a great impact on von Neumann, who developed out of this a
strong dislike, or even hatred, for all that communism stood for.[1]

After Bela Kun was defeated, a most remarkable period in the
history of science occurred in the fragments of the Austro-Hungarian Empire. Out of the repression and intellectual sterility that
were the hallmarks of Hapsburg rule there grew up suddenly most
remarkable groups of scientists of whom von Neumann and Eugene Wigner were examples in Hungary, Banach in Poland, and
Feller in Jugoslavia.[2] Prior to this period there had of course been
other great scientists in the Austro-Hungarian countries such as
Sigmund Freud, Georg von Hevesy, and Theodore von Karman, but
in general Central Europe was not an ideal climate under the Haps-

---

[1] The von Neumanns left Hungary as soon as they could, a month after Bela Kun
seized power. The Communist regime lasted 130 days; they returned almost two
months after he was driven out. In von Neumann's testimony at the hearing on
Oppenheimer he said, "I think you will find, generally speaking, among Hungarians
an emotional fear and dislike of Russia."

[2] From the early 1950s Feller had a considerable interest in computers and was
for many years a consultant to me and my successors at IBM, where he was of im-
measurable importance to the development of a mathematical sciences department
of first quality.

burgs and indeed both Hevesy and von Karman spent their intellectual lives outside Mittel Europa.

Whatever the reasons for the intellectual renascence, von Neumann was in all likelihood the greatest of all the remarkable figures produced during this period. This is a most difficult evaluation to make because the men of that period are by any standards outstanding. It has been more than fortunate for the United States that so many of the best of these men fled there to escape the intellectual, racial, and religious persecutions of the Nazi era.[3]

Von Neumann, who was known to almost everyone as Johnny —and to some as Jancsi—was so impressive in school that one of his teachers, Laszlo Ratz, persuaded his father to have him tutored privately in addition to his regular schooling.[4] Before he was 18 he published a paper with his tutor, M. Fekete, a well-known Hungarian mathematician. He attended the Lutheran gymnasium in Budapest from 1911 until his graduation in 1921. At this time his father made an extremely large gift to the school, which was one of the best in Hungary. It was very fortunate in having Ratz on its faculty; he was an excellent teacher, had a great influence on both von Neumann and Wigner, and finally became rector of the school.

Through the kindness of Carl Kaysen, Director of the Institute for Advanced Study, I had access to von Neumann's report cards issued while he was in this gymnasium. It is amusing to see his early strengths—he was recognized as the best mathematician the school produced—and his weaknesses. All his grades were A except for those in geometrical drawing B, writing B, music B, physical education C, and behavior sometimes A but more often B.

In 1921 he enrolled in the University of Budapest, but he spent the years 1921–1923 in Berlin where he came under the influence of Fritz Haber. From Berlin he went to the Swiss Federal Institute of Technology (Eidgenössische Technische Hochschule) in Zurich, where he had contacts both with Hermann Weyl, a superb mathematician and one of his future colleagues at the Institute for Advanced Study, and with George Polya, one of the greatest teachers of mathematics. He obtained a degree in chemical engineering

---

[3] The interested reader may wish to consult a charming book on some of these men. See Laura Fermi, *Illustrious Immigrants, The Intellectual Migration from Europe, 1930–1941* (Chicago, 1968).

[4] For biographical material see S. Ulam, "John von Neumann, 1903–1957," *Bull. Amer. Math. Society*, vol. 64 (1958), pp. 1–49; S. Bochner, "John von Neumann," National Academy of Sciences, *Biographical Memoirs*, vol. 32 (1958), pp. 438–457; or H. H. Goldstine and E. P. Wigner, "Scientific Work of J. von Neumann," *Science*, vol. 125 (1957), pp. 683–684.

at the Federal Institute in 1925; and the next year, on 12 March 1926 — at the age of twenty-two — he received his doctorate summa cum laude in mathematics with minors in experimental physics and chemistry from the University of Budapest!

Then in 1927 he became a Privatdozent in mathematics at the University of Berlin, where he stayed three years and established a world-wide reputation by his papers on algebra, set theory, and quantum mechanics. Ulam recounts how already by 1927, when von Neumann attended a mathematical congress in Lvov, Poland, he was pointed out to the students as a "youthful genius."

It is clear that by 1927 he was already recognized as a great mathematician, and after spending 1929 in Hamburg he was invited to Princeton University as a visiting lecturer in 1930. He stayed on as a visiting professor and in 1931 became a permanent professor. Then in 1933 he went to the Institute for Advanced Study, which was then housed in Princeton's Fine Hall, the mathematics building built by Veblen in honor of Dean Henry B. Fine. Earlier we recounted how von Neumann and Wigner came to the States at Veblen's invitation (above, p. 80).

The combination of the University's and the Institute's faculties of mathematics in one building constituted one of the greatest concentrations of leaders in mathematics and physics the world had ever seen. The only parallel was the great department at Göttingen, which was by this time (1933) already much on the decline; indeed, by then a Nazi was already head of the Mathematical Institute, and Courant, Landau, Emmy Noether, Bernays, Born, Franck, Weyl, and many others were soon to leave the Institute or had already done so. Otto Neugebauer was made head of the Institute and stayed in office for exactly one day. When he refused to take the loyalty oath required by the Nazis, he left Germany.[5] Hitler's rise to power was largely responsible for the great group of celebrities at Princeton. Not that all there were refugees, but many were. The time was now up for European dominance in mathematics and physics, a dominance that had been so strong that for many years the *Bulletin* of the American Mathematical Society regularly listed the courses of lectures to be given in Göttingen.

The great leader of German mathematics during the halcyon days of Göttingen was David Hilbert (1862–1943); he had a profound influence upon world mathematics and became during his lifetime the style-setter for the entire mathematical and theoretical physics worlds. This was an unparalleled accomplishment. Hil-

[5] C. Reid, *Hilbert* (New York, 1970).

bert's role in mathematics can perhaps best be summed up by mentioning his paper at the International Congress of Mathematicians in Paris in the year 1900. In this speech, the major address of the Congress, Hilbert undertook to formulate a set of 23 problems whose "solution we expect from the future." These problems were to be in a real sense a blueprint for modern mathematics. One of von Neumann's great accomplishments was to be a partial solution of the fifth problem. Even today the unsolved ones are still at the forefront.

It is against this background that we should understand some of what follows. Among the physicists who at various times during the 1920s were members of this magic group at Göttingen were Max Born and James Franck — permanently there — P. M. S. Blackett, Karl Compton, Paul Dirac, Werner Heisenberg, Pascual Jordan, Lothar Nordheim, Robert Oppenheimer, Wolfgang Pauli, Linus Pauling, and Eugene Wigner. During this same period von Neumann was also going to Göttingen to work with Hilbert on formal logics and physics. Miss Reid quotes Nordheim, in comparing these two men, as saying that Hilbert was "slow to understand" but that von Neumann had "the fastest mind I ever met." This was in 1924.

There were a number of standard anecdotes around the mathematical community to illustrate the fantastic speed of von Neumann. One of these he asserted was not true, but it is illustrative. Hermann Weyl is supposed to have given a preliminary lecture on the profundity of the next theorem he was going to prove indicating why the proof had to be very difficult. The next day he gave this lengthy and hard proof. At the end, the story goes, young von Neumann jumped up and said, "Would you be so good as to look over the following proof?" Whereupon he wrote a very few lines and gave an entirely novel and simple proof.

Another, true incident illustrating this speed occurred in Princeton. It was von Neumann's wont to keep an open door for all visitors to the Institute, and they usually came to see him for help when in mathematical trouble. Beyond anyone else he could almost instantly understand what was involved and show how to prove the theorem in question or to replace it by what was the true theorem. On this occasion, a young man stated his difficulty, von Neumann then gave the proof in detail on the blackboard, and the student nodded, thanked him, and left. The next Saturday night at von Neumann's party the same man approached von Neumann and said that he had forgotten the proof and would von Neumann mind repeating it. This von Neumann did standing in a crowded room.

In 1930 von Neumann married Marietta Kovesi, and in 1935 they had a daughter, Marina, who has had a distinguished career not only as a wife and mother but also as professor of economics at the University of Pittsburgh. Von Neumann's marriage ended in 1937, and in 1938 he married Klara Dan. She later became a programmer for the Los Alamos Scientific Laboratory and helped to program and code some of the largest problems done in the 1950s.

It is time now to return to von Neumann's early days with Hilbert. Under the latter's influence von Neumann embarked on a program which was to have a profound influence on his later work on computers and related topics. Hilbert had become strongly involved in a great program in the foundations of mathematics. He undertook this task because there was much concern at the turn of the century about the integrity of mathematics. For centuries it had been a basic creed of mathematicians that deductive reasoning when properly used could never lead to inconsistent results. Then Bertrand Russell and Alfred North Whitehead brought out their classic.[6] This linked back to Boole in the sense that he provided the genesis for the whole subject of mathematical logics. He had set up a formal machinery for expressing mathematical thoughts and this culminated in 1905 with a very exhaustive treatise by E. Schröder.[7]

Then the modern school started with the works of Frege and Peano which lead up to Whitehead and Russell. What they attempted was stated in 1903 by Russell: "The present work has two main objects. One of these, the proof that all pure mathematics deals exclusively with concepts definable in terms of a very small number of fundamental logical concepts, and that all its propositions are deducible from a very small number of fundamental logical principles. . . ." [8] This work was intended to annihilate both the positions of Hilbert and of L. E. J. Brouwer, a Dutch mathematician. Hilbert attempted to rescue mathematics "from the stickier quagmires of classical metaphysics"[9] by separating the ideas of mathematics from their meaning; concepts such as the positive integers were no longer to have any intuitive meaning drawn from experience but were to be viewed as abstract entities obeying certain formal laws and only those. This is known as *formalism*. To make this approach valid Hilbert and his colleagues, including

[6] *Principia Mathematica* (London, 1903).

[7] *Vorlesungen Über die Algebra der Logik*, 4 vols. (Leipzig, 1890, 1891, 1895, 1905).

[8] B. Russell, *Principles of Mathematics* (New York, 1950). The first edition appeared in 1903 and "most of it was written in 1900."

[9] E. T. Bell, *Development of Mathematics* (New York, 1945), p. 557.

von Neumann, embarked on a program to prove that all mathematics was consistent; to show that under their schema the old Greek ideal would reappear. There could be no contradictions, no inconsistencies. Hilbert undertook this program as his answer both to Russell-Whitehead on the one hand and Weyl-Brouwer on the other. This was his way of cutting mathematics loose from logics where Burali-Forti, Richard, and Russell had shown there were paradoxes to trap the unwary.

The group under Brouwer had gone on a very different tack. Brouwer, and later Weyl, refused "to regard a proposition as either true or false unless some method exists of deciding the alternative." [10] The method referred to must be expressed or utilized in a finite number of steps. Otherwise, in dealing with infinite sets Brouwer and Weyl rejected the law of the excluded middle which says that if a proposition is true its contradiction must be false. This law is familiar to all of us from plane geometry where we frequently made proofs by saying "suppose the contrary were true" and then reached an absurdity. This justified the proposition for us but not for Brouwer and Weyl. Their position rendered in doubt very large areas of mathematics and caused many mathematicians great anguish, among them Hilbert, who said: "I believe that as little as Kronecker was able to abolish the irrational numbers . . . just as little will Weyl and Brouwer today be able to succeed. Brouwer is not, as Weyl believes him to be, the Revolution — only the repetition of an attempted *Putsch*, in its day more sharply undertaken yet failing utterly, and now, . . . doomed from the start!" [11]

This is the scene upon which von Neumann came in 1924. In 1927 he published his famous paper on the problem of the freedom of mathematics from contradiction. It is worthy of note that von Neumann at that time conjectured that all of analysis can be proved consistent. A few years later, in 1930, another great young mathematician and logician, Kurt Gödel, showed that certain logical structures contain propositions whose truth is undecidable within the system, that not everything can be decided.[12] His proof was very direct. He in fact produced "a true theorem such that a formal proof of it leads to a contradiction." [13] The structures where this anomalous behavior can occur are not bizarre or pathological in

[10] Russell, *Principles*, p. vi.

[11] Quoted in Reid, *Hilbert*, p. 157.

[12] K. Gödel, "Über formal unentscheidbare Sätze der *Principia Mathematica* und verwandter Systeme," I, *Monatschafte für Mathematik und Physik*, vol. 38 (1931), pp. 173–198.

[13] Bell, *Development of Mathematics*, p. 576.

character. Even arithmetic cannot be shown to be contradiction-free in the Hilbert sense.

Von Neumann remarked that Gödel was the greatest logician since Aristotle and once told a charming tale about himself. To appreciate the story fully one should know that he was inclined to work on a particular problem as long as it went forward. At the end of a day's work he would go to bed and very often awaken in the night with new insights into the problem. In this case he was busily engaged in trying to develop a proof just exactly the opposite of Gödel's and was unsuccessful! One night he dreamed how to overcome his difficulty, arose, went to his desk, and carried his proof much further along but not to the end. The next morning he returned to the attack, again without success, and again that night retired to bed and dreamed. This time he saw his way through the difficulty, but when he arose to write it down he saw there was still a gap to be closed. He said to me, "How lucky mathematics is that I didn't dream the third night!"

It was his training in formal logics that made him very much aware of and interested in a result which foreshadowed the modern computer. This was contained in independent papers published by Emil L. Post and Alan M. Turing in 1936.[14] Post taught at City College of New York and Turing was an Englishman studying at Princeton University (1936–1938). Each of them conceived of what is now called an automaton and described it in similar, mechanistic terms. The men worked independently and in ignorance of each other. There is no doubt that von Neumann was thoroughly aware of Turing's work but apparently not of Post's. We shall revert to this topic again in its proper temporal order. At this time, however, it is just worth mentioning in passing that the Post-Turing work would have delighted Leibniz, since their automata surely carry out his dream of "a general method in which all truths of the reason would be reduced to a kind of calculation." Indeed, this is exactly what their automata do.

One of the interesting things Gödel did was to designate each provable theorem by a sequence of integers with a corresponding situation for remarks about the theorem. This provides a numerical algorithm for each theorem and puts us in the field of numerical computation.

Very early in von Neumann's career he displayed the interest he

---

[14] Post, "Finite Combinatory Processes — Formulation 1," *Journal of Symbolic Logic*, vol. 1 (1936), pp. 103–105; Turing, "On Computable Numbers," *Proc. London Math. Soc.*, ser. 2, vol. 42 (1936), pp. 230–265.

always had in the applications of mathematics. In 1927 and 1928, at about the same time as his work on proof theory, he published several papers on the mathematical foundations of quantum theory and probability in statistical quantum theory, which showed his profound understanding of physical phenomena. Indeed he was unsurpassed in his ability to understand completely very complex physical situations. Unlike many applied mathematicians who want merely to manipulate some equations given them by a physicist, von Neumann would go right back to the basic phenomenon to reconsider the idealizations made as well as the mathematical formulation.

He possessed along with all his other accomplishments a truly remarkable ability to do very elaborate calculations in his head at lightning speeds; this was especially noticeable when he would be making rough order of magnitude estimates mentally and would call upon an unbelievable wealth of physical constants he had available.

His great interest in the applications of mathematics was to become increasingly important as time went on, and by 1941 it had become his dominant interest. This was to have the most profound implications for the computer field in particular and for the United States in general. In this connection it is worth drawing attention to a paper von Neumann wrote in 1928 on game theory. This was his first venture in the field, and while there had been other tentative approaches—by Borel, Steinhaus, and Zermelo, among others —his was the first to show the relations between games and economic behavior and to formulate and prove his now famous minimax theorem which assures the existence of good strategies for certain important classes of games.[15] In the well-known book by von Neumann and Morgenstern it is stated that: "Our considerations will lead to the application of the mathematical theory of 'games of strategy' developed by one of us in several successive stages in 1928 and 1940–1941." [16]

This typifies something we find in all von Neumann's work: a very few motifs that constantly interweave and recur, usually in unexpected but profound ways; invariably they are aesthetically pleasing. Indeed, von Neumann was one of the greatest of all mathematical artists. He had a completely sure sense of what was elegant mathematically and was always watching for these aspects. It was never enough for him merely to establish a result; he had to

---

[15] "Zur Theorie der Gesellschaftspiele," *Math. Ann.*, vol. 100 (1928), pp. 295–320.
[16] *Theory of Games and Economic Behavior* (Princeton, 1944), p. 1.

do it with elegance and grace. He would often say to me while we worked on some topic, "Now here is the elegant way to do this."

One of the difficulties people experienced in listening to one of his lectures on mathematics was precisely its sheer beauty and elegance. They were often so taken in by the ease with which results were proved that they thought they understood the path von Neumann had chosen. Later at home when they tried to re-create it, they discovered not the magical path but instead a harsh, forbidding forest. Be it said in their defences that von Neumann usually chose a square area roughly two feet on a side on an enormous blackboard and seemed to play the game of seeing if he could confine all his writing to this tiny area. He did this by judicious use of an eraser and made it just barely possible for his auditors to take down what he had written.

It is interesting to contrast his style in lectures with his written one. His oral style was wonderfully clear and somehow very inspiring and uplifting. His written style was highly elegant, symmetrical, and in every detail complete but often lacking in indications as to why he chose as he did among various alternatives. It was these that he introduced into his oral expositions when he shared with his auditors his supreme insights. His written style was not heavy but had a certain inevitable complexity of thought that made it sometimes difficult to comprehend. Perhaps it derived from his familiarity with German, Latin, and Greek.

His oral expression on the other hand clearly derived from his absolute mastery cf idiomatic American English and his understanding of the American mind and style. The only difficulties he had in pronouncing English were those associated with "th" and "r"; but he had a delightful Hungarian accent and also certain fixed mispronunciations which he carefully preserved. One of the best of these was the way he pronounced "integer" as "integher." Once, in the author's hearing he said the word properly but then quickly corrected himself and again said it in his own style.

The story used to be told about him in Princeton that while he was indeed a demi-god he had made a detailed study of humans and could imitate them perfectly. Actually he had great social presence, a very warm, human personality, and a wonderful sense of humor. These qualities, together with his incredible mental capacity, made him a superb teacher. It has been said of him: "No appraisal of von Neumann's contributions . . . would be complete without a mention of the guidance and help which he so freely gave to his friends and acquaintances, both contemporary and younger than himself. There are well-known theoretical physicists who believe that they have learned more from von Neumann in personal conversations than from any of their colleagues. They value what they learned from him in the way of mathematical theorems, but they value even more highly what they learned from him in methods of thinking and ways of mathematical argument." [17] With real justice it can be said of him, in the words of Landor, that he "warmed both hands before the fire of life."

By the mid-1930s von Neumann had become deeply involved in the problems of supersonic and turbulent flows of fluids. "It was then that he became aware of the mysteries underlying the subject of non-linear partial differential equations. . . . The phenomena described by these non-linear equations are baffling analytically and defy even qualitative insight by present methods." [18] Thus by the beginnings of World War II von Neumann was one of the leading experts on shock and detonation waves and inevitably became involved with the Ballistic Research Laboratory, with the OSRD, with the Bureau of Ordnance, and with the Manhattan Project—all to their great good. It is not our place here to write anything like a biography of this great figure but rather to sketch in just those portions of his career that we need to explicate his role in the computer field. Instead of setting out all his activities in detail, perhaps we may be forgiven if we quote von Neumann himself speaking before the Special Senate Committee on Atomic Energy on his wartime activities and then copy Professor Bochner's brief chronology of von Neumann's career.

> Senator McMahon, Gentlemen:
> I assume that you wish to know my qualifications. I am a mathematician and a mathematical physicist. I am a member of the Institute for Advanced Study in Princeton, New Jersey. I have been connected with Government work on military matters for nearly ten years: As a consultant of Ballistic Research Laboratory of the Army Ordnance Department since 1937, as a member of its scientific advisory committee since 1940; I have been a member of various divisions of the National Defense Research Committee since 1941; I have been a consultant of the Navy Bureau of Ordnance since 1942. I have been connected with

[17] Goldstine and Wigner, "Scientific Work of J. von Neumann," p. 684.
[18] Ulam, *op. cit.*, pp. 7–8.

the Manhattan District since 1943 as a consultant of the Los Alamos Laboratory, and I spent a considerable part of 1943–45 there.[19]

JOHN VON NEUMANN: CHRONOLOGY [20]

1903   Born, Budapest, Hungary, December 28.

1930–33   Visiting Professor, Princeton University.

1933–57   Professor of Mathematics, Institute for Advanced Study, Princeton, N.J.

1937   Gibbs Lecturer, Colloquium Lecturer, Bôcher Prize, all in American Mathematical Society.

1940–57   Scientific Advisory Committee, Ballistics Research Laboratories, Aberdeen Proving Ground, Maryland.

1941–55   Navy Bureau of Ordnance, Washington, D.C.

1943–55   Los Alamos Scientific Laboratory (AEC), Los Alamos, N.M.

1945–57   Director of Electronic Computer Project, Institute for Advanced Study, Princeton, N.J.

1947   D.Sc. (hon.), Princeton University; Medal for Merit (Presidential Award); Distinguished Civilian Service Award, U.S. Navy.

1947–55   Naval Ordnance Laboratory, Silver Spring, Maryland.

1949–53   Research and Development Board, Washington, D.C.

1949–54   Oak Ridge National Laboratory, Oak Ridge, Tennesee.

1950   D.Sc. (hon.), University of Pennsylvania and Harvard University.

1950–55   Armed Forces Special Weapons Project, Washington, D.C.; Weapons System Evaluation Group, Washington, D.C.

1950–57   Member Board of Advisors, Universidad de los Andes, Colombia, South America.

1951–53   President, American Mathematical Society.

1951–57   Scientific Advisory Board, U.S. Air Force, Washington, D.C.

1952   D.Sc. (hon.), University of Istanbul, Case Institute of Technology, and University of Maryland.

1952–54   Member, General Advisory Committee, U.S. Atomic Energy Commission, Washington, D.C. (Presidential appointment).

1953   D.Sc. (hon.), Institute of Polytechnics, Munich; Vanuxem Lecturer, Princeton University.

1953–57   Technical Advisory Panel on Atomic Energy, Washington, D.C.

1955–57   U.S. Atomic Energy Commissioner (Presidential appointment).

---

[19] Statement of John von Neumann before the Special Senate Committee on Atomic Energy, Collected Works, vol. VI, pp. 499–502. He testified on 31 January 1946 concerning the pending legislation which created the Atomic Energy Commission.

[20] From Salomon Bochner, "John von Neumann," National Academy of Sciences, *Biographical Memoirs*, vol. 32 (1958), p. 447.

1956   Medal of Freedom (Presidential Award); Albert Einstein Com-
          memorative Award; Enrico Fermi Award.
1957   Died, Washington, D.C., February 8.

Academy memberships:
          Academia Nacional de Ciencias Exactas, Lima, Peru.
          Accademia Nazionale dei Lincei, Rome, Italy.
          American Academy of Arts and Sciences.
          American Philosophical Society.
          Istituto Lombardo di Science e Lettere, Milan, Italy.
          National Academy of Sciences.
          Royal Netherlands Academy of Sciences and Letters, Amster-
          dam, Netherlands.

How did von Neumann become involved in computers and computing? The answers to this question lie in his career as briefly
outlined in the last section and in a facet of his make-up we have
not yet mentioned. Along with all his other attributes, he had an
almost insatiable interest in new ideas — sometimes! At other times
— when, for instance, he was deeply committed to some intellectual
pursuit — he was completely impervious to new ideas. On these
occasions he would hardly bother to listen to the speaker. But in
general he was extremely receptive to new intellectual challenges,
and he always seemed to display a high degree of mental restless-
ness when he was between ideas. It seemed as if he was almost
always on the lookout for new fields to conquer and he probably
was. Naturally, those which appealed to him most were those that
involved his great interest in applied mathematics.

Thus we are not surprised to see him moving into hydrodynamics. This subject had several aspects which certainly must have
weighed strongly with him: it was and to some substantial extent
still is rife with difficulties of a very great mathematical sort; a
definitive advance in this field would have decisive implications
in both mathematics and theoretical physics; and finally hydrodynamics was and is enormously useful to the government. All
these reasons were important to von Neumann, and each must have
contributed to his decision to move into this field.

Perhaps it is worth saying a few words on each of these three
reasons. In the middle 1940s von Neumann and I were writing as
follows:

Our present analytical methods seem unsuitable for the solution
of the important problems arising in connection with non-linear
partial differential equations and, in fact, with virtually all types

of non-linear problems of pure mathematics. The truth of this statement is particularly striking in the field of fluid dynamics. Only the most elementary problems have been solved analytically in this field. Furthermore, it seems that in almost all cases where limited successes were obtained with analytical methods, these were purely fortuitous, and not due to any intrinsic suitability of the method to the milieu.[21]

An elaboration of the second reason occurs a little later in the same paper: ". . . that we are up against an important conceptual difficulty is clear from the fact that, although the main mathematical difficulties in fluid dynamics have been known since the time of Riemann and Reynolds, and although as brilliant a mathematical physicist as Rayleigh has spent the major part of his life's effort in combating them, yet no decisive progress has been made against them. . . ."[22]

To document the third reason it will perhaps suffice to quote Ulam: "Often he mentioned that personally he found doing scientific work there (Europe) almost impossible because of the atmosphere of political tension. After the war he undertook trips abroad only unwillingly."[23]

Von Neumann's knowledge of hydrodynamics was to be of inestimable value to the Los Alamos group which he joined as a consultant late in 1943. One of his earliest and most important contributions there was his work on implosions. To understand this it is perhaps desirable to say a little on what this is all about.

The big problem facing the physicists at Los Alamos was how to produce an extremely fast reaction in a small amount of the uranium isotope, U^{235}, or of plutonium so that a great amount of energy would be explosively released. This was quite another thing from the relatively slow build-up of neutrons in an atomic pile or reactor, this last having just been achieved on 2 December 1942. Many first-rate physicists were therefore proposing and examining various alternate schemes for achieving such fast reactions.

In principle, the idea was to start with material in noncritical form and somehow to transform it with extreme rapidity into a critical state. The need for speed was to prevent a very small explosion from occurring prematurely and blowing up the bomb. One scheme was to make two hemispheres of nuclear material,

---

[21] Goldstine and von Neumann, "On the Principles of Large-Scale Computing Machines," in von Neumann, Collected Works, vol. V, p. 2.

[22] *Ibid.,* pp. 2–3.

[23] Ulam, *op. cit.,* pp. 6–7.

each of which was subcritical but which when combined into a sphere would be critical. The two pieces were then assembled close together and at the requisite instant pushed together by means of an explosion of some conventional high explosive.

A much more sophisticated technique was to assemble a somewhat subcritical sphere of material and by conventional explosives around the sphere to compress it very rapidly with such great force that it became critical. It was this technique that was being experimented on by Seth Neddermeyer. Von Neumann in collaboration with Neddermeyer, Edward Teller, and James L. Tuck worked on this in detail. One of the big problems was the need to achieve a spherical shock wave that would push simultaneously on all points of the nuclear mass. If simultaneity were not achieved, the nuclear material would be extruded from low-pressure areas with a resultant loss of energy in the explosion. Tuck and von Neumann invented an ingenious type of high explosive lens that could be used to make a spherical wave. The idea was successful.

This was no small achievement but undoubtedly von Neumann's main contribution to the Los Alamos project was to lie in his showing the theoretical people there how to model their phenomena mathematically and then to solve the resulting equations numerically. A punched card laboratory was set up to handle the implosion problem, and this later grew into one of the world's most advanced and largest installations.

Von Neumann had an uncanny ability to solve very complex calculations in his head. This was a source of wonderment to mathematicians and physicists alike. It is possible to illustrate this quality of his by an amusing anecdote: One time an excellent mathematician stopped into my office to discuss a problem that had been causing him concern. After a rather lengthy and unfruitful discussion, he said he would take home a desk calculator and work out a few special cases that evening. Each case could be resolved by the numerical evaluation of a formula. The next day he arrived at the office looking very tired and haggard. On being asked why he triumphantly stated he had worked out five special cases of increasing complexity in the course of a night of work; he had finished at 4:30 in the morning.

Later that morning von Neumann unexpectedly came in on a consulting trip and asked how things were going. Whereupon I brought in my colleague to discuss the problem with von Neumann. We considered various possibilities but still had not met with success. Then von Neumann said, "Let's work out a few special cases." We agreed, carefully not telling him of the numerical work in the early morning hours. He then put his eyes to the ceiling and in perhaps five minutes worked out in his head four of the previously and laboriously calculated cases! After he had worked about a minute on the fifth and hardest case, my colleague suddenly announced out loud the final answer. Von Neumann was completely perturbed and quickly went back, and at an increased tempo, to his mental calculations. After perhaps another minute he said, "Yes, that is correct." Then my colleague fled, and von Neumann spent perhaps another half hour of considerable mental effort trying to understand how anyone could have found a better way to handle the problem. Finally, he was let in on the true situation and recovered his aplomb.

In any case, von Neumann had a profound interest and capability in numerical calculations, but his work in hydrodynamics would have been impossible without computers and computing. It is of course fortuitous that he linked Aberdeen and Los Alamos. It is precisely to this fortuity that we all owe so much.

Sometime in the summer of 1944 after I was out of the hospital I was waiting for a train to Philadelphia on the railroad platform in Aberdeen when along came von Neumann. Prior to that time I had never met this great mathematician, but I knew much about him of course and had heard him lecture on several occasions. It was therefore with considerable temerity that I approached this world-famous figure, introduced myself, and started talking. Fortunately for me von Neumann was a warm, friendly person who did his best to make people feel relaxed in his presence. The conversation soon turned to my work. When it became clear to von Neumann that I was concerned with the development of an electronic computer capable of 333 multiplications per second, the whole atmosphere of our conversation changed from one of relaxed good humor to one more like the oral examination for the doctor's degree in mathematics.

Soon thereafter the two of us went to Philadelphia so that von Neumann could see the ENIAC. At this period the two accumulator tests were well underway. I recall with amusement Eckert's reaction to the impending visit. He said that he could tell whether von Neumann was really a genius by his first question. If this was about the logical structure of the machine, he would believe in von Neumann, otherwise not. Of course, this *was* von Neumann's first query.

On 2 November 1944, my wife and I returned to Philadelphia and again took up residence there. This point in time formed the beginning of a long and very fruitful friendship and working relationship between ourselves and the von Neumanns that was to terminate only upon his untimely death.

To describe what happened next it is convenient to refer back to Babbage's Analytical Engine, to Stibitz's relay computers, to the Harvard-IBM computer, and to Post's and Turing's paper constructs. We have seen (above, p. 21) that Babbage's machine was conceived as being instructed in its tasks by a set of so-called operation cards strung together to describe the series of operations to be performed. Similarly, the other machines just mentioned all had paper tapes into which were punched holes that were a numerical code for the instructions to be effected. As early as November of 1943 Stibitz was describing how the orders were handled in a very small machine called a Relay Interpolator (RI), used for doing cubical interpolation, as follows: "To make the RI carry out a required computation, that computation is broken down into a succession of orders to the machine to memorize, read or write numbers, add, and so on. These orders are placed on a control tape with the aid of a device similar to the ordinary typewriter." [1]

Thus by 1944 there was general understanding among the group at the Moore School that the orders for a digital machine could be stored in numerical form on tape. This point is very important since it bears crucially on what happened next. My correspondence for the period August–October 1944 clearly indicates that as the two accumulator set-up came into operation the Moore School staff and I began to feel quite confident of our ability to construct the entire ENIAC as a reliably operating system without too many delays. In fact on 11 August 1944 I was optimistically writing that the ENIAC was to be ready about 1 January 1945, and that a room 20 feet by 40 feet should be provided. [2]

At the same time a general reassessment of the ENIAC's operational characteristics was being undertaken at the Moore School. This revealed a number of serious problems and deficiencies in the machine as an instrument for general-purpose scientific calculation. In particular I (and probably Cunningham as well) was

[1] G. R. Stibitz, Applied Mathematics Panel, NDRC, "A Statement Concerning the Future Availability of a New Computing Device," 12 November 1943.
[2] Memorandum, Goldstine to Simon, 11 August 1944.

bothered by the clumsiness of the mechanisms for programming the
ENIAC and by the small number — 20 — of electronically alterable
storage registers. We already had visualized doing much more
elaborate calculating than solving ballistic equations.

Thus on 11 August 1944 I was writing Col. Simon in this sense:

2. Due however to the necessity for providing the ENIAC in a
year and a half it has been necessary to accept certain make-shift
solutions of design problems, notably in the means of establishing connections between units to carry out given procedures and
in the paucity of high speed storage devices. These defects will
result in considerable inconvenience and loss of time to the
Laboratory in setting up new computing problems.

3. It is believed highly desirable that a new RAD contract be
entered into with the Moore School to permit that institution to
continue research and development with the object of building
ultimately a new ENIAC of improved design. . . .[3]

This memorandum makes very clear that conversations were
going on at the Moore School looking toward an amelioration of
the operational deficiencies of the ENIAC. It is evident that by mid-
August these conversations had progressed to the point where new
technological ideas had emerged. In fact there is an undated report
by Eckert and Mauchly, written that summer prior to 31 August,
when it was sent by Brainerd to me,[4] describing a device, called a
delay line, for increasing the storage capacity of the machine. This
device, discussed at length below, was to be crucial to the next
phase of development.

The 11 August memorandum was followed by extensive conversations at the Ballistic Research Laboratory, which resulted in a
meeting on 29 August 1944 of a committee known as the Firing
Table Reviewing Board and a recommendation to Simon backing
my request.[5] The minutes of that meeting show that von Neumann
was present. This makes it quite clear to me that von Neumann had
already visited the Moore School. My records indicate that I was
back on duty after my illness around 24 July 1944 and that I probably took von Neumann for a first visit to the ENIAC on or about
7 August. I recall that von Neumann's first visit was during the
two-accumulator test. My travel orders show that my first business

[3] Memorandum, Goldstine to Simon, Further Research and Development on
ENIAC, 11 August 1944.
[4] Letter, Brainerd to Goldstine, 31 August 1944.
[5] Memorandum, C. B. Morrey to L. E. Simon, 30 August 1944.

visit to Philadelphia after being released from the University of Pennsylvania Hospital was in the first week of August 1944. By the end of August I was urging development of a new electronic computing device that would:

a. Contain many fewer tubes than the present machine and hence be cheaper and more practical to maintain.

b. Be capable of handling many types of problems not easily adaptable to the present ENIAC.

c. Be capable of storing cheaply and at high speeds large quantities of numerical data.

d. Be of such a character that the setting up on it of a new problem will require very little time.

e. Be smaller in size than the present ENIAC.[6]

All this material makes clear that the thinking at the Moore School had advanced quite far by the end of August 1944. However, no real effort had gone into understanding how to store instructions or what they should be. The idea at that time was somehow to do this along the lines of the Stibitz machine, but nothing was formulated or thought out. Then Eckert came up with the idea that the delay line could be used for storage of information. This is in itself most important. It was fortunate that just as this idea emerged von Neumann should have appeared on the scene.

It was he who took the raw idea and perfected it. Starting in August 1944 von Neumann came regularly to the Moore School for meetings with Burks, Eckert, Adele Goldstine, Mauchly, and myself. Eckert was delighted that von Neumann was so keenly interested in the logical problems surrounding the new idea, and these meetings were scenes of greatest intellectual activity. Out of them arose very specific ideas on the types of mathematical problems needing solution and on the logical design of a new machine to handle them as well as on its engineering design.

Brainerd wrote in September:

The progress of work on the ENIAC has led to some rather extensive discussions concerning the solution of problems of a type for which the ENIAC was not designed. In particular, these discussions have been carried out with Dr. von Neumann. . . . Dr. von Neumann is particularly interested in mathematical analyses which are the logical accompaniment of the experimental work which will be carried out in the supersonic wind tunnels. . . .

[6] Draft memorandum, never sent, August 1944. This was my draft of the memorandum sent 30 August by Morrey to Simon.

It is not feasible to increase the storage capacity of the ENIAC
. . . to the extent necessary for handling non-linear partial differential equations on a practical basis. The problem requires an
entirely new approach. At the present time we know of two principles which might be used as a basis. One is the possible use of
iconoscope tubes, concerning which Dr. von Neumann has talked
to Dr. Zworykin of the R.C.A. Research Laboratories, and another of which is the use of storage in a delay line, with which we
have some experience. Such a line could store a large number of
characters in a relatively small space, and would . . . enable a
machine of moderate size to be constructed for the solution of
partial differential equations which now block progress in certain fields of research at the BRL.[7]

The details of the new machine — which was to be called EDVAC —
will be discussed later. Let us now return to von Neumann's collaboration with the Moore School group. The discussions and meetings are summarized in a first report on the new machine in March
of 1945 as follows:

The problems of logical control have been analyzed by means
of informal discussions among Dr. John von Neumann, . . . , Dr.
Mauchly, Mr. Eckert, Dr. Burks, Capt. Goldstine and others. . . .
Points which have been considered during these discussions are
flexibility of the use of the EDVAC, storage capacity, computing
speed, sorting speed, the coding of problems, and circuit design.
These items have received particular attention; . . . Dr. von Neumann plans to submit within the next few weeks a summary of
these analyses of the logical control of the EDVAC together with
examples showing how certain problems can be set up.[8]

In July Report No. 2 states that "discussions have been held at
regular intervals in the Moore School to develop a logical plan for
the EDVAC. These discussions continue. Dr. John von Neumann . . .

---

[7] Letter, Brainerd to Gillon, 13 September 1944. This letter also contained cost and
time estimates for the research and development work involved. The amount proposed was $105,600; the time was to start 1 January 1945 and to go on for one year
with work starting "on a small scale on October 1, 1944." On 18 September 1944
Gillon, in the Office of the Chief of Ordnance, issued an RAD Order to the Philadelphia Ordnance District for "Research and Experimental work in connection with
the Development of an Electronic Discrete Variable Calculator." The contract was
to be issued for the nine months beginning 1 January 1945, and the dollar amount
was as indicated above.

[8] J. P. Eckert, Jr., J. W. Mauchly, and S. R. Warren, Jr., PY Summary Report No. 1,
31 March 1945. PX was the Moore School accounting symbol for the ENIAC charges
and PY for the EDVAC.

has prepared a preliminary draft in which he has organized the subject matter of these discussions. This material has been mimeographed and bound. It consists of a 101-page report entitled *First Draft of a Report on the* EDVAC *by John von Neumann."* [9]

This work on the logical plan for the new machine was exactly to von Neumann's liking and precisely where his previous work on formal logics came to play a decisive role. Prior to his appearance on the scene, the group at the Moore School concentrated primarily on the *technological* problems, which were very great; after his arrival he took over leadership on the *logical* problems. Thus the group tended to split into the technologists — Eckert and Mauchly — and the logicians — von Neumann, Burks, and I. This was a perfectly natural division of labor, but the polarization was to become increasingly severe as time went on and was finally to disrupt the group. At the same time, tensions were also developing between the university's and Moore School's administration on the one hand and Eckert and Mauchly on the other. All these conflicts play a role in the story and will be discussed later.

Before describing von Neumann's contributions to the EDVAC — Electronic Discrete Variable Calculator — we need to explain what a delay line is, and how it is used in an electronic computer.

These devices can be built in a number of ways, but the type relevant here is the so-called ultrasonic delay line. As a general class they were developed so that electrical signals could be delayed by given predetermined amounts of time. The ultrasonic ones are very useful for achieving long delays, i.e. of several milliseconds.

These ultrasonic devices operate by transforming the electrical signal to be delayed into an ultrasonic signal in some fluid and then transforming it back again to an electrical one. The delay comes in from the fact that transmission of signals through fluids is very slow as compared to that of electricity through a wire. The speed of such a signal in mercury, for example, is 1,450 meters per second, whereas the speed in a wire of an electrical signal is comparable to that of light, 3×10^8 meters per second. Thus, by suitably fixing the length of the container of fluid, a preassigned time delay can be achieved.

The first such device was built at the Bell Telephone Laboratories by William B. Shockley, one of the co-inventors of the transistor. It used a mixture of water and ethylene glycol for the fluid. The second was built by Eckert and his associates at the Moore School

[9] Eckert, Mauchly, Warren, PY Summary Report No. 2, 10 July 1945.

for the Radiation Laboratory at MIT and its properties studied during the summer of 1943.[10] Then a group was set up at Radiation Laboratory to exploit these devices for radar applications.

As we have said, these delay lines require among other things that an electrical signal be transformed into a sonic one and conversely. This may sound quite bizarre but is a very usual thing. In fact, for very low frequencies—less than a 100 kiloherz per second —loudspeakers are examples of the first transformation and microphones of the latter. For the present application, though, extremely high frequencies are involved, and such devices are not satisfactory. Instead, the so-called piezoelectric effect is utilized.

This property—the word piezo derives from the Greek word *piezein,* to press—was discovered in 1880 by Pierre and Jacques Curie. They showed that a mechanical change in shape of a suitably cut quartz crystal would result in an electric charge being formed in the crystal, and, conversely, that an electrical charge being imposed on such a crystal would result in a mechanical change in the crystal. Such crystals can then be used to transform electrical signals into mechanical ones and vice versa. If these signals are superimposed on a suitably determined carrier, the crystals can vibrate at very high frequencies—millions of cycles per second. The way these crystals were used was to attach them to either end of a tube filled with mercury. When an electrical pulse was impressed on the crystal at the initial end of such a tube, the crystal vibrated and thereby sent a sonic wave down the mercury at a speed of 1,450 meters per second. Upon arrival at the other end, the sonic disturbance compressed the terminal crystal which then emitted an electrical disturbance, mirroring the input signal but delayed by $d/1,450$ seconds where d is the length in meters of the tube. Thus a tube 1.450 meters (= 4 feet 9 inches) long will give a delay of 1 millisecond.

How can such a device be used for storing information? Visualize a delay line as in Fig. 13. Further suppose the input and output

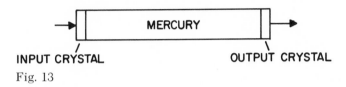

Fig. 13

[10] A. G. Emslie, H. B. Huntington, H. Shapiro, and A. E. Benfield, "Ultrasonic Delay Lines II," *Journal of the Franklin Institute,* vol. 245 (1948), pp. 101–115; see also pp. 1–23.

wires are joined together. Then in principle any pattern of sonic disturbances in the mercury will reappear periodically. This follows since upon reaching the output crystal the sonic pattern will become transformed into the corresponding electrical pattern and will then reappear at the input crystal where it will be transformed back into the original sonic pattern. The period of the phenomenon will of course be determined by the length of the column of mercury, as we saw earlier.

However, the idealization we have just made is unrealistic since friction both in the wires and in the mercury will stop the device from ever operating. To remedy this loss of energy a black box containing the order of ten vacuum tubes is needed to retrieve and reshape the pattern, as indicated in Fig. 14. To quantise what we have

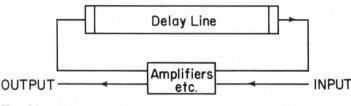

Fig. 14

described, let us suppose the binary digit 1 is represented by a pulse 0.5 microseconds wide and a 0 by no pulse. A tank 1.45 meters long can contain a pattern of 1,000 binary digits and hold it as long as the power stays on.

Let us understand the significance of this compared to what was done in the ENIAC. There a binary digit could be stored using a flip-flop, i.e. at the cost of a pair of vacuum tubes; however, since such a pair came in one glass envelope, let us say one binary digit could be stored at the cost of one vacuum tube. In the case of the delay line 1,000 binary digits could be stored at a cost of 10 vacuum tubes or less. Thus the cost of storing a binary digit dropped from 1 to 1/100 vacuum tube under the new technology. Moreover by means of suitable vacuum tube circuits called gates the recirculation can be interrupted to permit information to be read out or new information to be fed in as desired. Therefore, a pattern may be changed at will. This, then, was Eckert's great new technological invention.

In the ENIAC it was possible to store only 20 words because of the clumsy way in which binary digits—or bits, to use John Tukey's word—were stored, i.e. in vacuum tubes. For the proposed new machine using the mercury delay line it was decided tentatively to increase the storage capacity to about 2,000 words, i.e. by a factor of a hundred, thereby taking advantage of the hundred-fold drop in the cost of storage.

It is somewhat irrelevant but amusing to recount at this point a story about Alan Turing, the logician we mentioned earlier. In late 1946 or early 1947 Turing visited von Neumann and me for several weeks, during which time he argued that the mercury delay line used as a memory could not work. His argument was based on various signal-to-noise ratio considerations and seemed most convincing. In fact, he persuaded us, but fortunately experiment and experience proved him wrong.

We are now in a position to give a preliminary account of von Neumann's contributions to the EDVAC. They were contained in the 101-page draft of a report mentioned earlier which was issued by the Moore School on 30 June 1945.[11] This report represents a masterful analysis and synthesis by him of all the thinking that had gone into the EDVAC from the fall of 1944 through the spring of 1945. Not everything in there is his, but the crucial parts are. He stated that he wrote the report "in order to further the development of the art of building high speed computers and scientific as well as engineering thinking on this subject as widely and as early as feasible." [12] In a sense, the report is the most important document ever written on computing and computers.

Eckert and Mauchly in writing a progress report on the EDVAC had this to say: "During the latter part of 1944, and continuing to the present time, Dr. John von Neumann, consultant to the Ballistic Research Laboratory, has fortunately been available. . . . He has contributed to many discussions on the logical controls of the EDVAC, has proposed certain instruction codes, and has tested these proposed systems by writing out the coded instructions for specific problems. . . . In his report, the physical structures and devices . . . are replaced by idealized elements to avoid raising engineering problems which might distract attention from the logical considerations under discussion." [13]

Von Neumann was the first person, as far as I am concerned, who understood explicitly that a computer essentially performed logi-

[11] J. von Neumann, *First Draft of a Report on the* EDVAC (Philadelphia, 1945).
[12] Sworn deposition, von Neumann, notarized 8 May 1947.
[13] Eckert and Mauchly, "Automatic High-Speed Computing, A Progress Report on the EDVAC," 30 September 1945.

cal functions, and that the electrical aspects were ancillary. He not only understood this was so but he also made a precise and detailed study of the functions and mutual interactions of the various parts of a computer. Today this sounds so trite as to be almost unworthy of mention. Yet in 1944 it was a major advance in thinking. In a letter to von Neumann I said: "All . . . have been carefully reading your report with the greatest interest and I feel that it is of the greatest possible value since it gives a complete logical framework for the machine." [14]

There was an imperative need precisely at this time for a complete analysis of how a computer operates when viewed as a logical mechanism. This need developed from two new factors: the speed of the electronic device and the invention of the delay line as a storage device.

In electromechanical devices such as the Harvard-IBM machine there were 72 storage registers for numerical information, and the programs were stored in paper tapes. The speed of operation of the tapes was more or less conformed to that of the counter wheels making up the registers. There was therefore no unbalance between the numerical storage organs and those for instructional storage. In the case of an electronic machine the instructions had to be handled at the same electronic speed as the numbers or else there would have been a total unbalance. Thus a rethinking of the storage requirements both for numbers and instructions was essential. This, among other things, is what von Neumann accomplished.

Prior to von Neumann people certainly knew that circuits had to be built to effect the various arithmetic and control functions, but they concentrated primarily on the electrical engineering aspects. These aspects were of course of vital importance, but it was von Neumann who first gave a logical treatment to the subject, much as if it were a conventional branch of logics or mathematics. His account shows how he impressed his stamp on the field:

> 1.1 The considerations which follow deal with the structure of a *very high speed automatic digital computing system,* and in particular with its logical control. . . .
>
> 1.2 An *automatic computing system* is a (usually highly composite) device which can carry out instructions to perform calculations of a considerable order of complexity — e.g., to solve a non-

[14] Letter, Goldstine to von Neumann, 15 May 1945.

linear partial differential equation in 2 or 3 independent variables numerically.

The instructions which govern this operation must be given to the device in absolutely exhaustive detail. They include all numerical information which is required to solve the problem. . . . These instructions must be given in some form which the device can sense: punched into a system of punch cards or on teletype tape, magnetically impressed on steel tape or wire, photographically impressed on motion picture film, wired into one or more fixed or exchangeable plugboards. . . . All these procedures require the use of some code, to express the logical and the algebraical definition of the problem under consideration. . . .

Once these instructions are given to the device, it must be able to carry them out completely and without any need for further intelligent human intervention. At the end of the required operations the device must record the results again in one of the forms referred to above. The results are numerical data. . . .[15]

He then went on to an enumeration of the organs of the machine:

2.2   First: Since the device is primarily a computer, it will have to perform the elementary operations of arithmetics most frequently. These are addition, subtraction, multiplication and division: $+, -, \times, \div$. It is therefore reasonable that it should contain specialized organs for just these operations.

It must be observed, however, that while this principle as such is probably sound, the specific way in which it is realized requires close scrutiny. . . . At any rate a *central arithmetical* part of the device will probably have to exist and this constitutes *the first specific part: CA.*

2.3   Second: The logical control of the device, that is, the proper sequencing of its operations, can be most efficiently carried out by a central control organ. If the device is to be *elastic,* that is, as nearly as possible *all purpose,* then a distinction must be made between the specific instructions given for and defining a particular problem, and the general control organs which see to it that these instructions—no matter what they are—are carried out. The former must be stored in some way—in existing devices this is done as indicated in 1.2—the latter are represented by definite operating parts of the device. By the *central control*

[15] Von Neumann, *First Draft,* pp. 1–2.

we mean this latter function only, and the organs which perform it form *the second specific part: CC.*

2.4   Third: Any device which is to carry out long and complicated sequences of operations (specifically of calculations) must have a considerable memory. . . .

(b)   The instructions which govern a complicated problem may constitute considerable material, particularly so, if the code is circumstantial (which it is in most arrangements). This material must be remembered. . . .

At any rate, the total *memory* constitutes *the third specific part of the device:* M.

2.6   The three specific parts CA, CC (together C), and M correspond to the *associative* neurons in the human nervous system. It remains to discuss the equivalents of the *sensory* or *afferent* and the *motor* or *efferent* neurons. These are the *input* and *output* organs of the device. . . .

The device must be endowed with the ability to maintain the input and output (sensory and motor) contact with some specific medium of this type (cf. 1.2): That medium will be called the *outside recording medium of the device:* R. . . .

2.7   Fourth: The device must have organs to transfer . . . information from R into its specific parts C and M. These organs form its *input,* the *fourth specific part: I.* It will be seen, that it is best to make all transfers from R (by I) into M and never directly into C. . . .

2.8   Fifth: The device must have organs to transfer . . . from its specific parts C and M into R. These organs form its *output, the fifth specific part: O.* It will be seen that it is again best to make all transfers from M (by O) into R, and never directly from C. . . .[16]

We can see from the little quoted above how von Neumann gave a logically complete analysis of the structure of the EDVAC in his report. This was his first major contribution. We need not argue here whether others could or would have done this without him. The fact is that he did it. All his training in formal logics fitted him to do it, and it unequivocally bears the mark of his genius.

There has been much, quite bitter controversy about the attribution of credits for the various ideas connected with the EDVAC. It is therefore proper at this point to set out the facts as seen at the time by me.

[16] *First Draft,* pp. 3–7.

The discussions on the logical structure of this machine were very extensive and served to focus sharp attention on a myriad of problems. The whole matter was then synthesized by von Neumann in his *First Draft*. Eckert and Mauchly writing at the time said: "Those who have contributed in this way are Dr. Arthur Burks, of the Moore School Staff, Capt. H. H. Goldstine, Army Ordnance, and especially Dr. John von Neumann, consultant to the Ballistic Research Laboratory." [17] Actually, many engineering matters were also threshed out in these meetings, as we shall see.

To illustrate how the discussions went—even when von Neumann was away at Los Alamos he participated by mail—let me quote from a letter from von Neumann to me:

> Here are some further small items on which I wanted to write to you, namely:
>
> I want to add something to our discussion of feeding data into the machine, and getting results out. As you may recall, in feeding in, two types of numbers occurred: Binary integers x, y which denote positions in the memory: binaries ξ. . . . In printing only the ξ occur.
>
> Clearly the ξ ought to be typed (by a human operator) and absorbed by the machine as decimal numbers, and also printed as such. So we need here decimal $\rightarrow$ binary and binary $\rightarrow$ decimal conversion facilities.
>
> As to the x, y we had doubts. Since they have logical control functions, it is somewhat awkward to have to convert them. I argued for having x, y always in binary form, but we finally agreed that binaries are hard for handling and remembering by humans.
> . . .
>
> I think that we overlooked an obvious solution: That is, to handle x, y outside the machine in the octal (base 8) system. . . .
>
> Does my suggestion seem reasonable to you and to the others?
>
> There exist two recent (1944) midget pentodes which may be of interest to us: 6AK5 and 6AS6. . . .
>
> Both have sharp cutoff on the control grids. 6AK5 has inner connection between suppressor and cathode, 6AS6 however brings the suppressor out separately and has a sharp cutoff on the suppressor too: $-15v$ for $+150v$ on the screen. . . .
>
> I am continuing working on the control scheme for the EDVAC and will definitely have a complete writeup when I return. I am

---

[17] Eckert and Mauchly, "Automatic High-Speed Computing . . . ," Acknowledgements.

also working on the problem of formulating a two-dimensional, non-stationary hydrodynamical problem for the ENIAC. . . .[18]

To indicate further how things were done at this juncture here are excerpts from my reply:

> We are enclosing some further tube characteristics in which you will be interested. As you will notice, one . . . is the 6AS6 which you mentioned. . . . We have also ordered the 6AK5. . . .
>
> We all agree that the use of the octal system for handling the pairs x, y . . . is a very sound solution. . . .[19]

Or again:

> The contents of this letter belong, of course, into the manuscript, and I will continue the manuscript and incorporate these things also after I get it back from you — if possible with comments on both items from you. . . .[20]

These few illustrations should serve to show the true situation: all members of the discussion group shared their ideas with each other without restraint and therefore all deserve credit. Eckert and Mauchly unquestionably led on the technological side and von Neumann on the logical. It has been said by some that von Neumann did not give credits in his *First Draft* to others. The reason for this was that the document was intended by von Neumann as a working paper for use in clarifying and coordinating the thinking of the group and was not intended as a publication. (In fact, on 25 June 1945 copies were distributed to 24 persons closely connected to the project.) Its importance was so clear however that later as its fame grew many outsiders requested copies from the Moore School or me. Through no fault of von Neumann's the draft was never revised into what he would have considered a report for publication. Indeed, not until several years later did he know that it had been widely distributed.

Hopefully having laid that ghost to rest, we return to von Neumann's contributions. First, his entire summary as a unit constitutes a major contribution and had a profound impact not only on the EDVAC but also served as a model for virtually all future studies of logical design. Second, in that report he introduced a logical notation adapted from one of McCulloch and Pitts, who used it in a

[18] Letter, von Neumann to Goldstine, 12 February 1944. The dating by von Neumann is erroneous; it should be 1945. This is evident both from the substance of the letter and from the date of the reply.

[19] Letter, Goldstine to von Neumann, 24 February 1945.

[20] Letter, von Neumann to Goldstine, 8 May 1945.

study of the nervous system.[21] This notation became widely used, and is still, in modified form, an important and indeed essential way for describing pictorially how computer circuits behave from a logical point of view.[22]

Third, in the famous report he proposed a repertoire of instructions for the EDVAC, and in a subsequent letter he worked out a detailed programming for a *sort and merge* routine. This represents a milestone, since it is the first elucidation of the now famous stored program concept together with a completely worked-out illustration.

Fourth, he set forth clearly the serial mode of operation of the modern computer, i.e. one instruction at a time is inspected and then executed. This is in sharp distinction to the parallel operation of the ENIAC in which many things are simultaneously being performed.

Fifth, he gave clear indications that some modification of the so-called iconoscope or television camera tube would be a valuable memory device. This is a foreshadowing of the project at the Institute for Advanced Study which was to be the prototype for the computers of yesterday, today, and perhaps tomorrow.

These are the things von Neumann did. They have left their mark on the whole computing world. Thus, for example, the designers of the EDSAC, the first machine in the world to use the stored program, say: "Several machines working on the same principles as the EDSAC are now in operation in the United States and in England. These principles derive from a report drafted by J. von Neumann in 1946 [*sic*]. . . . It is found that machines designed along the lines laid down in this report are much smaller and simpler than the ENIAC and at the same time more powerful. The methods by which programs are prepared for all these machines are, as might be expected, similar, although the details vary according to the different codes used. Anyone familiar with the use of one machine will have no difficulty in adapting himself to another." [23]

To recapitulate: It is obvious that von Neumann, by writing his report, crystallized thinking in the field of computers as no other

[21] W. S. McCulloch and W. Pitts, "A Logical Calculus of the Ideas Immanent in Nervous Activity," *Bull. Math. Biophysics*, vol. 5 (1943), pp. 115–133.

[22] See, e.g., Hartree, *Calculating Instruments and Machines* (Urbana, 1949), pp. 97–110.

[23] M. V. Wilkes, D. J. Wheeler, and S. Gill, *The Preparation of Programs for an Electronic Digital Computer* (Cambridge, Mass., 1951). This machine was the first electronic one to be put in operation after the ENIAC. It was built for the Cavendish Laboratory in Cambridge, England.

person ever did. He was, among all members of the group at the
Moore School, *the* indispensable one. Everyone there was indispensable as regards some part of the project—Eckert, for example,
was unique in his invention of the delay line as a memory device—
but only von Neumann was essential to the entire task.

The thorny question of exact authorship of each idea will never
be resolved satisfactorily because almost all ideas were generated
in a joint mode. In fact von Neumann once summarized the situation in this regard very well by saying:

> There are certain items which are clearly one man's . . . the
> application of the acoustic tank to this problem was an idea we
> heard from Pres Eckert. There are other ideas where the situa-
> tion was confused. So confused that the man who had originated
> the idea had himself talked out of it and changed his mind two or
> three times. Many times the man who had the idea first may not
> be the proponent of it. In these cases it would be practically im-
> possible to settle its apostle.[24]

Let us now return to our temporal sequence and bring the story up
to date. At the beginning of September 1944 the construction of
virtually all parts of the ENIAC was well underway, and it was
generally felt that the problem had decisively shifted from a
developmental to a constructional phase. Indeed, this is why there
was so much emphasis just at this moment on the EDVAC. The
design engineers had little to do at this time on the ENIAC while
those engaged in the actual construction were fully occupied.

It is interesting to see now how the shortcomings of the ENIAC
were acknowledged at that time. In writing to Gillon, who was
then in the Pacific Theatre of Operations, I said:

> To illustrate the improvements I wish to realize, let me say
> that to solve a quite complex partial differential equation of von
> Neumann's . . . the new Harvard IBM will require about 80
> hours as against ½ hour on ENIAC of which about 28 minutes will
> be spent just in card cutting and 2 minutes for computing. The
> card cutting is needed simply because the solution of partial
> differential equations requires the temporary storage of large
> amounts of data. We hope to build a cheap high-speed device for
> this purpose. . . . The second major improvement . . . can again
> be illustrated on the Harvard machine. To evaluate seven terms

---

[24]Remarks from Minutes of Conference held at the Moore School of Electrical
Engineering on 8 April 1947 to discuss Patent Matters.

of a power series took 15 minutes on the Harvard device of which 3 minutes was set-up time, whereas it will take at least 15 minutes to set up ENIAC and about 1 second to do the computing. To remedy this disparity we propose a centralized programming device in which the program routine is stored in coded form in the same type storage devices suggested above. The other crucial advantage of central programming is that any routine, however complex, can be carried out whereas in the present ENIAC we are limited.[25]

This letter is interesting in that it shows how far developed thinking was at such an early stage. Recall that during the two prior months the two-accumulator test was the all-consuming interest of the group. Furthermore, at that time von Neumann had only just become acquainted with the project. This is an excellent example of the speed of his mind. In that same letter I said: "Now that we seem to be on the fairway as far as development goes, I feel it most important to make plans for further improvements to realize in a second machine the highly important features that seemed too difficult in the first model. Von Neumann, Eckert and I have formulated quite definite ideas along these lines. . . ."

Even during the latter part of August there was great progress on the ideas leading to the EDVAC. In an earlier letter to Gillon I wrote:

Von Neumann is displaying great interest in the ENIAC and is conferring with me weekly on the use of the machine. He is working on the aerodynamical problems of blast. . . . As I now see the future course of the ENIAC, there are two further directions in which we should pursue our researches. After talking to S. B. Williams of Bell Telephone, I feel that the switches and controls of the ENIAC now arranged to be operated manually, can easily be positioned by mechanical relays and electromagnetic telephone switches which are instructed by a teletype tape. . . . In this manner tapes could be cut for many given problems and reused whenever needed. Thus we would not have to spend valuable minutes resetting switches when shifting from one phase of a problem to the next. The second direction to be pursued is of providing a more economical electronic device for storing data than the accumulator. Inasmuch as the accumulator is so powerful an instrument, it seems foolish to tie up such tools merely to hold numbers temporarily. Eckert has some excellent ideas on a very cheap device for this purpose.

[25] Letter, Goldstine to Gillon, 2 September 1944.

The new torque amplifiers [this of course is a statement about the differential analyzer] are installed at Philadelphia on the integrators and also on the input tables. . . . The analyzer is operating . . . between three and four times the maximun speed ever before used with excellent results. . . .[26]

Notice that in the fortnight between the two letters the idea of the stored program seems to have evolved. Indeed, in the September letter the concept already appears in quite modern guise, whereas in the August one the author was trying to evolve an emendation of the ENIAC's decentralized controls to make it a little more useful.

All this must make clear that the fall of 1944 was perhaps the most eventful time in the intellectual history of the computer. All evidence from the correspondence of the period certainly bears this out. It also helps reinforce the comments in the previous section about the general impossibility and impracticability of sorting out authorship of ideas.

Before describing in detail the logical structure of the EDVAC and the stored program concept, the subject of the next section, we should perhaps mention a few other items.

Brainerd was so involved with administrative details in connection with the ENIAC that Prof. S. Reid Warren, Jr. was appointed to head the EDVAC project. Typical of the sorts of things happening at that time are three memoranda of the period. In April 1945 Brainerd assigned Burks to be in charge of writing the ENIAC reports and placed under him Adele Goldstine and Prof. Harry Huskey. They were charged with the tasks of producing an operating manual, a complete technical report, and a maintenance manual.[27] It is clear from the tone of Brainerd's memorandum that there were great pressures. It says: "In addition, Dr. Burks is requested to discuss with me the status of the report writing work at least once a week."

In May Brainerd wrote to a visitor from the British Scientific Office, John R. Womersley, apologizing for his delay in sending reports on the ENIAC.[28] Womersley's visit was interesting because it opened the door for modern computation to go abroad. As we shall see later, Womersley was the first in a long line of visitors; originally they came from England, then from France and Sweden.

[26] Letter, Goldstine to Gillon, 21 August 1944.

[27] Memorandum, Brainerd to Dr. Burks, Mr. Eckert, Mrs. Goldstine, Captain Goldstine, Dr. Huskey, 27 April 1945.

[28] Letter, Brainerd to Womersley, 9 May 1945.

Brainerd instituted new working rules in September to ensure that "work on the ENIAC itself [will] proceed from 8:30 in the morning to 12:30 the following morning, each day except part of Saturday and all day Sunday." [29]

It was during the spring of 1945 that the problem of where to house the ENIAC at Aberdeen began to come to the fore. In order to ensure an orderly transfer of the equipment, contract W-18-001-ORD 335 (816) was entered into in May of 1945 between the University of Pennsylvania and Aberdeen Proving Ground to move the ENIAC into the Ballistic Research Laboratory.

In the summer of 1944 I had written Simon calling his attention to the need for a room about 20 feet by 40 feet to house the machine. At first Simon proposed space in the then recently completed Wind Tunnel Building at Aberdeen. This space turned out to be inadequate, and Simon thereupon initiated a request to erect a suitable structure to house all the computing equipment of the Ballistic Research Laboratory: this included about 100 people, a set of IBM machines, the differential analyzer, the expected ENIAC and EDVAC as well as the two Bell Relay Computers.

By the spring of 1945 the Office of the Chief of Ordnance had agreed to the erection of a Computing Annex to the Ballistic Research Laboratory, and Dean Pender was requested to have his people prepare plans and specifications for locating the ENIAC in this new location. As will be seen later, this was done but much later than originally planned—not because of delays in the operability of the ENIAC but because it was not in the public interest to have the only operative electronic computer disassembled during a critical time in the life of the United States. This will be discussed in the proper place in this history.

At this same time the Moore School received the power supply transformers and chokes which had been built by an outside contractor for the ENIAC. These devices, which were essential for providing electrical power at the proper voltages, had been ordered some time before. When they arrived in the spring of 1945, all were found to be completely defective, and the entire lot was rejected.[30]

[29] Memorandum, Brainerd to Captain Goldstine, Messrs. Burks, Chu, Eckert, Sharpless, Shaw, 8 September 1945.

[30] Letter, Lt. Col. C. H. Greenall to Goldstine, 20 April 1945. This letter has as an enclosure a memorandum showing the details of the tests. An idea of the size of the equipment in question is indicated by the fact that it cost well over $10,000. Fortunately for the project, the Moore School had also ordered a duplicate set from a very small firm as a "back-up" or auxiliary set. This order was filled satisfactorily,

A number of other projects were being brought into the fore-ground by the impending completion of the ENIAC. For one thing there was, as seen by me, a real need for a permanent cadre of operating personnel for the machine. To this end I had John V. Holberton assigned to take charge of the operations of the ENIAC upon its completion: "Effective 1 June 1945 . . . [he] will re-port to Capt. H. H. Goldstine." [31] I also had two enlisted men, Cpl. Irwin Goldstein and Pfc. Homer Spence, both of whom had some technician's training in electronics, assigned to work with the ENIAC engineering group. It was the intention that they should become the service or maintenance men for the machine after it was turned over to the government. As it happened, Goldstein left, but Spence stayed on and later became an important member of the computing group at the Ballistic Research Laboratory.

To support Holberton on the programming side I assigned six of the best computers to learn how to program the ENIAC and report to Holberton. They were the Misses Kathleen McNulty, who subsequently married Mauchly; Frances Bilas, who married Spence; Elizabeth Jennings who married a Moore School engineer; Elizabeth Snyder, who married Holberton; Ruth Lichterman; and Marilyn Wescoff.

Also, a $45,000 contract was entered into between the Army and the Moore School providing a considerable amount of test equip-ment for the ENIAC and also ensuring the coordination of the IBM equipment with that machine.[32]

Thus the problems of housing and staffing the ENIAC upon its completion were handled. These preparations were quite satisfac-tory, and there was an adequate staff to maintain and operate the ENIAC both in Philadelphia and in Aberdeen after it was moved. This was particularly important since the end of the war had a very strong centrifugal effect on the staff.

One last problem was solved by establishing a committee at the Ballistic Research Laboratory under the chairmanship of L. S. Ded-erick, one of the associate directors, and including Haskell Curry, one of Hilbert's last Ph.D. students, and Derrick Henry Lehmer, a

---

and the equipment was installed so that little real delay occurred. To give a feeling for the physical dimensions of the apparatus we might mention that the iron needed for the transformer laminations weighed over two tons. (See letter, Goldstine to V. W. Smith, 20 February 1945.)

[31] Memorandum, Lt. Col. J. M. Shackelford, Organization of the Computing Branch, 31 May 1945.

[32] Letter, T. E. Bogert to R. Kramer, 18 June 1945.

well-known computer and table compiler, who, with his wife Emma, pioneered in the use of computers for probing number theoretical problems. This committee took over the ENIAC from the Moore School and supervised its usage by the Ballistic Laboratory staff.

Chapter 8

The Structure of

the EDVAC

We can now describe in some detail the logical structure of the EDVAC. It was the immediate antecedent of the Institute for Advanced Study computer, which was to be the prototype for the systems of today, and is thus important historically. In 1945 however these things were as yet no more than dreams of the future. But von Neumann's *First Draft* was to be a blueprint for a whole line of machines starting with the EDSAC at the Cavendish Laboratory in England and continuing through a number of other delay line machines.

This machine, as we saw earlier in the quotation from the *First Draft,* had five principal organs: an *arithmetical* one capable of performing the elementary operations of arithmetic, a *central control* for executing the instructions it received, a *memory* for storing both the numerical as well as the instructional information for a given problem, and finally *input* and *output* units for intervening between the mechanism and the human. In connection with these organs recall that Babbage recognized explicitly two of these: his store and his mill (above, p. 21). The store was the memory in our parlance, and the mill the arithmetic organ. Apparently neither he nor Lady Lovelace seem to have made explicit the need for a control or input-output organs.

Many different machine organizations and architectures are possible with the five organs of von Neumann's account, and we shall see that there was a great difference between the EDVAC and the Institute for Advanced Study machine. Let us therefore indicate a few of the salient characteristics of the EDVAC, as then envisioned, without going into too many details.

There were several aspects of the ENIAC which were particularly worrisome: it had too many vacuum tubes — over 17,000 — for comfort, too small a memory for numbers, and a highly cumbersome decentralized control as well as awkward means for setting up a new problem.

In a previous section we showed how the delay line provided a means for storing a number at about 1/100 the cost in vacuum tubes

of the ENIAC circuits. This was a first major step in bringing down the
tube count. Another was to abandon the highly parallel operation of
the ENIAC where many additions and a multiplication all proceeded
in parallel. Thus, instead of 20 accumulators (they were of course
adders) plus a multiplier and separate divider-square rooter, von
Neumann's account envisages one arithmetic organ capable of just
one operation at a time. The basic idea here was that the electronic
technique is potentially so fast one can afford to abandon all paral-
lelism and still obtain enough speed. This postulate has underlain
virtually all machine developments ever since and is only in the
last few years being questioned.

This notion of serial operation was pushed to its absolute limit in
the EDVAC design. Not only was there to be a complete serial character to the execution of operations, but each operation itself, in so
far as possible, was also to be broken down into its atomic parts and
each of these executed serially. Thus when two n-digit numbers
were to be added, the operation was performed by adding them a
digit at a time as we humans do instead of treating all digits in
parallel. This serial procedure is of course slower than the parallel
one, but it requires much less equipment. Again, the slow-down
was acceptable because of the extremely high unit speeds achievable by electronics.

Moreover, on this score it is clear that the delay line lends itself
very well to a highly serial mode of operation. If the individual
digits comprising a number occur in serial order in a delay line,
they naturally will emerge a digit at a time at the output of the line
ready to be processed serially as indicated above. To make this
more obvious suppose that we have two delay lines, I and II, each
containing a separate number, and that we wish their sum formed
and put into a third line, III.

This is schematized in Fig. 15. The adder is supposed to be
capable of adding three digits at a time: the corresponding digits of

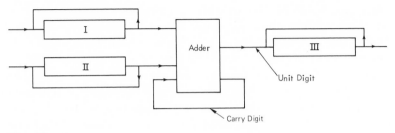

Fig. 15

the numbers in I and II and the carry digit from the previous addition. Thus, in order to add 14 and 18, first we add 4 and 8, obtaining a unit digit of 2 and a carry of 1; then we add the next digits, 1 and 5, together with the carry of 1.

How large is such an adder? This brings us into the next simplification: von Neumann's decision to press for the binary representation of numbers. In that number system the only digits are 0 and 1, and every number is an array of 0's and 1's. Thus, the number representable decimally as

$$141 = 1 \times 10^2 + 4 \times 10^1 + 1$$

is representable binarily as

$$128 + 8 + 4 + 1 = 1 \times 2^7 + 0 \times 2^6 + 0 \times 2^5 + 0 \times 2^4 + 1 \times 2^3 + 1 \times 2^2 + 0 \times 2^1 + 1$$

$$= 10001101$$

This representation is considerably longer than the corresponding decimal one but much simpler in the sense that there are only two kinds of digits – 0 and 1 – instead of ten kinds in the decimal representation. The addition of binary digits is very simple: $0 + 0 = 0$, $0 + 1 = 1$, $1 + 0 = 1$, $1 + 1 = 10$. The last one is merely the statement that one plus one is two, which is represented in the binary form as $1 \times 2^1 + 0$ or in shorthand as 10 (one zero). A simple adder of the sort suggested above can probably be made with less than twenty vacuum tubes.

How about multiplication? How complex is it in this milieu of serial operation and binary numbers? To answer these questions we should first write down the multiplication table for binaries: $0 \times 0 = 0$, $1 \times 0 = 0$, $0 \times 1 = 0$, $1 \times 1 = 1$. This is all there is to it. Instead of the 100-entry table of our childhoods for decimals here we find only a 4-entry table. This great compression of information is exactly mirrored in the electronic circuitry. To do a multiplication of two ten-decimal digit numbers requires of course 100 individual products plus the accompanying additions. The same numbers each require about 32 binary digits to represent them; thus, about 1,000 individual steps are now required. Thus we see that although there is a great simplification of the complexities of arithmetic by going from decimal to binary representations for numbers, it is achieved at the price of a considerable increase in the number of steps. This price is however very cheap under the electronic dispensation but is impossibly expensive under the electromechanical one. Since a vacuum tube circuit could operate in 1945 in times of the order of a microsecond—a millionth of a second—it is not disturbing to have to do 1,000 of these time units. Indeed 1,000 microseconds is only a millisecond—a thousandth of a second. Thus the extreme simplicity of binary arithmetic made it the "medicine of choice" for the EDVAC and for all modern machines as well. Its advantages more than make up for the increase in the number of steps.

In his account von Neumann estimated the memory needs of the EDVAC based on some heuristic appraisals of hydrodynamical calculations, and proposed a memory of between 2,000 and 8,000 numbers of 32 binary digits each. We noted earlier that a delay line of about five feet in length could store 30 such numbers at a cost of around 10 vacuum tubes; hence, a 2,000 to 8,000 word memory would cost from 700 to 2,800 vacuum tubes. The other parts of the system would require considerably less than 1,000 tubes. These rough arguments convinced von Neumann—correctly—that "the decisive part of the device, determining more than any other part its feasibility, dimensions, and cost, is the memory." [1]

It is now time for us to say a few words about the control of the machine and the instructions it could execute. A general-purpose device is by its very nature capable of doing a great many things, but it must somehow be instructed as to which particular task it is supposed to do at any given time. Thus a desk computing machine is such a device, but it cannot do any task without a human to decide upon and control its sequence of operations. The electromechanical machines, such as Stibitz's, were instructed by paper tapes on which were impressed instructions coded in numerical form. These devices contained an organ that could read these instructions serially and execute them. The speed balance between the relay registers that constituted the memory for such machines and these paper tapes was relatively good; both operated in the same speed range.

Obviously a paper tape would never do in a machine with an electronic memory. The imbalance in speed between electronic circuits and electromechanical ones is so severe that a machine of such type would be completely mismatched, so that the electronic portions would almost never operate while waiting for the tapes to advance. The speed of such a device would then be determined by that of the tape, and all advantages of the electronic discipline would be lost.

[1] *First Draft*, p. 67.

To get around this difficulty the ENIAC was modeled upon the old-style IBM machines with all instructions prewired in. As we said earlier, this was workable but clumsy. In casting around for a better way to handle the problem Eckert and Mauchly first thought of a rotating magnetic device "wherein the successive digits of a number were transmitted in timed sequence from magnetic storage or memory devices through electronic switches to a central electronic computing circuit and similarly returned to magnetic storage." [2] This device is of course the magnetic analog of Atana- soff's rotating capacitor drum and no doubt derived from it. How- ever, the supersonic delay line was superior to both and displaced them. It permitted the storage of both numbers and instructions in the inner memory so that again a speed balance was regained. An- other type memory was discussed by von Neumann in some detail and investigated by him in talks with Vladimir Zworykin of RCA: the iconoscope, the camera tube of television. We discuss this de- vice in detail later when we come to the Institute for Advanced Study phase of our account.

What are the instructions like, and how are they coded into nu- merical form? To discuss these questions we need to make two assumptions: first, that the locations in the memory are numbered sequentially; second, that there is an organ in the machine, the control, which normally goes through the memory in this sequen- tial order looking for, interpreting, and executing the instructions. Suppose that the arithmetic organ contains three delay lines each capable of storing a number. The operations of addition, subtrac- tion, multiplication, and division all are of the form: $x \, O \, y = z$, where O is $+$, $-$, $\times$ or $\div$. There is then one delay line for each of the x, y, and z. Typical orders are these: transfer a number from a spe- cific location in the memory into the arithmetic organ; add the con- tents of the delay lines in the arithmetic organ; transfer a number from the arithmetic organ into a specific memory location; take one or the other course of action depending on the sign of a number in the arithmetic organ; etc. We do not need to describe the orders in more detail beyond noting that each instruction or order consists of two parts: a location either in the memory or the arithmetic organ, and a statement of which operation is to be executed. The locations in the memory can be enumerated 0, 1, 2, . . . , and the operations can also. Thus, an instruction can be expressed as a number which can therefore be stored, just as is any other numeri- cal material, in the memory.

[2] Eckert and Mauchly, "Automatic High-Speed Computing," p. 2.

In von Neumann's *First Draft* he preceded each number by a digit 0 and each instruction by a digit 1. Then the central control was so designed that "as a normal routine . . . [it] should obey the orders in the temporal sequence, in which they naturally appear. . . . There must, however, be orders available, which may be used at the exceptional occasions . . . to transfer its connection to any other desired point. . . ." [3] By May of 1945 he was making some emendations and elaborations in the repertoire of instructions for the EDVAC and mentioned a program he had been working on for meshing and sorting of data.[4] In that letter he sums up the sorting situation thus:

> At any rate the moral seems to be that the EDVAC with the logical controls as planned for "mathematical" problems, and without any modifications for "sorting" problems, is definitely faster than the IBMs on sorting. The factor seems to be 1–3, according to the problem — it may be improved somewhat, but hardly beyond 6. Since the IBMs are really very good in sorting, and since according to the above sorting can be meshed with the other operations of the EDVAC without human intervention or need for additional equipment, etc., the situation looks reasonably satisfactory to me. On really big sorting problems it will, of course, be necessary to consider the interaction of the EDVAC and the magnetized tape which it produces and which may be fed back into it. . . . I think, however, that it is legitimate to conclude already on the basis of the now available evidence, that the EDVAC is very nearly an "all purpose" machine, and that the present principles for the logical controls are sound.

At some time in this period von Neumann made important changes in his ideas on the logical structure of the EDVAC. These changes appear implicitly in a manuscript in my hands which is, as far as I know, the first stored program written. It is a meshing and sorting routine. There is an excellent analysis both of the changes in the EDVAC and of this program in a paper by Knuth.[5]

One final remark is worth making at this point. The design of the EDVAC was already so sophisticated by this time that the use of magnetic tape as an input-output medium was firmly entrenched in the

[3] *First Draft*, p. 86.
[4] Letter, von Neumann to Goldstine, 8 May 1945.
[5] D. E. Knuth, "Von Neumann's First Computer Program," *Computing Surveys*, vol. 2 (1970), pp. 247–260.

planners' minds. This machine was by mid-1945 very advanced in its design over the ENIAC, which of course had not yet been completed. Indeed the EDVAC's design is quite modern even today. The improvements since then are, with a few exceptions, rather refinements than great new insights. We shall say more on this later.

During this period very many visitors came to the Moore School and left to spread the word about the new world of the computer. These people were of course all connected with the war effort but virtually all were so connected only for the duration. They had university positions and by this date were already conscious of the fact that the end of the war was in sight. They were the cadre around which the early computer people were formed.

Perhaps the most important of the domestic visitors were various members of the Institute for Advanced Study, the NDRC's Applied Mathematics Panel, MIT's Servomechanism Laboratory, RCA's Research Laboratory in Princeton, and Los Alamos's Theoretical Physics Division.

From the Institute came James W. Alexander, Oswald Veblen, and John von Neumann. We have already discussed the latter two in some detail. Alexander was a most distinguished topologist who as a lieutenant in Ordnance had worked closely with Veblen and Kent during World War I and had done a great deal for the Air Force in World War II as a civilian. They, along with Marston Morse, helped materially at the proper time to make the Institute for Advanced Study the next world center for computers. This was to happen in 1946 as the Moore School's position changed. But more of this later.

From NDRC came a number of very important mathematicians whose interest in the ENIAC and EDVAC projects was to be of inestimable importance to the computer field. Perhaps foremost among them was Mina Rees, president of the Graduate School of the City University of New York and a University of Chicago Ph.D. in Mathematics, who was at this period executive assistant to Warren Weaver, the chief of the Applied Mathematics Panel.[1] Through her thoughtful and forward-looking efforts, the Office of Naval Research played a key role in the post-war computer world. It was probably

[1] Weaver himself is a figure of first importance to science policy and has had a major influence on it for almost fifty years. The interested reader may wish to read his entertaining memoirs, *Scene of Change: A Lifetime in American Science* (New York, 1970).

she who involved her superior at ONR, Dr. Emanuel R. Piore, in this field. He played a key role in making that agency into what it became: for many years the most intelligent and stable funder of science in the United States. As Chief Scientist of ONR he was instrumental in shaping scientific policy in this country and abroad. It is certain that he very early recognized the importance of the computer to science and gave Dr. Rees his full support. Then in 1956 he joined IBM as Director of Research. Since then he has, in addition to many detailed accomplishments, served admirably as that company's "scientific conscience," a task he is uniquely qualified to fill. It was he who was to take up the early interests of the Watsons in scientific computing and help give them tangible expression. It is perhaps noteworthy that Piore and Weaver first became acquainted at the University of Wisconsin in the early 1930s when Piore was a young instructor in physics and Weaver was chairman of the Mathematics Department.

From MIT came Jay W. Forrester, who with Gordon Brown founded the Servomechanism Laboratory at MIT in 1940. Forrester is today best known for his absolutely fundamental invention of the coincident-current magnetic core memory that was one of the basic technological discoveries in the entire computer field — more on this later. The laboratory at MIT did very important work on fire control and shipboard radar systems. Then in 1944 Forrester in a pioneering spirit undertook, under a contract with the Navy Office of Research and Inventions, the extremely difficult and challenging task of building a computer system to simulate an airplane's performance with the pilot as a part of the system. As originally planned, the device would have had all the controls available to a pilot, and for his purposes would give in "real" time all the responses that the plane would have. Forrester envisioned this computer as a means of simulating engineering changes in a plane without the expense, time loss, and possible dangers inherent in building such a modified plane. He originally thought in terms of an analog computer, but after surveying the field and talking with various members of the Moore School group he changed his mind.[2] In the fall of 1945 he described his project as one of designing

. . . a computer to solve design problems arising in aircraft research. The problem requires the simultaneous solution of

[2] See S. B. Williams, *Digital Computing Systems* (New York, 1959), p. 10; or B. V. Bowden, *Faster than Thought* (London, 1953), p. 177.

many differential equations and the techniques visualized by the University of Pennsylvania on their EDVAC computer show considerable promise.

Your office authorized our visiting the University . . . last week to obtain information and it would be most helpful if the recently published report on the EDVAC system could be made available to us. We would appreciate your sending this as soon as possible. . . .[3]

Much of interest and one thing vital to the computer field flowed out of Forrester's work. His work in many ways was fundamental to MIT: through Forrester's leadership that institution resumed a leading role in the computing field. He founded and headed from 1951 through 1956 its Digital Computer Laboratory.

From the RCA group came V. K. Zworykin and Jan Rajchman whom we mentioned earlier in passing. Zworykin was one of the great inventors of the electronics industry; two of his most important inventions are the iconoscope, the camera tube of television, and the electron microscope. Rajchman, like Forrester, had been a pioneer in the computing field; in fact, he independently conceived of the magnetic core for storage of information and has made a number of inventions of other storage devices, one of which will figure in what we write about the Institute for Advanced Study. From the beginning these men, together with a colleague of theirs, Richard L. Snyder, were much interested in the technological problems and challenges of the digital computer field. As mentioned earlier, they even designed a digital fire control system for Frankford Arsenal before the ENIAC project. In a weekly progress report to Simon I noted: "The RCA Company has become interested in digital computing devices and has agreed to assist in the development of special tubes for memory purposes for use in the EDVAC."[4] However, this relationship did not flourish until the

[3] Airmail, Special Delivery letter, Forrester to Gillon, 13 November 1945. The authority for me to send the von Neumann report is contained in letter, J. J. Power to Goldstine, 20 November 1945. Forrester's first visit to the Moore School took place on 7 November 1945. In addition to Forrester, Perry Crawford of Special Devices, Navy Dept. came along. Von Neumann and I were also present. (See letter, Forrester to R. C. Mork, 31 October 1945.) This visit had been preceded on 9 October 1945 by a visit to the Moore School by Lts. Mork and Sidney Sternberg of the Navy's Office of Research and Inventions. (See teletype clearance, 8 October 1945, Power to Goldstine.) This is the predecessor office to the Office of Naval Research.

[4] Goldstine, Progress Report on ENIAC and EDVAC Projects for Week ending 17 November 1945. See also letter, Warren to Goldstine, 10 October 1945 regarding the loan of certain RCA tubes to the EDVAC project.

Institute for Advanced Study computer was initiated, since the
EDVAC adopted the acoustic delay line technology.

Finally, from the Theoretical Physics Division at Los Alamos
came two young theoretical physicists, Stanley P. Frankel and
Nicholas Metropolis. They were to have the honor of running the
first problem on the ENIAC. Very early von Neumann realized the
supreme importance that machine could have for Los Alamos,
especially for studying the feasibility of various ideas arising in
that laboratory. Accordingly, he urged the theoreticians there to
examine the ENIAC with a view to using it for their calculations.
They agreed with him and plans were initiated to do a very large
scale calculation as a test of the feasibility of a process that was
then felt to be of great importance to Los Alamos.

I agreed with von Neumann on using the ENIAC for this purpose
and, after a considerable amount of discussion, persuaded Brainerd that the Los Alamos problem should be used as the first problem to be run on the ENIAC. Brainerd had originally wanted a series
of problems, ranging from very simple to very complex ones, run
for the "shake-down cruise" of the ENIAC. He however acquiesced
graciously after he realized the urgency of the Los Alamos problem
and succeeded in persuading Dean Pender to go along with him.
All this was quite complex because I was not "cleared" by Los
Alamos and had to have faith in von Neumann's basic assertion as to
the importance of the problem; the equations were not classified so
I could at least understand the mathematical character of the
problem. I then had to pass along my faith to Brainerd without telling him or the dean more than that several scientists working for
the War Department would come and run a problem on the ENIAC
and that no details of the problem or the men could be discussed!

In any case Frankel and Metropolis arrived in the summer of
1945 to learn about the ENIAC. My wife and I spent considerable
time in teaching them how to program the machine; and they made
several trips between Los Alamos and Philadelphia in this connection.[5] From then on until the end of November Frankel and Metropolis were to work on the preparation of their problem for the ENIAC.
This problem was of great importance, since it was to test out a
dramatic new idea for Los Alamos and an equally dramatic one for
the Moore School, the ENIAC.

---

[5] Letter, Goldstine to Metropolis and Frankel, 28 August 1945. By this date we had
already become friends since the salutation reads "Dear Nick and Stan." It is clear
from the letter that they had recently left Philadelphia to return to Los Alamos but
would return "within a couple of weeks."

The calculation, which was a very large one, was successful, and in the early spring of 1946 Dr. Norris Bradbury, a well-known physicist who succeeded Robert Oppenheimer as Director of the Los Alamos Scientific Laboratory—or P.O. Box 1663, Santa Fe, New Mexico, as it was styled in the early days—was writing to Major General Gladeon M. Barnes and Gillon thanking them for the successful use of the ENIAC. He said:

> The calculations which have already been performed on the ENIAC as well as those now being performed are of very great value to us. . . . The complexity of these problems is so great that it would have been almost impossible to arrive at any solution without the aid of the ENIAC. We are extremely fortunate in having had the use of the ENIAC for these exacting calculations.
>
> I should like also to offer our congratulations for the successful development of so valuable an instrument. Our experience in this project has supplied a number of excellent illustrations of the dependence of modern pure and applied physics on advanced calculational techniques. It is clear that physics as well as other sciences will profit greatly from the development of such machines as the ENIAC.
>
> I should like also to express my thanks for the direct cooperation and sustained interest of Captain Goldstine and of Eckert and Mauchly as well as of the rest of the engineers and operators.[6]

While we have jumped ahead in time, it seems reasonable to do so in order that the reader may have some feeling of the excitement and tension that was building up in the summer and fall of 1945 around the ENIAC as it neared completion. We will return to this subject again in its proper place.

During the critical spring and summer of 1945 the NDRC was becoming much interested in the electronic computer, and Warren Weaver requested von Neumann to write a report on computing machines, including both the ENIAC and EDVAC.[7] Weaver furthermore suggested that the report be issued with a "Restricted" classification. Since both the ENIAC and EDVAC projects had been classified as "Confidential," this required a change of classification. Simon was agreeable to this provided that the Patent Branch of the

---

[6] Letter, Bradbury to Barnes and Gillon, 18 March 1946.
[7] Letter, Weaver to Col. Ritchie, 8 March 1945.

Office, Chief of Ordnance was.[8] Finally, after much correspondence the Ballistics Branch of the Office, Chief of Ordnance wrote to Weaver giving its approval for an Applied Mathematics Panel report by von Neumann with a classification of "Restricted." [9] The report by von Neumann was never completed, although it proceeded very far in draft form. For some reason unknown to me, he never carried it to the stage where he was satisfied.[10]

In the approval from the Ordnance Office to Weaver on 8 May 1945 for a report on computing devices it was stated that "The Moore School will be authorized to collaborate in preparation of the report. . . ." Weaver now asked Brainerd to prepare a report. After some debate within the ENIAC staff over the authorship of the document, a compromise was reached, and the list of authors as it appears on the report is Eckert, Mauchly, Goldstine, Brainerd.[11]

Both Gillon and I were eager for information on these machines to reach the scientific community as soon as possible, since we realized the great significance these devices would have. We therefore did all in our power to lower the classification of the projects and to encourage von Neumann to write his report. As a matter of fact, so well had word on the ENIAC and EDVAC projects "leaked" to the scientific community that the number of visitors both domestic and foreign had reached such proportions that there was concern about the delays these visits were causing.[12]

---

[8] First Indorsement 23 March 1945 to basic letter, Proposed NDRC Report on Computing Devices, 13 March 1945. This indorsement, written by me, said, ". . . the NDRC report mentioned in basic communication is highly desirable at the present time, and it is further very desirable that the ENIAC and EDVAC machines be included."

[9] Letter, Col. J. H. Frye to Weaver, 8 May 1945.

[10] It was from this draft that von Neumann lectured in May of 1946 before an advisory panel of the Office of Research and Inventions. A. H. Taub, in a footnote to a paper on which von Neumann and I collaborated, "On the Principles of Large-Scale Computing Machines" (von Neumann, Collected Works, vol. V, pp. 1–32), states this paper was the basis for those lectures and that it further contained material to be found in another report by the same authors. Actually this is incorrect. The facts are these: The lectures were based on von Neumann's unpublished manuscript. The article appearing in his Collected Works was first written by us in November of 1946, the first draft being dated 5 November 1946, and was prepared as an invited paper for the editor, L. R. Ford, of the American Mathematical Monthly. Four or five years later the author, with von Neumann's concurrence, started to prepare a revised joint paper and to this end added some parts of the material mentioned by Taub. However, pressure of other work prevented its publication in any form.

[11] Description of the ENIAC and Comments on Electronic Digital Computing Machines, AMP Report 171.2R, Moore School of Electrical Engineering, University of Pennsylvania, distributed by the Applied Mathematics Panel, National Defense Research Committee, 30 November 1945.

[12] On 26 June 1945, I informed Dean Pender that no further clearances would be

It was also true that various commercial concerns were desirous of publicizing their contributions to the ENIAC. Thus, for example, the International Resistance Company, of which Dean Pender was one of the founders, had asked and received permission to state in a prospectus that "the Army has an electronic calculating machine with more than 15,000 tubes." [13]

A meeting was held in the fall of 1945 at the Ballistic Research Laboratory to consider the computing needs of that laboratory "in the light of its post-war research program." The minutes indicate a very great desire at this time on the part of the leaders there to make their work widely available. "It was accordingly proposed that as soon as the ENIAC was successfully working, its logical and operational characteristics be completely declassified and sufficient publicity be given to the machine . . . that those who are interested . . . will be allowed to know all details. . . ." [14]

During this same eventful spring and summer a number of British visitors came to the Moore School, and from these visits stemmed the computerization of Great Britain. Specifically, they resulted in the development and construction of machines at the National Physical Laboratory, Teddington—the ACE, Automatic Computing Engine; at Cambridge University—the EDSAC, Electronic Delay Storage Automatic Calculator; and at the University of Manchester —MADM, Manchester Automatic Digital Machine. The principal visitors were John R. Womersley, who headed the Mathematical Division of the Teddington Laboratory, Douglas R. Hartree, and L. J. Comrie, all of whom we have met before. In addition to Womersley, two others connected with the British Post Office Research Station also came. [15]

The developments at Teddington were initially under the direction of Alan Turing from 1945 until July 1947 when he left to return to Cambridge University. Soon thereafter he went to Manchester and became part of a major project there, which we shall speak of again.

Womersley was the first of the Britishers to visit. He stayed in the States from February to April 1945, at which point he went home to become head of the Mathematical Division of the National Physi-

---

granted to visit the ENIAC. Memorandum, Goldstine to Pender, Clearance to visit ENIAC. This policy was not very actively carried out, but it did serve to keep down the numbers of visitors to some extent.

[13] Memorandum, Goldstine to Pender, 16 July 1945.

[14] Minutes, Meeting on Computing Methods and Devices at Ballistic Research Laboratory, 15 October 1945.

[15] Letter, Hartree to Goldstine, 23 August 1945.

cal Laboratory and to build an electronic computer there based on what he had learned in the States. He visited me at the Moore School on 12–14 March 1945 and at that time became acquainted with modern developments.[16] It is much to his credit and to the far-sightedness of his superior, Sir Charles G. Darwin, that a machine development started so promptly in England. In addition to Turing, Womersley was able to bring Harry D. Huskey over in 1946 to work on his machine. It was Huskey who wrote the engineering manual for the ENIAC. He also brought in J. H. Wilkinson, who is now a world leader in numerical analysis.

The ACE was planned by Turing and bore his mark. As Huskey wrote: "Emphasis on a simple machine led to more complicated systems of coding than were used in any of the computers being designed in this country." [17] There were several versions of this machine on paper. Copies of the first of them were brought over by Hartree and given to Caldwell and me in 1946 together with a third version.[18] The logical complexity of the ACE is not surprising since Turing had a preference for this type of activity to engineering. The type of complexity Turing proposed, while attractive in some respects, did not in the long run flourish and selection weeded it out.

It was while Turing was at Teddington that he visited von Neumann and me at Princeton. The correspondence of the period shows he was there for about a fortnight in the latter part of January 1947. Turing's plan was first to build a pilot machine, Pilot Ace. He was successful in this, and it was demonstrated in the fall of 1950 after he had left Teddington. It had been started under Sir Charles Darwin's directorship and finished under that of Sir Edward Bullard, with the immediate supervision of Womersley.

Womersley's successor, Dr. E. T. Goodwin, wrote Turing's mother: ". . . at about the time Alan took his sabbatical year at Cambridge, it was decided to produce a smaller version which would be entitled the Pilot Ace. Though the basic ideas . . . were largely Alan's you will understand that the detailed arrangement was decided by others. This machine did magnificent work for four

[16] Teletype, Gerber to Philadelphia Ordnance District, 10 March 1945. This message grants clearance for Womersley to visit Goldstine. He was sent a copy of von Neumann's *First Draft* as soon as it was available.

[17] H. D. Huskey, "Electronic Digital Computing in England," *Math. Tables and Other Aids to Computation*, vol. 3 (1948–49), pp. 213–225.

[18] A. M. Turing, "Proposals for the Development in the Mathematics Design of an Automatic Computing Engine (ACE)," first version; and "Circuits for the ACE," third version. This material may be found in a letter, Goldstine to Col. G. F. Powell, 27 February 1947.

or five years. . . . The original Pilot Ace was eventually given to the Science Museum to use as an exhibit. . . ." [19]

The full scale version of ACE was eventually completed in 1958 by the Control Mechanisms and Electronics Division of the National Physical Laboratory, and Dr. A. M. Uttley, the then superintendent of the division said, "Today, Turing's dream has come true." [20]

But all this takes us far ahead of our story. Let us therefore return to Philadelphia in the summer of 1945. Douglas Hartree first came there shortly before the Fourth of July as a visitor sponsored by the British Commonwealth Scientific Office and immediately became deeply interested and involved in the ENIAC and in the plans for the EDVAC. In fact, as soon as he returned home he used all his good offices to push England into the electronic computing field. In this he was very successful. In a later section we shall describe in a little more detail the developments there.

In addition to these activities the Committee on Mathematical Tables and Other Aids to Computation of the National Research Council under the leadership of Prof. R. G. Archibald, its chairman, called a conference on computing devices to be held 29–31 October 1945 at the Massachusetts Institute of Technology and Harvard. The organizing committee consisted of L. C. Comrie, chairman; S. H. Caldwell, vice-chairman; Howard Aiken; Derrick Lehmer; J. C. Miller, University of Liverpool; G. R. Stibitz; I. A. Travis, Moore School; and John Womersley. [21] In part the meeting was to celebrate the public announcement of the new MIT Differential Analyzer. Brainerd wrote Gillon asking approval to discuss the ENIAC and EDVAC projects at this meeting and received the latter's approval: "I think it would be helpful for you to give a coherent story on these developments." [22] To allay Gillon's concerns about achieving proper publicity only after the ENIAC was successfully operating, Brainerd corresponded with Caldwell, who reassured him that "The conference . . . will be entirely without

[19] Sara Turing, *Alan M. Turing* (Cambridge, 1959), p. 84.

[20] *Ibid.*, p. 89. Uttley is a distinguished engineer who had done considerable research on electronic computers while he was at the Telecommunications Research Establishment, Malvern, of the Ministry of Supply. His machine was designed after a 1948 visit to von Neumann and me in Princeton. It was known as the T.R.E. High-Speed Digital Computer.

[21] Announcement by the National Research Council, Program Conference on Advanced Computation Techniques, Massachusetts Institute of Technology and Harvard University, 29–31 October 1945.

[22] Letters, Brainerd to Gillon, 29 September 1945 and Gillon to Brainerd, 4 October 1945.

publicity and attendance to it by invitation only." [23] Brainerd, Eckert, and Mauchly went as Moore School representatives; von Neumann and I attended as government representatives. There was a presentation of the Moore School's work on the ENIAC and the EDVAC, made in four parts by Brainerd, Eckert, Mauchly, and me.

At the same time I was searching for a means to carry on the development of computers into the post-war world. To this end I canvassed a variety of places without success. The solution that was finally settled on was the result of many soul-searching conversations between von Neumann and me. There were several great problems: Were computers an essential part of the peacetime scientific world? How could research on them be funded? Where was there a congenial atmosphere to undertake this work?

Both of us answered the first question affirmatively virtually as a matter of faith. And we both felt confident that we could somehow out of our friendly relations with the governmental establishment find the money to continue our work. The site for the work was more of a problem. The Institute for Advanced Study had been originally conceived as a place to do theoretical studies and had no experimental tradition or facilities. It was therefore not *a priori* a likely place for the work. Eventually, however, these objections proved, at least for a temporary project, not insurmountable and the project went there. Much more on this later.

Before going forward with our story it is perhaps necessary now to digress and discuss the various tensions and centrifugal forces that have been alluded to before. They eventually were to reduce considerably the Moore School's contribution in the emergent computer field.

The contract between the University of Pennsylvania and the government was typical of research and development contracts of that era. As far as patents were concerned, the contractor had two options available to him: either he could take out the patents and grant the government various royalty-free licenses, or the government would take over the task of patent preparation for him. In either case, "Title to the inventions will remain in the inventors, and an appropriate license to the Government will be executed." [24]

[23] Letter, Caldwell to Brainerd, 8 October 1945.

[24] Letter, Col. E. H. Herrstrom to Pender, Patent Application on ENIAC, 23 December 1944. The university opted for the latter course. (Herrstrom headed the Patent Branch, Office of the Chief of Ordnance.)

Most universities in this era were quite naive about business matters, and the University of Pennsylvania was no exception. Its officials never bothered to consider how they were going to get their engineers to execute the appropriate licenses, as required under the contract. Today these matters are well understood; there are clear policies, and a condition of employment for scientific and engineering employees is that they execute legal papers assigning their prospective patents to their employers. At the University of Pennsylvania this had not been done.

Many universities turn patents over to foundations who collect the royalties and use them for the common good of the institutions. An example is the University of Wisconsin Alumni Foundation, which uses such funds to underwrite university research projects; its funds come largely from a process to introduce vitamin D into milk and from an improved rat poison.

The University of Pennsylvania had in those days a vague policy of permitting each employee who requested it all rights to his inventions. This was not an automatic procedure and required a petition by the employee to the Board of Trustees. There was much confusion in the Moore School as to who was entitled to be considered an inventor. All were working to help defeat the enemies of the United States. They had put all patent considerations very much into the background. There was also a belief at that time, which is expressed in a Moore School document of the period, that "anyone in the project who had worked for Eckert and Mauchly, even though they might have contributed to the program in general, were probably not entitled to patents since it is the custom that the originator of an idea usually holds all patentable ideas which might originate in the group working under him." [25] I do not comment on the validity of this view since it is a legal one, and I am certainly not competent to discuss it. Furthermore, the whole question of the patentability of the ENIAC and who were the inventors is now *sub judice;* it is therefore improper and indeed impossible at this time to do more than to record the situation as it developed. The final word on this subject cannot be written for some time.

Now the problem of patents on the ENIAC and then a little later on the EDVAC was to have an explosive impact on the University of Pennsylvania. As far back as November 1944 Dean Pender was writing to Dr. George W. McClelland, the then President of the

[25] J. Warshaw, Report on trip to Patent Branch, Legal Division, Office of the Chief of Ordnance, 3 April 1947.

University, asking for a clarification of the university's patent policies. Dr. McClelland responded saying that the Executive Committee had taken no action on the matter.[26] However, after much discussion between Brainerd and Pender on one side and Eckert and Mauchly on the other, Eckert and Mauchly wrote President McClelland asking for rights to the inventions made by them in the course of work on Contract W-670-ORD-4926. McClelland wrote them in March of 1945 granting them this right, waiving the university's right to a patent assignment with one stipulation: he provided that they grant the United States Government a non-exclusive, royalty-free license and the university such a license with the further right to sub-license "any established eleemosynary institution to build and to use such devices for essentially non-commercial and non-profit purposes." [27]

However, a month earlier Eckert and Mauchly had had assurances that President McClelland would act favorably on their request. They accordingly had hired an attorney to assist the Ordnance lawyers in preparing the necessary applications. As might be imagined, there was very great heat generated over this entire question of patents. It served to cleave Eckert and Mauchly apart from the university and Moore School officials, and it created tensions between Pender and Brainerd and finally between Eckert and Mauchly on the one hand and Gillon and me on the other. The reasons for the last rift had to do with publicity. Gillon and I were very anxious to declassify ENIAC and EDVAC and give them wide publicity throughout the scientific community. We did not however want to hurt Eckert and Mauchly in the process. Thus in November of 1945 Gillon and I were corresponding on this subject. Gillon wrote: "On the protection of Mauchly and Eckert, how much time do they require to file and how will that change our present publicity plans?" [28] Eckert was loath to have any publicity until he

[26] J. Warshaw, Chronological Résumé of Available Correspondence and Memoranda Pertaining to General University Patent Policy and to ENIAC and EDVAC Patent Proceedings, Compiled 14 March 1947.

[27] Letter, McClelland to Mauchly, 15 March 1945. This letter is the university's official response to a request to Dean Pender from Eckert and Mauchly dated 9 March 1945. Pender forwarded their letter to McClelland on 10 March with a favorable recommendation. The original letter to be sent by President McClelland did not give the university the right to give a free sub-license to other educational institutions. I proposed the formula regarding eleemosynary institutions that was in the final letter.

[28] Letter, Gillon to Goldstine, 6 November 1945, in response to letter, Goldstine to Gillon, 2 November 1945. The latter letter contained as an enclosure a letter from the attorney for Eckert and Mauchly indicating various serious patent consequences of public disclosures of the ENIAC and EDVAC.

and Mauchly had filed their patent applications. Indeed, Eckert "thinks we should say nothing . . . but simply attend as auditors." [29]

A further exacerbation of the problem had to do with authorships. Both Eckert and Mauchly were much offended that Brainerd had been asked by Weaver to write the NDRC report, and Mauchly was upset by the fact that Brainerd and Eckert had originally been invited to the MIT conference and he had not. While each of these crises was eventually smoothed over, each served to deepen the rift that was rapidly developing between Eckert and Mauchly on one side and Brainerd and Pender on the other. It should be said in connection with the above-mentioned grievances that Eckert and Mauchly felt with some justice that no one in the Moore School administration had any deep technical understanding of the ENIAC. There was truth in this. The way the dean organized things, Brainerd was so deeply immersed in all the administrative details of the research commitments of the Moore School that he did not have the time or strength to follow in detail the ENIAC project. This was in part why Prof. S. Reid Warren, Jr. was put in charge of the EDVAC project. Warren's participation was also not as technically deep as he would have liked to have made it. However, he also became immersed in paper work to the exclusion of the technical work.

It is my considered opinion that the lack of technical participation by the senior staff of the Moore School was real—probably unavoidable as the school was then structured—and ultimately led to its loss of leadership in the field it had pioneered. It is also my opinion that vigorous technical participation by members of the faculty of the Moore School would have resulted in a further cleavage, as we shall see from the relations of von Neumann and Eckert and Mauchly.

Von Neumann's keen participation and leadership of the logical design work on the EDVAC became a source of substantial conflict between him and me on one side and Eckert and Mauchly on the other. While these matters did not come to a head until later, perhaps it is as well to finish off the subject at this time.

As we mentioned above, when the discussions leading up to von Neumann's *First Draft* had taken place it had been against a background of complete mutual openness and desire to produce the best possible ideas. Later it turned out that Eckert and Mauchly viewed themselves as the inventors or discoverers of all the ideas and concepts underlying the EDVAC. This view was strenuously

[29] Letter, Warren to Brainerd, 9 October 1945.

opposed by von Neumann and me. A meeting on this subject took place in Washington in March 1946, attended by us, Col. Gillon, and two representatives of the Patent Branch. As a result of this meeting von Neumann submitted material stating that he had contributed three ideas to the EDVAC. As formulated by the lawyers it stated his claims as:

1. A new code for enabling the operation of the EDVAC.
2. The serial performance or progression through the system of the various arithmetical operations required for the solution of a whole problem.
3. The use of the "iconoscope" [by which was understood the electron beam oscilloscope substantially as used in video systems] as a memory device.[30]

Finally, after considerable acrimony a meeting was held on 8 April 1947 to try to resolve the problem relating to the EDVAC. This was attended by Dean Pender and several of his associates, Eckert and Mauchly, von Neumann and me, Dederick and an associate, as well as representatives of the Legal Branch of the Ordnance Department.[31] The upshot of the meeting was that von Neumann's *First Draft* was treated by the Ordnance lawyers as a publication in the strict legal sense. This meant that the distribution given to that report had placed its contents in the public domain, and hence anything disclosed therein became unpatentable. The Ordnance lawyers thereupon withdrew from the task of preparing patents on the EDVAC work in behalf of Eckert and Mauchly. At the meeting von Neumann and I proposed sorting out those ideas which could be attributed to specific people and agreeing to joint patents on the balance. But no agreement could be reached on either procedural or substantive points.

While the placing of the EDVAC report in the public domain was very satisfactory to both von Neumann and me, it ended our close relations with Eckert and Mauchly. There is much correspondence on the whole controversy, but since I was a participant it is probably best if some future historian analyzes the material objectively.

[30] Informal Report in re Disclosure of John von Neumann's *First Draft of a Report on the* EDVAC; written shortly after 2 April 1946.

[31] Minutes of Conference held at the Moore School of Electrical Engineering on 8 April 1947 to discuss Patent Matters.

By the fall of 1945 the ENIAC was fast approaching completion. Gillon, who had been on an assignment in the Pacific, returned and Simon wrote Pender, saying: "No doubt you remember the stimulus for the new computing devices came largely from Gillon. The supervision of the contracts came to me because of Colonel Gillon's departure. . . . Neither Dr. Dederick nor I really get the time . . . in fact I would really like to see the entire supervision of the contracts passed back to the Research Division of the Ordnance Office so that Gillon would control it and the Ballistic Research Laboratory would merely be consulted as to its needs and wishes." [1]

The ENIAC contract was due to expire on 30 September 1945, but it was clear that it could not be completed by that date. The executive vice-president of the university accordingly requested an extension of the contract until 31 December 1945.[2] Irrespective of all these administrative problems the finishing touches were being put on the ENIAC, readying it for its "maiden voyage." Late in October I notified Frankel and Metropolis: "It is not expected ENIAC will be ready for your problem on 7 November as was planned. It is felt however that middle of November is about the time at which your problem can be handled." [3]

On 15 November I was writing Dederick: "It looks at the present time as if the ENIAC will be ready for testing on or about this coming Monday. It is our plan to run the Los Alamos problem at that time as a test of the workability of the machine." [4] This letter was apparently a response to one of Dederick's in which he felt that the first problem "should be that of firing tables." [5]

In a progress report to Simon I stated that "there will be about three more weeks of construction work needed to complete the

[1] Letter, Simon to Pender, 20 September 1945.

[2] W. H. DuBarry to Philadelphia Ordnance District, 26 September 1945. A rather grudging acceptance of this request is contained in a response dated 15 October 1945. Various other letters passed back and forth and by 26 October 1945 a more satisfactory extension was worked out.

[3] Teletype, Goldstine to Frankel and Metropolis, undated.

[4] Letter, Goldstine to Dederick, 15 November 1945.

[5] Dederick to Goldstine, 15 November 1945.

divider and the test equipment for the ENIAC. . . . It is expected that the Los Alamos problem will be used for this purpose, and to this end that project has transferred 1,000,000 IBM cards to the Moore School." [6]

This was an absolutely frenetic period with myriad things to be done and minutiae to be attended to. I will always remember the all-day and almost all-night character of the work for the next several months. Both my wife and I shared in this, and in retrospect it seems like an absolutely wonderful but rugged experience.

Exactly when the Los Alamos problem was put on the ENIAC is not clear from the correspondence. However, on 23 November I was writing to Edward Teller saying, "Nick tells me that Dr. Fermi was interested in seeing the ENIAC and EDVAC but was unable to come here last Saturday." [7] From this it seems clear that Frankel and Metropolis were in Philadelphia reviewing details of their problem with my wife and me prior to putting the problem on the machine. [8]

The Los Alamos problem was classified as far as the underlying physical situation was concerned but not as regards the numerical or mathematical form of the equations to be solved. This policy of keeping the numerical equations unclassified was a wise one that was long maintained. It made possible doing calculations for Los Alamos without obtaining clearances for any of the personnel involved and without having to maintain elaborate security measures in the ENIAC room itself.

The service log kept for the ENIAC shows that Problem A was in some state of running on 10 December 1945. This must have been highly tentative, but by 18 December the log states, "check E, runs after this were terribly wrong in the E numbers." [9] This run was a part of the Los Alamos problem. Each day thereafter this problem was on the ENIAC, and the run was being used not only to obtain results for Los Alamos but also to find "bugs" in the machine.

There is a certain amount of amusement to be had from reading through the various entries in the log such as a pathetic note by me on 23 December 1945 at 3 P.M. saying "Steam pipe going thru

[6] Progress Report on the ENIAC and EDVAC Projects for week ending 3 November 1945.

[7] Letter, Goldstine to Teller, 23 November 1945.

[8] Reply Affidavit of Mrs. Adele K. Goldstine, United States Patent Office, in re application of John P. Eckert, Jr. and John W. Mauchly, Serial No. 757,158, Filed 26 June 1947. Petition for the Institution of Public Use Proceedings.

[9] Service Log, p. 50; the E numbers were the left-hand column of tabulation. There is also a terse entry in the log between 8 January and 9 January 1946 saying "End of first month."

window ventilator broken. Capt. Ryan and Brainerd notified." Or 25 December 1945 at 9:30 P.M.: "Heavy rain and melting snow leaking into 2nd floor . . ."; at 3 A.M.: "About five men still working, mopping up water and emptying buckets which catch drip. . . ." [10]

Early in December Gillon, Feltman, and I met at Aberdeen to plot out a strategy for giving "suitable publicity for the ENIAC, which is now in its final testing stage." [11] Various recommendations emerged from this discussion which were promptly implemented. It was decided that the "ENIAC should be declassified as soon as possible with only such material as is necessary for patent security continuing under secrecy classification"; that the Public Relations Office of the War Department should be involved in the publication of the ENIAC; that articles should be written for various professional journals; that a dedication ceremony be held at the University of Pennsylvania, with mid-January picked tentatively as the date; and that eminent scientists from the United Nations, interested in computing, be invited to the ceremony.

The results of this meeting were then discussed with Brainerd and Pender, and they enthusiastically began to carry out their part of the plan. The Army assigned two officers, Major Thomas G. Gentel and Captain John Slocum of the Publication Branch of the War Department's Public Relations Office, to help me arrange for the ceremonies from the Army point of view. One of the tasks involved was the declassification of the ENIAC. This was done by an action of the Ordnance Committee. An item in its minutes reads: "The design details and circuits . . . will remain in the confidential category. All other phases of the project, including general principles of design and operational and functional characteristics of the equipment are hereby unclassified." [12]

A Moore School committee was set up consisting of Brainerd, chairman, Warren, Weygandt, and me, ex officio. [13] It is clear that the committee planned a preliminary press conference and briefing for 1 February and a formal dedication on 15 February. These were carried out as we shall see.

[10] Service Log, pp. 56 and 60.
[11] Minutes of Meeting on ENIAC and EDVAC at Ballistic Research Laboratory, 7 December 1945.
[12] Col. S. B. Ritchie to Secretary of the Ordnance Technical Committee, Reclassification of the Project for Development of the Electronic Numerical Integrator and Computer (ENIAC), 17 December 1945. It is Item 29904 and was read for record on 20 December 1945 before the Ordnance Committee.
[13] Memorandum, Brainerd to Messrs. Warren, Weygandt, Capt. Goldstine, 18 January 1946, and Memorandum, Pender to Goldstine, Committee on Arrangements, 1 February and 15 February 1946.

At the press conference Major General Gladeon M. Barnes, the head of Research and Development Service of the Office of the Chief of Ordnance, Eckert, Mauchly, Brainerd, and I addressed the reporters. Arthur Burks, assisted by Kite Sharpless, was to conduct the formal ENIAC demonstration, which consisted of the following five simple problems to illustrate the speed and capability of the machine. These were

1. 5,000 additions in one second.
2. 500 multiplications in one second.
3. Generation of squares and cubes.
4. Generation of a sine and cosine table, to be tabulated.
5. A modification of the E-2 ENIAC run as an illustration of a long and complicated calculation.[14]

All sorts of plans were put on foot for scientific as well as popular articles on the machine as well as for radio coverage. Most of these came off rather well after much travail.[15]

The formal dedication ceremony itself was a considerable success. A dinner was held in the university's Houston Hall, at which scientific celebrities were present and at which President McClelland presided. The principal speaker for the occasion was Dr. Frank B. Jewett (d. 1949), the then president of the National Academy of Sciences. It was particularly fitting that he was the speaker, since he had a long and distinguished career starting with a position as research assistant to Michelson in 1901–02. He had also held many important posts in the worlds of the university, industry, and government. These included being a life member of the Massachusetts Institute of Technology Corporation, first president and then Chairman of the Board of the Bell Telephone Laboratories, as well as a member of the National Defense Research Committee. General Barnes carried out the actual dedication; after giving a short speech, he pressed a button that turned on the ENIAC. The guests then proceeded to the Moore School to see the actual demonstration.

One part of Barnes' speech is perhaps worth quoting since it showed the vision of the future that even then was had by some.

---

[14] ENIAC Guides for Press Day, 1 February 1946 (rehearsal to be held January 30). The so-called E-2 run was part of the Los Alamos problem.

[15] See, e.g., Burks, "Electronic Computing Circuits of the ENIAC," op. cit.; Goldstine and Goldstine, "The Electronic Numerical Integrator and Computer (ENIAC)," op. cit.; Brainerd, "Project PX – The ENIAC," The Pennsylvania Gazette, vol. 44 (1946), pp. 16–32.

In the meantime the ENIAC will provide the means of extending the frontiers of knowledge with all that implies for the betterment of mankind. It will provide exceptionally powerful means for facilitating the acquisition of basic scientific knowledge through fundamental research, which will in turn permit the replenishment of the reservoir of scientific knowledge. It will additionally provide the means of circumventing the serious limitations of analytical methods or analogy devices currently being encountered in attempting to solve the important problems of industrial technology, and finally it provides the firm bases for developing the tools of the future in man's endless search for scientific truths.

In conclusion, I should like to indicate that this important milestone on the highroad of scientific advancement has been reached none too soon. . . . Within the limitations imposed by the requirements of national security every effort will be made to permit the great potential usefulness of this great scientific tool to be realized as broadly as possible. . . .

By pressing the prepared switches on this control box I will symbolically initiate the solution of the first problem on the ENIAC formally dedicating the machine to a career of scientific usefulness.[16]

The impressario at the demonstration was Burks, who performed superbly. Just before the guests arrived a malfunction developed in the multiplier, which Burks detected, localized, and repaired in a very few minutes. The actual preparation of the problems put on at the demonstration was done by Adele Goldstine and me with some help on the simpler problems from John Holberton and his girls.

Since there has been some controversy about the training of this group and about the people who prepared the test problems for the demonstrations, perhaps this is as good a place as any to make clear what the facts actually were. Holberton and his group had been assigned the responsibility under my direction of becoming the programming staff for the ENIAC when it was turned over by the Moore School to the government. They took up their task in July of 1945, as has been recounted earlier. They were trained largely by my wife, with some help from me. Although this has been disputed, the facts are very clear. The only persons who really had a completely detailed knowledge of how to program the ENIAC

[16] Barnes, Dedication Address, 15 February 1946.

were my wife and me. Indeed, Adele Goldstine wrote the only manual on the operation of the machine.[17] This book was the only thing available which contained all the material necessary to know how to program the ENIAC and indeed that was its purpose. It is also relevant to remark that programming the ENIAC was at this stage, at very least, an exceedingly difficult and arcane art form. Reviewers of the book emphasized that "Because of the vast wealth of detail presented, however, it seems unlikely that the report will attract many thorough readers outside of the group actually engaged in performing manipulations on the ENIAC. . . . A brief examination of the report will make it clear that the operator is able to convert the machine to his purpose only at the cost of much tedious preliminary labor." [18]

The programming of the simpler problems for the demonstration were done by Holberton and his people, but the main calculation and the interrelationship between the various problems were prepared solely by my wife and me. This was a detailed trajectory calculation which was run for the dedication ceremony on 15 February and for the open house for students and faculty members of the University of Pennsylvania on the next day. There is a page of programs in the log dated 30 January, 12:05 A.M., and signed by Adele Goldstine showing how she and I had revised the demonstration set-ups as regards the Initiating Unit and Master Programmer. There are detailed instructions left indicating how Burks was to proceed. Moreover under the date of 1 February, 12:30 A.M., I wrote: "All problems O.K. Do not touch any switch settings or program lines." There are also detailed instructions from my wife to Holberton.[19]

Again, under date of 13 February at 2 A.M. I noted that "all problems except trajectory are O.K." and then in an entry which must mean 12:30 P.M. that same day I jubilantly noted "Demonstration problems O.K.!!" [20] Indeed, I well recall the kindness and humor of Dean Pender that night. He suddenly appeared very late in the evening with a bottle of bourbon in hand and gave it to my wife to help sustain us in our effort.

Finally, it might be pointed out that the Moore School gave me full responsibility for the demonstration "including all aspects.

[17] Adele K. Goldstine, *Report on the* ENIAC, *Technical Report I*, 2 vols., Philadelphia, 1 June 1946.

[18] J. T. Pendergrass and A. L. Leiner in *Mathematical Tables and Other Aids to Computation*, vol. 4 (1948–49), pp. 323–326.

[19] Log, pp. 101–105.

[20] Log, p. 109.

Burks, Sharpless, and other engineers associated with the demonstration report to Goldstine." [21]

I need say only a little more about the ENIAC to complete my story
of that system. There has been considerable legal interest in exactly
when the ENIAC became operational, and while I should not
attempt to answer legal questions in a history, I can illuminate the
factual situation.

At least my wife and I were certainly of the opinion that the
ENIAC was running satisfactorily prior to the formal dedication.
Neither we nor the responsible officials of the university or the
Army would have participated in the demonstrations on 1, 15, and
16 February had there been any doubt. Late in December I wrote
that the failure rate of tubes was less than one a day and that in
four previous days there had not been a single such failure.[22] Again,
early in January I wrote to Womersley saying the "ENIAC has been
running about 1,000 hours at the present time and has been performing quite decently. We put on . . . a problem involving almost
all the facilities . . . and are using this . . . as a test. By means of it,
we have located a small number of badly soldered joints and other
small defects. In general, however, the main problem with the
machine is vacuum tube failure. Our average is about one tube per
day failure, and I estimate that it takes about one hour per day to
cure the trouble. The longest we ran without tube failures was 120
hours." [23]

By early January Dean Pender stated in a message to the university's assistant comptroller that the "testing of the ENIAC has been
rather thoroughly mixed up with the initial operation. . . . The
introduction of a relatively complicated practical problem into the
test period has obliterated the line between testing and operation.
Despite the fact that the ENIAC has not yet been formally accepted
. . . Captain Goldstine has assured me he would approve bills for
operating expenses of the ENIAC beginning December 1st, 1945." [24]

During the demonstration period there was little point in the
Los Alamos people, Frankel and Metropolis, staying in Phila-
delphia, but by February 1946 Metropolis and Prof. Anthony
Turkevich, now a distinguished physicist at the University of

---

[21] Moore School Memorandum, Staff Responsibility for ENIAC Dedication, 15
February 1946; it is dated 11 February 1946.
[22] Letter, Goldstine to Simon, 26 December 1945.
[23] Letter, Goldstine to Womersley, 8 January 1946.
[24] Memorandum, Pender to Murray, Moore School-EO Account, 9 January 1946.

Chicago, were at the Moore School setting up the Los Alamos problem.[25] As we noted earlier this problem was run satisfactorily and Dr. Norris Bradbury, the director of Los Alamos, had written Barnes and Gillon on 18 March 1946 thanking them for the successful use of the machine.[26]

Moreover on 4 February both Frankel and Metropolis wrote a formal letter to me telling me officially that they were taking up positions at the University of Chicago and asking whether that institution could rent time on the ENIAC from the government.[27] This complies with Gen. Barnes' statement in his dedication remarks quoted above that "every effort will be made to permit the great potential usefulness of this great scientific tool to be used as widely as possible." Both Gillon and I were very much for this idea and I wrote encouraging them.[28] It is noteworthy that already at this stage both Fermi and Teller were very much interested in the idea of electronic computation. Somewhat later Fermi invited me to Chicago to meet with him and Prof. Samuel Allison to discuss the details of electronic computers. This was a most fascinating conversation for me; it revealed the tremendous powers of concentration Fermi could bring to bear on a problem.

In any case, the Army agreed to allow university scientists to use the ENIAC free of charge, and a number of problems were run under this arrangement. Profs. Hans Rademacher and Harry Huskey did computations of tables of sines and cosines to study the way round-off errors develop in numerical calculations (15–18 April 1946). Frankel and Metropolis did an extensive calculation of the so-called liquid drop model of fission (15–31 July 1946), a model developed by Niels Bohr and John Wheeler in 1939.[29] Hartree did a very large calculation of the behavior of boundary layers in compressible flows. (There is contention about the dates of this calculation, but Hartree stated that he did "the first stage of the work on it while I was in Philadelphia this summer." [30]) Prof. A. H. Taub, then of Princeton University, and Adele Goldstine did a calculation on

[25] Letter, Goldstine to Frankel, 19 February 1946.

[26] Letter, Bradbury to Barnes and Gillon, 18 March 1946.

[27] Letter, Frankel and Metropolis to Goldstine, 4 February 1946.

[28] Letter, Goldstine to Frankel, 19 February 1946.

[29] Frankel and Metropolis, "Calculations in the Liquid-Drop Model of Fission," *Physical Review*, vol. 72 (1947), pp. 914–925.

[30] Hartree, *Calculating Machines, Recent and Prospective Developments and Their Impact on Mathematical Physics* (Cambridge, 1947). This was the lecture he gave in 1946 upon taking up his chair as Plummer Professor of Mathematical Physics in succession to the late Sir Ralph Fowler, one of the very great mathematical physicists and a man who did fundamental work in ballistics as well.

reflection and refraction of shock waves (3–24 September 1946).[31] A calculation on thermodynamical properties of gases was done for Dean J. A. Goff of the University of Pennsylvania (7–18 October 1946).[32] Prof. D. H. Lehmer, one of the mathematicians at Aberdeen for the duration and a great pioneer along with his wife on the application of computers to number theory, did a most interesting study of certain primes [33]; and the first numerical calculations of the weather were undertaken by von Neumann and his associates on the ENIAC.[34] This is by no means a complete list of problems done for university people, but it is representative and serves to indicate the enlightened policy of making the ENIAC available to reputable scientists, a policy initiated by Gillon and Simon and carried on by their successors.

Of course, many other problems were done on the ENIAC for Aberdeen itself, a number of large ones for the Atomic Energy Commission, and several for the U.S. Bureau of Mines. It is not our aim here to enumerate these; complete lists exist at Aberdeen.[35]

During 1947 von Neumann realized that the lack of a centralized control organ for the ENIAC was not an incurable deficiency. He suggested that the whole machine could be programmed into a somewhat primitive stored program computer. He turned the task over to Adele Goldstine, who worked out such a system and passed it along to Richard Clippinger, who was then the head of the Computing Laboratory at the Ballistic Research Laboratories and is also a mathematician of note. He and his associates made certain emendations and put the final touches on the idea, and on 16 September 1948 the new system ran on the ENIAC. Although it slowed down the machine's operation, it speeded up the programmer's task enormously. Indeed, the change was so profound that the old method was never used again.

Hartree's period in Philadelphia can be dated easily. On 30 July he wrote a "Dear Herman and Adele" letter from England saying "I got off all right on Saturday 20th." (Letter, Hartree to H. H. and A. K. Goldstine, 30 July 1946.)

[31] A. H. Taub, "Reflection of Plane Shock Waves," *Physical Review*, vol. 72 (1947), pp. 51–60.

[32] For the details, see Letter, Irven Travis to G. J. Kessenich, 18 November 1946.

[33] D. H. Lehmer, "On the Factors of $2^n \pm 1$," *Bull. Amer. Math. Soc.*, vol. 53 (1947), pp. 164–167.

[34] J. G. Charney, R. Fjörtoft, and J. von Neumann, "Numerical Integration of the Barotropic Vorticity Equation," *Tellus*, vol. 2 (1950), pp. 237–254.

[35] W. Barkley Fritz, A Survey of ENIAC Operations and Problems: 1946–1952, Ballistic Research Laboratories, Memorandum Report No. 617, August 1952. This is a description and bibliography of all unclassified work done on the ENIAC during the cited period.

The system as worked out by Adele Goldstine provided the ENIAC programmer with a 51-order vocabulary. This was modified to 60 orders by Clippinger and then later to 92 orders.[36]

The ENIAC was formally accepted by the government on 30 June 1946. The date of 30 June is quite a reasonable one since the ENIAC was complete and the final reports were dated 1 June 1946; the actual mailing date of most copies was 6 June. The delay in getting these reports written lay mostly in the very large amount of material that had to be written. In all, there were five volumes in the report, the first two of which were, respectively, an operating and a maintenance manual for the ENIAC. The remaining volumes were entitled *Technical Description of the ENIAC*, Parts I and II. Volumes 3 and 4 comprised Part I and were written by my wife. They contained 301 pages plus 122 figures. Volume 5 comprised Part II and was written by Huskey. It contained 163 pages plus 12 figures. Thus, there was much to be prepared.

The 30 June date is contained in a letter from the Philadelphia Ordnance District to the Chief of Ordnance which says: "This machine has now been completed and will be accepted by this District as of 30 June 1946." [37]

The delay in moving the ENIAC to Aberdeen can be attributed to two factors. For one thing, the construction of the building into which the ENIAC was to be moved was halted late in the fall of 1945. In May of 1946 Simon wrote to say it was nearing completion and that the ENIAC would be moved into it in the fall.[38] Thus the Ballistic Research Laboratory was not itself ready to receive the machine for some time. Furthermore, Hartree, von Neumann, and I were much concerned that the "main computation workhorse for the solution of the scientific problems of the Nation" [39] should not be put out of commission. This point was particularly urgent in early 1946 since there were very pressing calculational problems that the atomic energy people needed solved. Accordingly, we persuaded Simon to delay moving the machine.

In the event the ENIAC was "turned off" on 9 November 1946 to

[36] Description and Use of the ENIAC Converter Code, Ballistic Research Laboratories, Technical Note No. 141, November 1949.

[37] Letter, Welch to Ritchie, 25 June 1946. I believe that on 30 June the Philadelphia District did accept the machine and then transferred responsibility to Aberdeen. The letter requests "a shipping order confirming the transfer of property accountability to Aberdeen Proving Grounds without physical movement of the machine." A 26 July date for formal acceptance is claimed by some, but it is contained in a document written many years later and may be erroneous.

[38] Letter, Simon to Goldstine, 9 May 1946.

[39] M. H. Weik, "The ENIAC Story," *Ordnance* (1961).

ready it for the move to Aberdeen; it was not started up again until 29 July 1947 and thus was inoperable for a nine-month period. It then became once again a most useful instrument, and continued operating until 11:45 P.M. on 2 October 1955. It was then disassembled, and a part of it forms part of a most interesting exhibit at the Smithsonian Institution in Washington, D.C.

Part Three

Post–World War II:
The von Neumann Machine
and the Institute for
Advanced Study

As mentioned earlier, it was clear to me by the summer of 1945 that the development of computers had to continue and in a more normal peacetime mode. I therefore had a number of conversations with friends to gain some feeling for what was possible and desirable. Among those I talked to was John Kline, chairman of the mathematics department at the University of Pennsylvania and Secretary of the American Mathematical Society. Kline was both a wise and kindly gentleman, and he gave me much good advice.

Of course, the most obvious choice for a location for this work was the University of Pennsylvania, but unfortunately this was not a good time in the life of that institution for innovation. Perhaps institutions as well as people can become fatigued, or perhaps it is the fatigue of the leaders of an institution that gives the place itself that feeling. In any case, the University of Pennsylvania had to go through seven difficult years before a new president, Gaylord Harnwell, appeared on the scene and revitalized the school. Moreover, the tensions in the Moore School arising from the ENIAC and EDVAC projects were already being felt.

It is not easy even today to analyze exactly why the Moore School became unsuitable for this type of project. However, some of the factors can be distinguished. First perhaps may be the failure on the part of the senior staff of the Moore School — Brainerd excepted — to appreciate the overwhelming importance the computer was to have to electrical engineering as a discipline. It is my surmise that while these professors saw that the school had done an excellent piece of war work, they did not see the implications of these devices for the future. As we have already mentioned, none of them was close technically to the situation, but all saw or heard the fireworks arising over intellectual credits and patents. Perhaps, therefore, they were "turned off" by the disruptive aspects that they witnessed.

It may also be that the dean felt intuitively that a major effort in computers was too large and disruptive a program for the orderly life of a school dedicated to teaching the young. He never stated this explicitly, however, and he did see to it that the Moore School ultimately finished the EDVAC. This effort was made not without discord, and it probably was with a considerable sense of relief that the machine was finally accepted by the Ballistic Research Laboratories in 1950, moved to Aberdeen, and put into successful operation.[1]

To my knowledge neither Eckert nor Mauchly was offered a professorship at the university, nor was any other member of the ENIAC-EDVAC staff. It is probably a fair evaluation that the Moore School was tired of electronic computers and their concomitant strains on human nerves and material resources. It is only fair however to point out a very real effort by Warren and others on the faculty to reorganize and modernize the Moore School. This was initiated by Dean Pender and bore fruit in that a considerable revitalization, enlargement, refurbishment of the Moore School took place not long after the war ended.[2] Fittingly, an annex to one of the university's buildings bears Pender's name.

Perhaps it was the strong impetus at the Institute for Advanced Study, more than any other reason, which served to end the hegemony of the Moore School in the field. Von Neumann's return to the Institute certainly meant the loss to the Moore School of the greatest thinker of our times on computers and computing. The great contributions in the computer field by the Moore School were two in number: the brilliant conception and execution of the ENIAC and the equally brilliant conception of the EDVAC. The former was independent of von Neumann, but the latter was already highly dependent on him, as we have seen, and his departure was a great loss. Next, the departures of Eckert and Mauchly in 1946 to form their own company, Electronic Control Co., and of Burks, my

[1] The first successfully operating machine of the binary, serial type using supersonic delay lines was not the EDVAC but the EDSAC, the Electronic Delay Storage Automatic Calculator, designed and built at the University Mathematical Laboratory, University of Cambridge, England, by M. V. Wilkes and his colleagues. It was publically displayed at a meeting held there 22–25 June 1949, to mark the completion of the project. See M. V. Wilkes, "Progress in High-Speed Calculating Machine Design," *Nature*, vol. 164 (1949), pp. 341–343.

[2] Warren's creative proposals were contained in a memorandum entitled "Outline of a Research Program for the Moore School of Electrical Engineering," written shortly prior to 15 December 1945. He specifically urged a "particularly strong effort . . . to acquire funds (for general computing projects) . . . because the Moore School has been so active in this field during the past eleven years." He envisioned a dual computing group: one side to develop and build machines, and the other to use such devices for solving research problems. One of his most important proposals was that the School should hire a business manager to handle financial matters, so that the director of a project could function solely as the technical director.

wife, and me to go to Princeton with von Neumann caused further depletion of the staff.

The Moore School had a resurgence in the summer of 1946. From 8 July to 31 August, under the direction of Carl C. Chambers, the school ran a special course entitled "Theory and Techniques for Design of Electronic Computers." The course was sponsored by the Army's Ordnance Department and the Navy's Office of Naval Research. The preface to a book by Wilkes and his colleagues acknowledges the value of the course and also provides a link to the next stage at the Institute: "We are deeply conscious of the debt we owe to our colleagues engaged on other projects, especially to those who were instructors of a course attended by one of us at the Moore School of Electrical Engineering, University of Pennsylvania, in 1946, and to Dr. J. von Neumann and Dr. H. H. Goldstine of the Institute for Advanced Study, Princeton, whose privately circulated reports we have been privileged to see."[3]

Whatever the reasons, the leadership of the Moore School was almost at an end. Kline helped me make a highly tentative approach to the American Philosophical Society on behalf of the university, but this was not fruitful. After many long conversations between von Neumann and me on the subject of who would carry on computer development, the idea gradually became clear to both of us that if it were to continue, *we* would have to be the ones to do it. As mentioned earlier, the Institute for Advanced Study did not seem the place to undertake this sort of work because of its small size and because of its aversion to experimental as opposed to theoretical work. Moreover, at about this time the then president of the University of Chicago, Robert M. Hutchins, was making a great and successful effort to revitalize that university. Many of the outstanding members of the faculty were quite old and most had retired or left. Hutchins therefore gave *carte blanche* to his Dean of Physical Sciences, Walter Bartky, to hire a group of eminent scientists. In this effort Bartky was very successful in general; but in the case of von Neumann, whom he approached, he did not succeed. In November of 1945 von Neumann went to Chicago to tell this to Hutchins. By this time he had persuaded Frank Aydelotte, Director of the Institute for Advanced Study, to involve the Institute in a computer project. In this James Alexander, Marston Morse, and especially Oswald Veblen were quite helpful.

[3] M. V. Wilkes, D. J. Wheeler, and S. Gill, *The Preparation of Programs for an Electronic Digital Computer, With Special Reference to the* EDSAC *and the Use of Library Subroutines* (Cambridge, Mass., 1951), preface.

As the plan was first conceived, the Institute for Advanced Study, Princeton University, and the Radio Corporation of America were jointly to collaborate on the project; and as early as November of 1945 von Neumann was writing to Aydelotte, Dean Hugh S. Taylor of the Princeton University Graduate School, and Elmer W. Engstrom, vice-president for research of RCA, outlining to them the purposes of the project.[4] The plan set down by von Neumann, which mirrored rather closely his *First Draft* report, was not followed in a number of significant respects. However, it contained one very visionary paragraph regarding visual displays. Such a device was actually built in the 1950s at the Institute, and the principle has since become fundamental in many machines. It is therefore worth recalling his words here: "In many cases the output really desired is not digital (presumably printed) but pictorial (graphed). In such situations the machine should graph it directly, especially because graphing can be done electronically and hence more quickly than printing. The natural output in such a case is an oscilloscope, i.e. a picture on its fluorescent screen. In some cases these pictures are wanted for permanent storage (i.e. they should be photographed); in others only visual inspection is desired. Both alternatives should be provided for." [5]

At first both von Neumann and I were very dubious about the possibility of securing financial support from the Army because of its commitments to the Moore School. Indeed, in the minutes of a meeting on publicizing the ENIAC written by me it was stated that the Office of the Chief of Ordnance should enter into a contract for one dollar with the Institute for Advanced Study, so that it could receive copies of all reports, drawings, etc.[6] Instead, we turned to RCA and the Rockefeller Foundation for support, and at first all looked well in these quarters. The RCA support was generously and fully forthcoming in the form of storage tube developments (more on this later), but the Rockefeller help did not materialize. Then we approached the Navy, but an impasse was reached on the question of who would have title to the machine that was to eventuate. Next a joint funding arrangement with the Army Ordnance Department, the Navy Office of Research and Inventions, the Radio Corporation of America, and Princeton University was envisaged.[7]

[4] Von Neumann, Memorandum on the Program of the High-Speed Computer Project, 8 November 1945.

[5] *Ibid.*, p. 3

[6] Minutes of Meeting on ENIAC and EDVAC at Ballistic Research Laboratory, 7 December 1945.

[7] Von Neumann, Report on Computer Project, 16 March 1946.

This scheme partly succeeded and partly floundered. In the end, the funding for the mathematical and meteorological work came from the Office of Naval Research; for the engineering and logical design and programming it came originally from Army Ordnance, then from a tri-service contract which for a while was supplemented by the Atomic Energy Commission. The key names in the government who should be mentioned are Mina Rees, Joachim Weyl, and Charles V. L. Smith of ONR; Paul Gillon, Sam Feltman, and later George Stetson of Ordnance; Merle Andrew of the Office of Air Research; and Joseph Platt and Holbrook MacNeille of the Atomic Energy Commission.

Through all these manifold twistings and turnings, Veblen, who was a trustee of the Institute, stood firmly behind von Neumann, and Aydelotte never wavered in his backing of the project. Here again Veblen was to make one of those crucial contributions of his to science just when it was most needed. It is doubtful to me whether von Neumann would have persevered in trying to convince the Institute to take on this project if it had not been for Veblen's courage, foresight, and unlimited patience. In the end these triumphed, and a project to develop and build a computer was initiated at the Institute for Advanced Study under von Neumann's direction.

Aydelotte was absolutely influential in pushing the project. He said to his Board: "I think it is soberly true to say that the existence of such a computer would open up to mathematicians, physicists, and other scholars areas of knowledge in the same remarkable way that the two-hundred-inch telescope promises to bring under observation universes which are at the present moment entirely outside the range of any instrument now existing." [8] He recommended the Board underwrite this project to the extent of $100,000. He went on to say:

> Such an electronic computer could be built for a total cost of about $300,000. It would be a new instrument of research of unprecedented power. Curiously enough the plan to such a machine is partly based on what we know about the operation of the central nervous system in the human body. This means, of course, that it would be the most complex research instrument now in existence. It would undoubtedly be studied and used by scientists from all over the country. Scholars have already ex-

[8] Minutes of Regular Meeting of the Board of Trustees, Institute for Advanced Study, 19 October 1945.

pressed great interest in the possibilities of such an instrument and its construction would make possible solutions of which man at the present time can only dream. It seems to me very important that the first instrument of this quality should be constructed in an institution devoted to pure research though it may have many imitations devoted to practical applications.

Aydelotte's Board approved his recommendation, and he continued to give real direction to the Institute's effort to initiate the project. A Committee on the Electronic Computer was formed with him as chairman and Engstrom, Taylor, Veblen, and von Neumann as members. It held its only meeting on 6 November 1945. At that time Aydelotte told the members about the backing by the Institute Trustees and of RCA's promise of a similar amount. Taylor pledged that the university "would make a contribution in the form of the services of members of the Faculty." [9] Furthermore, the committee took the momentous step of stating that "It was the sense of the Committee that we were justified in beginning work on the project immediately on the basis of the funds secured. It was therefore agreed, on the suggestion of Dr. Engstrom, that a committee should be formed to advise Professor von Neumann, this committee to consist in the first instance of Dr. von Neumann as Chairman, Professor J. W. Tukey, and Dr. V. K. Zworykin. It was understood that the committee should be free to coöpt additional members who might contribute to the discussion of various aspects of the problem." Finally, it was agreed that von Neumann would undertake to prepare a statement of objectives. This is undoubtedly the 8 November memorandum previously cited.

On 12 November von Neumann's committee held its first meeting in Zworykin's office at RCA. The committee was already enlarged to include, in addition to Tukey and Zworykin, G. W. Brown, J. A. Rajchman, A. W. Vance of RCA, and me. It worked very hard and by 29 November had already held its fourth meeting.

The members not already met in these pages consisted of John W. Tukey, a well-known statistician on the faculty at Princeton with close connections with the Bell Telephone Laboratories; George W. Brown, also a well-known statistician, who was then a colleague of Rajchman and is now chairman of the Department of Business Administration at the University of California at Los Angeles; and A. W. Vance, an electrical engineer, a colleague of

---

[9] Minutes of the Meeting of the Committee on the Electronic Computer held in the Director's office, Institute for Advanced Study, Tuesday, 6 November 1945.

Rajchman at RCA. Vance did not have the leisure to join the project: the minutes of the first meeting indicate he "will be in a consulting capacity, with further participation contingent on future contract commitments." [10] Actually, he became deeply involved in a large analog computer project and did little more on this one.

It was stated at the first meeting on 12 November that "H.H.G. may be released by the Army in the not too distant future; it is hoped that at that time H.H.G. will become associated with the Institute for Advanced Study." It was also indicated that "J.A.R. and G.W.B. will be available full time or nearly full time for E.C. after January 1, 1946."

On 27 November 1945, von Neumann wrote me a formal offer, saying: "The duties of your office will be connected with the automatic, high speed computer project undertaken jointly by the Institute, the Radio Corporation of America, and Princeton University, and they will consist of participating in planning the device, in working on its mathematical theory, and in assisting me in coordinating the various phases of this project, which will be worked upon by several different groups." [11] This then was for me the guarantee that the computer would live and develop further toward maturity. I formally accepted the offer on 25 February 1946, after the initial operation and dedication frenzy died down at the Moore School.[12]

Earlier in November I already knew of the Institute's commitment, mentioned above, under which the trustees pledged $100,-000 toward the total expense—a not inconsiderable sum in those days, when the total cost of the EDVAC was to be $467,000. I was therefore busy discussing this with my friends both at the Moore School and elsewhere in order to create an engineering staff. In particular I discussed the situation relative to the Institute with Burks, Eckert, and C. Bradford Sheppard, an excellent engineer on the EDVAC project. Burks was much interested, received an offer, and accepted on 8 March 1946. Eckert was for a long time torn between academic life and a business of his own; and Sheppard had thrown in his lot with Eckert. In the end, as we have seen, Eckert decided to embark on a career as co-owner with Mauchly of

[10] Minutes of E. C. Meetings, 12, 19, 21, 29 November 1945. These meetings were largely tutorial in nature and had the essential purpose of acquainting the RCA group with the essentials of the ENIAC and EDVAC. After these four meetings, the members met more informally but in the early days perhaps weekly to keep abreast of each other's work.

[11] Letter, von Neumann to Goldstine, 27 November 1945.

[12] Von Neumann, Report on Computer Project, 16 March 1946.

the first commercial enterprise in what is now a multi-billion dollar industry.

In October 1946 Eckert and Mauchly set up a partnership, the Electronic Control Co., in Philadelphia and almost at once started the design and development of a mercury delay line machine, BINAC (Binary Automatic Computer), for the Northrop Aircraft Co. It was not dissimilar to the EDVAC in design philosophy.[13] It was the first American machine built after the ENIAC. The machine became operational in August 1950.

Soon after setting up their partnership Eckert and Mauchly reorganized and formed Eckert-Mauchly Computer Corporation. The new company was successful in obtaining a contract with the National Bureau of Standards to build a machine designed on EDVAC principles for the Bureau of the Census. The machine, UNIVAC (Universal Automatic Computer), was started about August 1947 and was operational in March 1951.

With the dedication of the ENIAC, the computer revolution swept very rapidly throughout the United States and Western Europe — with such speed, indeed, that it was obvious to everyone that there was an extremely large but latent worldwide need for computers. This narrative will focus principally on developments at the Institute for Advanced Study, but those of us working there were of course not unaware of computer activity in other places or of its significance. A few comments may be in order at this point.

As far as Great Britain is concerned, there is no doubt that Douglas Hartree made the big initial contribution by using his prestige with the government and university communities to press for machine design there. Originally, as mentioned earlier, he first came to the States for a short visit. He made such an excellent impression that I persuaded Gillon to invite him back for an extended stay in the spring-summer of 1946.[14] He came on 20 April and left on 20 July 1946.[15] He was of great help to the ENIAC group coming when he did, because he helped keep up the morale and intellectual tone of the ENIAC operating staff, headed then by Dederick and under him Holberton, just when they most needed such bolstering: the scientists at the Ballistic Research Laboratory for the "duration" were now streaming back to their universities, with a conse-

---

[13] See A. A. Auerbach, J. P. Eckert, Jr., R. F. Shaw, J. R. Weiner, and L. D. Wilson, "The BINAC," *Proceedings of the IRE*, vol. 40 (1952), pp. 12–29.

[14] Letter, Goldstine to Hartree, 19 February 1946.

[15] Letter, Gillon to Commanding General, Aberdeen Proving Ground, Temporary

quent sense of depression in those who remained. This depression lifted in part due to Hartree's stimulation.

Another who significantly aided computer developments in England was Sir Geoffrey Taylor, who came to the ENIAC dedication with von Neumann. G. I. Taylor is one of England's most distinguished mathematical physicists as well as being a man who knew even then how important computers would be. Several years ago a former colleague, Dr. Bruce Gilchrist and his wife, friends of mine since the early days at the Institute for Advanced Study, were on an airplane flight from London to New York sitting next to an elderly English gentleman. In chatting with him somehow my name was mentioned, whereupon the old gentleman – none other than Sir Geoffrey – recalled meeting me at the ENIAC dedication.

Hartree wrote a number of popular articles on electronics in English journals such as *Nature* and lectured on the subject extensively. As we mentioned earlier, he did a great deal to influence Wilkes at Cambridge. In fact, as early as January 1946 he was writing as follows: "Which reminds me that I have had an enquiry from M. V. Wilkes, acting director of the Mathematical Laboratory at Cambridge, asking if I could give him information on electronic computing machine developments in U.S.A." [16] He also interested a colleague, Maxwell H. A. Newman, F.R.S., Professor of Pure Mathematics at the Victoria University in Manchester, in computers. Newman was "interested in developing an electronic machine for non-numerical mathematics, based on the logical rather than the arithmetical side. . . ." [17] This interest of Newman's was very serious. He sent one of his young lecturers, Dr. David Rees, over to the Moore School course in 1946, and later that year came himself to the Institute. Hartree wrote, "Newman tells me he hopes to leave as soon as possible after October 21st." [18] Newman also secured the services of an absolutely first-class engineer, Frederic C. Williams, F.R.S., who took a chair in electrotechnics at Manchester in 1947, coming there from the Telecommunications Research Establishment at Malvern, where he had been associated with Uttley.

Appointment of Professor D. R. Hartree, 17 April 1946; Letter, Goldstine to Gillon, 13 April 1946; Letter, Goldstine to Hartree, 11 April 1946; Letter, Hartree to Goldstine, 6 April 1946.

[16] Letter, Hartree to Goldstine, 19 January 1946.

[17] Letter, Hartree to Goldstine, 2 March 1946.

[18] Letter, Hartree to Goldstine, 30 July 1946. Newman's term at the Institute was from 24 October through 27 December 1946.

Newman also brought Turing in as associate director of the project in 1948.

While Williams was at Malvern, he had begun to study the storage of pulses on the face of a cathode-ray tube—the details will be described later. This study was to lead to results which influenced the computing field profoundly. Prior to then, experiments in this line had been undertaken at the Radiation Laboratory in Cambridge, Massachusetts. This type of device was being tried as an alternate to the delay line for radar application. (It was being developed by E. F. MacNichol, assisted by P. R. Bell and H. B. Huntington. They worked under the general direction of Britton Chance who later went to the Johnson Foundation of the Hospital of the University of Pennsylvania. The mercury delay line work was done by D. Arenberg and Huntington.[19]) Williams and his colleague T. Kilburn, writing in 1952, said: "Another American source reported that the observed phenomena were quite unsuitable as the basis for a computing machine store. The reason given was that the memory was too transient, and thus the record could be read only a very few times before it faded away." [20] One of Williams' greatest accomplishments was to understand that if information can be stored for very short times, it can be stored indefinitely. He realized the importance of the basic idea, understood how to exploit it, and set up a project at Malvern in 1946 to explore this technology. He was successful for reasons we shall discuss below in some detail, and the so-called Williams tube was to be perhaps the best memory device available until the invention of magnetic core. This latter displaced these tubes, and they are now only curiosities. However, they made possible a whole generation of machines. In 1950 several developments that were independent of but parallel to Williams' project took place in the United States.[21]

Perhaps Newman's most important accomplishment in the computer field—apart from sponsoring Williams—was his vision

[19] Letter, von Neumann to H. P. Luhn, 20 May 1953. See also B. Chance, F. C. Williams, V. W. Hughes, D. Sayre, and E. F. MacNichol, Jr., *Waveforms*, chap. 9 (New York, 1949).

[20] F. C. Williams and T. Kilburn, "The University of Manchester Computing Machine," *Review of Electronic Digital Computers*, Joint AIEE-IRE Computer Conference (1953), pp. 57–61. See also F. C. Williams and T. Kilburn, "A Storage System for Use with Binary Digital Computers," *Proc. IEEE*, vol. 96, pt. 2, no. 81 (1949), pp. 183–200.

[21] J. P. Eckert, H. Lukoff, and G. Smoliar, "A Dynamically Regenerated Electrostatic Memory System," *Proc. IRE*, vol. 38 (1950), pp. 498–510; J. W. Forrester, "High-Speed Electrostatic Storage," in *Proceedings of a Symposium on Large-Scale Digital Calculating Machinery* (Cambridge, Mass., 1948; see below, n. 28).

that the computer could be used as a fundamental tool for heuristic investigations in areas of pure mathematics where the burden of algebraic calculation was too great for the human. He wrote that "with the development of fast machine techniques mathematical analysis itself may take a new slant, apart from the developments that may be stimulated in symbolic logic and other topics not usually in the repertoire of engineers or computing experts; and that mathematical problems of an entirely different kind from those so far tackled by machines might be tried, e.g. testing out, the 4-colour problem or various theorems on lattices, groups, etc., for the first few values of n." [22] This is the first instance, as far as I know, of the use of the computer to gain heuristic insights. Turing had great interest in this subject in early days and perhaps derived it from his conversations with Newman. In any case, he wrote a paper on "Intelligent Machinery" as early as 1947, while at the National Physical Laboratory and another one in 1950.[23]

Swedish scientists and Government have been much interested in digital computers since the times of Scheutz and Wiberg. Not long ago Professor Arne Beurling, now of the Institute for Advanced Study but formerly of the University of Uppsala, told me of trying to interest a leading firm in Sweden in building a computer in 1945. The suggestion turned out to be a little premature. But in July 1946, as soon as publicity on the ENIAC appeared and had been digested in Sweden, a charming gentleman, Professor Stig Ekelöf of the Chalmers Institute of Technology in Gothenburg, visited me in Princeton. We went together to Philadelphia to see the ENIAC and possibly Ekelöf attended the Moore School course.

He became quite excited about the idea of an electronic computer and wrote me:

> However I am afraid I am still on the same level as the old man in those years when the electric light began, who understood everything quite well except one thing — how the oil could pass through the fine wires!!! . . .
>
> There is an enormous interest in Sweden right now for these machines. If the ENIAC were for sale and if it were a good idea to buy it (which I understand it is not!) then we would certainly have the money available. Probably some young people will be sent out shortly to acquaint themselves with this new field.[24]

[22] Letter, M. H. A. Newman to von Neumann, 8 February 1946.
[23] Sara Turing, *Alan M. Turing*, bibliography.
[24] Letter, Ekelöf to Goldstine, 9 November 1946.

In a few years these things came to pass in Sweden and will be described in due course. Indeed a very substantial interest developed in that country which resulted in several machines being built both in the university and industrial communities.

The French also were soon interested in the field. During September 1946 Louis Couffignal of the Laboratoire de Calcul Mécanique of the Institut Blaise Pascal visited me in Princeton and went to see the ENIAC in Philadelphia.[25] Couffignal and his associates designed an electronic machine which was described in a 1951 article.[26]

After the war the Swiss, like the Germans, came under the influence of a quite remarkable German relay computer designer, Konrad Zuse, who had designed and built several such devices during the war years. Zuse and another engineer, Gerhard Overhoff, went into business together in 1943. Overhoff had been chief of the department in the Henschel Flugzeugwerken, A.G., Berlin, where remote controlled rockets were tested. He therefore realized the great value there would be in having an automatic machine to help in this testing. Zuse had already been at work, and they became partners.

The two men had both worked at Henschel, where they became friends in 1941. Zuse, however, spent his spare time in developing his ideas on relay calculators. After many years of hard work he completed a first model and obtained the fundamental support which enabled him to carry on further research and development.

The second machine Zuse built apparently appealed to the Nazis and the Deutschen Versuchsanhalt für Luftfahrt (German Research Institute for Aerodynamics) expressed considerable interest in this device. Then in early 1943 the German Air Ministry placed an order with Henschel for a general-purpose relay machine. Henschel put Zuse in charge and appointed Overhoff as the technical and fiscal coordinator. It also gave Zuse an order for two more special-purpose machines for its own aerodynamic calculations. All went well — from their point of view — and Zuse established his own company in the latter part of 1943. His firm was called Zuse Apparatebau and was quite small, employing about 15 men. The scientific staff consisted — in addition to Zuse and Overhoff — of a mathematician and two engineers. He was assisted both by the Air Ministry and by Henschel. His three machines were finished before the war was

[25] Letter, Goldstine to von Neumann, 20 September 1946.
[26] Couffignal, "Report on the Machine of the Institut Blaise Pascal," *Math. Tables and Other Aids to Computation*, vol. 5 (1951), pp. 225–229.

ended and operated successfully.[27] After the war he continued his very successful career. In many ways he was the premier builder of electromechanical computers.

Somewhat later virtually every country of Western Europe became active in the field, and I have provided a brief survey of these developments in the Appendix. They are out of place here, and it is time to go back to the early days in Princeton to pick up our thread again.

The great difficulty in writing a history of computers is that about 1950 scientists in virtually every highly civilized country realized the great need for these instruments. This caused a great awareness of the scientific importance of computers to the university and government community, and the earliest and boldest developments are in general in the university rather than in the industrial world. Indeed, at this stage of the field's development most industrialists viewed computers mainly as tools for the small numbers of university or government scientists, and the chief applications were thought to be highly scientific in nature. It was only later that the commercial implications of the computer began to be appreciated.

As a result, the seminal ideas developed largely in the university world, with the help of far-sighted government support. It is here we must look for the basic notions which underlay the computer revolution. They were disseminated to the world by means of a number of large scientific meetings and by reports written for the government and given wide circulation by various governmental agencies. Notable among the meetings were several organized by Aiken at Harvard, which were widely attended.[28] Perhaps the most noteworthy of the reports was a series of papers that emanated from the Institute for Advanced Study starting in mid-1946 — much more on these later.

[27] This material was extracted from an Interrogation Summary entitled Automatic Calculating Machine and was prepared by Capt. Robert E. Work, Air Division, Headquarters United States Forces in Austria, Air Interrogation Unit, 8 November 1946. See also W. deBeauclair, *Rechnen mit Maschinen* (Braunschweig, 1968). This is a generally good account, complete with pictures, of many early machines. The reader is also referred to Zuse's autobiography, *Der Computer — mein Lebenswerk* (Munich, 1970).

[28] See, e.g., *Proceedings of a Symposium on Large-Scale Digital Calculating Machinery, Jointly Sponsored by the Navy Department Bureau of Ordnance and Harvard University at the Computation Laboratory, 7–10 January 1947* (Cambridge, Mass., 1948). At this meeting there were 157 university, 103 government, and 75 industry attendees, the largest single industrial group being 9 from IBM. There were papers presented by 24 university, 11 government, and 7 industry people. Scientists from Belgium, Great Britain, and Sweden were also present.

The staffing of the Institute project really began in March 1946 when I arrived full time in Princeton. The man who was for a number of years to be the chief engineer, Julian Bigelow, accepted the Institute's offer on 7 March with the intention of coming full time to Princeton as soon as he could free himself from his duties. These consisted of responsibilities to the Fire Control Division of NDRC as well as its Applied Mathematics Panel. "He has a profound interest in automatic computing and control which is clearly a very important asset in this work. Part of his war work was done in one group with Norbert Wiener from M.I.T. whose ideas in this field are very significant." [1]

Arthur Burks accepted von Neumann's offer on 8 March with the intention of joining full time in May or June, and in the interim he was to be a part-time consultant. The reason he could not start sooner was a teaching responsibility he had in Philadelphia. His stay at the Institute was brief; he left in the fall of 1946 when he received a call from the University of Michigan to teach in the philosophy department there. However, he did return for the summers of 1947 and 1948 and made most important contributions. He is one of those rare people with a real understanding of mathematics, philosophy, and engineering and their interrelationships.

James H. Pomerene, who was then an engineer with the Hazeltine Co. of New York, accepted an offer on 9 March to start in April. He is an excellent engineer who became the chief engineer of the Institute project when Bigelow went on a Guggenheim fellowship. He remained in this capacity until nearly the end of the project, when he went to IBM. There he has had a distinguished career as a machine designer.

Ralph J. Slutz joined the project in June coming from Division 2 of NDRC. He is a physicist with an excellent understanding not only of physics but also of electronics. He was with the project until

---

[1] Von Neumann, Report of Computer Project, 16 March 1946. This was apparently a memorandum to Aydelotte bringing him up to date on project developments.

he left in 1948 to join the National Bureau of Standards group in Boulder, Colorado.

Next Willis H. Ware, a friend of Pomerene's from Hazeltine and a man who has since headed the computer science work at the Rand Corporation, joined the project. He is today a leader in the field and one of the American experts on the Soviet computer industry.

These were the starting engineering and logical design groups. During the first months of the project two very different but equally important tasks confronted the two groups. For one thing, a logical design for the machine had to be articulated, and this was done by Burks and me in collaboration with von Neumann and assists from John Tukey. The other task was to create and organize a laboratory in an institution with no experimental facilities or shops whatsoever. Both tasks were done with great speed. Let us now however concentrate on the design problem since the results of that work were to be the basic ideas underlying essentially all modern machines. Out of this arose the concept today popularly and fittingly known as "the von Neumann machine."

It was very early contemplated that the project would be a multi-faceted one embracing engineering, formal logics, logical design, programming, mathematics, and some significant revolutionary application. This latter evolved in 1948 as numerical meteorology. All this was carried out with signal success and was the reason that the Institute project assumed the fundamental role it did. It truly embraced all aspects of the field as then understood. Indeed, it dimensioned the field.

In the memorandum of 8 November 1945 von Neumann had written:

> (1) The purpose of this project is to develop and construct a fully automatic, digital, all-purpose electronic calculating machine. . . . Furthermore while the machine is to be definitely digital in character, it is important to provide it with some continuous variable organs, essentially as alternative inputs and outputs. . . .
>
> (2) The overall logical control of the machine will be effected from the memory . . . by orders formulated in a binary digital code. . . . These orders form a system which gives the machine a very considerable flexibility. It is expected that it will be able to deal efficiently and at extremely high speeds with very wide classes

of problems. . . . In this sense the machine is intended to be an all-purpose device. . . .

(3) A machine as described . . . will certainly revolutionize the purely mathematical approach to the theory of non-linear differential equations. It will also (practically for the first time!) permit extension of . . . compressible hydro- and aerodynamics . . . as well as to the more complicated problems of shock waves. It is likely to extend quantum theory to systems of more particles and more degrees of freedom than one could heretofore. . . . It may render a computational approach to the decisive phases of (incompressible) viscous hydrodynamics possible: To the phenomena of turbulence and to the more complicated forms of boundary layer theory. It is likely to make the theories of elasticity and plasticity much more accessible than they have been up to now. It will probably help a great deal in 3-dimensional electrodynamical problems. It will certainly remove many very critical bottlenecks in the computing approach to ordinary and electron optics. It may be useful in stellar astronomy. It will certainly open up a new approach to mathematical statistics: the approach by computed statistical experiments. . . .

Apart from these things, however, such a machine, if intelligently used, will completely revolutionize our computing techniques, or, to formulate it more broadly, the field of approximation mathematics. Indeed, this device is likely to be in any objective sense at the very least 10,000 times faster (actually probably more) than the present human computer-and-desk multiplying machine methods. However our present computing methods were developed for these, or for still slower (purely human) procedures. The projected machine will change the possibilities, the difficulties, the emphases, and the whole internal economy of computing so radically, and shift all procedural options and equilibria so completely, that the old methods will be much less efficient than new ones which have to be developed. These new methods will have to be based on entirely new criteria of what is mathematically simple or complicated, elegant or clumsy.

The development of such new methods is a major mathematical and mathematical-logical program. Certain phases of it can be visualized now, and will necessarily influence the scheme and the control arrangements of the projected machine. Further investigations will have to be carried out in parallel with the development and construction of the machine. However, the main work

will have to be done when the machine is completed and available, by using the machine itself as an experimental tool. It is
to be expected that the future evolution of high-speed computing
will be decisively influenced by the experiences gained and the
interpretations and theories developed in this stage of the
project.[2]

Within a very short time the dimensions broadened even more
than was then envisioned by von Neumann.

The so-called Electronic Computer Project of the Institute for
Advanced Study was undoubtedly the most influential single
undertaking in the history of the computer during this period.
Clearly, the pioneering thrust came from that Institute project,
and to it may be attributed many of the ideas that were, and still
are, basic to the entire field. The project consisted of (1) engineering, (2) logical design and programming, (3) mathematical, and
(4) meteorological groups. To each of these fields it made fundamental contributions. The first group was under the initial direction of Bigelow and then later of Pomerene; the second and third
groups were under my direction; the fourth group was directed by
Dr. Jule G. Charney, a first-rate meteorologist. The Director of the
Project was von Neumann, and I was the Assistant Director. In
July 1952 Bigelow and I were made permanent members of the
Institute, and a little later Charney also received a long-term
appointment.

By the latter part of June 1946, a report was issued by Burks, von
Neumann, and me as "the first of two papers dealing with some aspects of the overall logical considerations arising in connection
with electronic computing machines."[3] Actually, the second paper
appeared in three separate parts, and a fourth was planned but
never completed because of other pressures. These parts started
appearing in April 1947.[4] They were sent as they appeared to the

[2] Von Neumann, Memorandum on the Program of the High-Speed Computer,
8 November 1945.

[3] Burks, Goldstine, and von Neumann, *Preliminary Discussion of the Logical
Design of an Electronic Computing Instrument,* Princeton, 28 June 1946. This
report was very widely circulated, and on 2 September 1947 a second edition was
issued which contained a considerable expansion of the earlier material. It has been
reprinted many times, most recently as a chapter in C. G. Bell and A. Newell,
Computer Structures: Readings and Examples (New York, 1971).

[4] Goldstine and von Neumann, *Planning and Coding Problems for an Electronic
Computing Instrument,* Part II, vol. 1, 1 April 1947; vol. 2, 15 April 1948; vol. 3,
16 August 1948. These papers have also been extensively reprinted.

United States Patent Office and to the Library of Congress with an affidavit from the authors asking that the material be placed in the public domain.[5]

This series was in many ways the blueprint for the modern computer. It contained a very detailed discussion of how a computer should be organized and built as well as how to program it. The papers describe the so-called von Neumann machine, a structure which is, with small modifications, the one still used in virtually all modern computers. Because of the critical importance of these papers to the field, we should now pause in our temporal account to discuss some part of their contents.

It is of some interest perhaps to read what Paul Armer, one of the leaders in the computer science field, wrote about this work. He said:

> Who invented stored programming? Perhaps it doesn't matter much to anyone other than the principals involved just who gets the credit—we have the concept and it will surely stand as one of the great milestones in man's advance. . . . The leading contenders are the authors of the paper reprinted here and the group at the University of Pennsylvania led by J. Presper Eckert and John Mauchly. Others undoubtedly contributed, not the least of whom was Babbage. . . .
>
> Nevertheless the paper reprinted here is the definitive paper in the computer field. Not only does it specify the design of a stored program computer, but it anticipates many knotty problems and suggests ingenious solutions. The machine described in the paper (variously known as the IAS, or Princeton, or von Neumann machine) was constructed and copied (never exactly), and the copies copied. . . .
>
> At the time the paper was written, the principle of automatic calculation was well established (with Harvard's Mark I) as was the great advance gained by electronics (with ENIAC). The jump from that state-of-the-art to the detail of their paper is difficult to measure objectively today. . . .[6]

[5] See, e.g., Deposition by Burks, Goldstine, and von Neumann, dated June 1947. This states that the first edition of the *Preliminary Discussion* dated 28 June 1946 was widely circulated—about 175 copies—both in the United States and abroad, and that the authors "desire that any material contained therein which might be of patentable nature should be placed in the public domain and that the report itself should be regarded exactly as is any other scientific publication."

[6] *Datamation* reprinted parts of the Burks, Goldstine, and von Neumann paper in vol. 8 (1962) with an introduction by Armer, parts of which are quoted here.

Any device which lays claim to being a general-purpose computing machine must clearly contain certain units: it must have one where arithmetic processes can be performed; one where numbers can be stored; one where the set of orders or instructions that characterize the particular problem being solved can be stored; one that can automatically execute these instructions in the correct sequential order; and one that makes connection with the human operator. Conceptually, we have distinguished two different sorts of memory or storage: storage of numbers that are used or produced during a calculation and storage of instructions. As we indicated for the EDVAC, if the instructions are reduced to a numerical code and if the control can in some manner distinguish an order from a number, then one memory organ will suffice for both. We discuss below both how a numerical code was developed and also how the machine was able to differentiate between numbers and orders.

To sum up, there were four main organs contemplated for the machine: an arithmetic unit, a memory, a control, and an input-output. Let us now consider some of their properties.

The arithmetic organ—in modern parlance the central processing unit—was the part of the system in which the arithmetic operations viewed as elementary by the machine were to be executed. There were not—and indeed there cannot be—any fixed rules as to what these are. But the elementary processes, whichever ones are chosen, are those that are wired in. Other operations can be built out of them by properly arranged sequences of instructions. To illustrate by an extreme example, one could in principle build an arithmetic organ having only the fundamental operations of *and*, *or*, and *not* (actually two suffice) and then as Boole and his successors showed build up the usual arithmetic operations out of these as properly arranged sequences of these basic logical ones. In fact, such an approach would be impractically slow. A less extreme illustration would be to omit either multiplication or division or both and have only addition and subtraction as basic operations. The multiplication could be built up out of a sequence of properly ordered additions and similarly for division.

In general, the inner economy of the arithmetic unit is determined by a compromise between the desire for speed of operation—a non-elementary operation will generally take a long time to perform, since it is constituted of a series of orders given by the control—and the desire for simplicity, or cheapness, of the machine. (Actually these remarks are somewhat too simplistic. One can in fact build a machine with operations that are programmed out of a special memory and still achieve reasonable speeds. For our purposes it is however as well to ignore this possibility.)

The memory organ has always been a complex one; above all the others, this unit, since the ENIAC days, has caused the biggest technological problems. The user always wants an indefinitely capacious memory, and the engineer has never been able to meet this need as elegantly as either would like. This was recognized from early times, and a compromise was agreed upon. There was to be a hierarchy of memories consisting of an innermost one of electronic speed with a capacity "of about 4,000 numbers of 40 binary digits each," a secondary one "of much larger capacity on some medium such as magnetic wire or tape," and a tertiary one on paper tape or punch cards. The reality was an inner one of 1,000 numbers and, for a long while, no secondary one. Finally however a magnetic drum was inserted instead of magnetic wire or tape.

The concept of a hierarchical memory is still a fundamental one today and always will be. Fundamentally, the user of a machine wants, and indeed needs from a mathematical point of view, an indefinitely large memory capable of operating at the highest speeds. This is of course impossible to attain in the real world so he must accept a compromise: a hierarchical one in which the successive stages are more capacious but slower in speed. If the sizes and speeds of the various stages are properly arranged, a very reasonable overall balance can be achieved for a large class of problems.

The control organ was perhaps most mysterious in the mid-1940s. The work of Post and Turing made it very clear that from the point of view of formal logics there was no problem to devise codes which were "in abstracto adequate to control and cause the execution of any sequence of operations which are individually available in the machine and which are, in their entirety, conceivable by the problem planner." The problem is of a practical nature and is closely allied to that connected with the choice of the elementary operations in the arithmetic organ. The code for a machine is in reality the vocabulary or totality of words or orders that the machine can "understand" and "obey." The designer needs to compromise between his desire for simplicity of the equipment required by the code and the applicability and speed with which the really important problems can be solved.

These things are still true today. The designer still must make the same compromise between the size of the machine's vocabulary, the amount of equipment (more properly its cost) that is required to implement this vocabulary, and the speed that results on important problems.

What needs to be in a minimal code? First, of course, there must be orders for each of the elementary arithmetic operations; this goes almost without saying since otherwise there would be no way to utilize those operations. Second, there must be orders for allowing intercommunication between the memory and arithmetic organ. Moreover, since the memory contains two kinds of information, numbers and orders, it is plausible that there need to be two kinds of transfers from the arithmetic unit to the memory: total and partial substitutions. The former are obvious; a number is transferred into the memory. The latter is more subtle, but it is what helps give a computer its great power as well as its great simplicity. This order allows the device to alter its own instructions, and this ability has been shown to be crucial. We discuss the point later.

There is another type of transfer order that is of great significance. These are the so-called transfer of control orders. There are two kinds: unconditional and conditional. To understand what is involved let us realize that the various locations in the memory can be enumerated in a serial order; these may be termed memory location-numbers. Having said this, we next decree that although we need to get numbers from any part of the memory we can be more methodical with orders. The control organ is so constructed that normally it proceeds seriatim through the memory, i.e. after the order at location-number n has been executed, it normally moves to $n + 1$. For an unconditional transfer the control is shifted to a preassigned location-number, irrespective of where it was. For a conditional one, this shift takes place conditioned by the sign of some number. The transfer of control orders are the way of causing the control to behave abnormally.

This sounds very complex but is not. The real importance of an automatic computer, as Babbage and Lady Lovelace knew, "lies in the possibility of using a given sequence of instructions repeatedly, the number of times being either preassigned or dependent upon the results of the computation." [7] At the conclusion of this iteration, a quite different sequence of orders will need to be executed. It is therefore necessary at the end — or beginning — of such a sequence of orders to introduce an order that can decide whether the iteration is complete and, if it is, cause the next sequence to be executed.

[7] *Preliminary Discussion*, p. 3.

Finally, there need to be instructions that integrate the input-output organ into the device.

This then is a brief description of the order code for the Institute for Advanced Study machine. (We say more on it later.) Fundamentally, though, we have done more: we have described the basic vocabulary that must underlie any modern computer. The newer codes are very much richer than was the one used at the Institute. It contained about two dozen words in its vocabulary, whereas modern machines may have two hundred or more. In the past, however many words there might have been in a machine's vocabulary, a description of a particular problem was essentially an essay written in the pidgin English understood by the machine. Nowadays this is no longer true in quite this literal sense, and we will discuss later the role of computer languages. Let us now resume dimensioning the Institute computer.

What about the number system for the machine? "In spite of the longstanding tradition of building digital machines in the decimal system, we feel strongly in favor of the binary system for our device." [8] There are several reasons for this choice, the outstanding ones of which are these: the greater simplicity and speed with which the elementary operations can be performed; the fact that electronic circuitry and technology tends to be binary in character; and the fact that the control portions of a computer are not arithmetical but rather logical in nature — logics is a binary system. Recall that in binary multiplication the product of a digit in the multiplier by the multiplicand is either the multiplicand or null, according to whether the digit is one or zero. This process is extremely easy to construct as against the decimal process where there are 10 possible values.

It used to be argued in the mid-1940s that the reason for adopting the decimal system for computers was that the problem of conversion from binary to decimal and vice versa was a major one. In the *Preliminary Discussion* we pointed out that this is not so. There are simple, unambiguous ways to make such conversions involving only very simple arithmetic operations. Indeed, in volume 1 of Part II, *Planning and Coding*,[9] von Neumann and I programmed these conversions and showed that: "The binary-decimal and the decimal-binary conversions can be provided for by instructions which require 47 words. . . . The conversions last 5 msec and 5.9 msec, respectively." (A msec = a millisecond = 1/1000 second.)

[8] *Ibid.*, p. 7.   [9] *Op. cit.*, pp. 133ff.

While it is not our idea here to develop detailed mathematical techniques for calculating, it may be worthwhile to give a brief indication of the true simplicity of the process of conversion from one number system to another. (We shall ignore a variety of details that are not strictly relevant to our point of view.) Suppose that a is a given positive binary number lying between 0 and 1, $0 \leq a < 1$. Let us further suppose that its decimal representation is

$$z_1 \ldots z_n = z_1/10 + z_2/10^2 + z_3/10^3 + \cdots + z_n/10^n.$$

Then $10a$ is a number consisting of an integral and a fractional part; the integer part is clearly z_1. This suggests an algorithm:

$$a_1 = a,$$
$$a_{i+1} + z_{i+1} = 10a_i; \; 0 \leq a_{i+1} < 1, \; z_{i+1} = 0, 1, \ldots, 9; \; i = 0, \ldots, n - 1.$$

In words this says start with $a = a_1$, multiply it by 10 and express the result as an integer plus a fractional part; the integer is z_1, the first digit in the decimal representation of a; the fractional part will be called a_2; it is the rest of the decimal representation of a_1. Again multiply this, a_2, by 10 and again express it as an integer z_2 plus a fraction a_3. Repeat this as many times as desired. Let us illustrate with the conversion of the binary number $.1011 = 1/2 + 1/8 + 1/16 = 11/16$. We multiply by 10 and find $110/16 = 6 + 14/16$. Thus to one decimal place $11/16 = 6/10$. Next we operate on $14/16$ by multiplying by 10 and find $140/16 = 8 + 12/16$, so to two decimals $11/16 = 6/10 + 8/100$. Again, $120/16 = 7 + 8/16$, and the third decimal place is 7. Let us calculate one more: $80/16 = 5 + 0/16$, and our fourth and last decimal place is 5. That is, $11/16 = .6875$.

Now there are certain problems connected with the conversion from binary to decimal in a binary machine which we have slurred over, but they are trivial. For example, how does one express a decimal digit in a binary device? This is done by expressing the 10 decimal digits (0 to 9) thus:

0000	0	0100	4	1000	8
0001	1	0101	5	1001	9
0010	2	0110	6		
0011	3	0111	7		

Next, is there an easy way to multiply by 10? This is done by noting that $10 = 8 + 2$ and that a multiplication by 2 is a shift left of one place and by 8 of three places.

The conversion from decimal to binary is equally simple.[10]

The EDVAC development was, as we saw earlier, predicated on a memory system, the supersonic delay line, that had a serial quality to it. The digits appeared seriatim at the end of each delay line, and this more or less fixed the organization of the machine. The Institute development all along contemplated an alternate mode of organization. Von Neumann and I felt that there was an important class of devices in which it was at least conceptually possible to avoid the need for a serial mode of operation. In the *Preliminary Discussion* the authors stated the following:

> We must therefore seek out some more fundamental method of storing electrical information than has been suggested above. One criterion for such a storage medium is that the individual storage organs, which accommodate only one binary digit each, should not be macroscopic components, but rather microscopic elements of some suitable organ. They would then, of course, not be identified and switched to by the usual macroscopic wire connections, but by some functional procedure in manipulating that organ.
>
> One device which displays this property to a marked degree is the iconoscope tube. In its conventional form it possesses a linear resolution of about one part in 500. This would correspond to a (two-dimensional) memory capacity of $500 \times 500 = 2.5 \times 10^5$. One is accordingly led to consider the possibility of storing electrical charges on a dielectric plate inside a cathode-ray tube. Effectively such a tube is nothing more than a myriad of electrical capacitors which can be connected into the circuit by means of an electron beam.
>
> Actually the above-mentioned high resolution and concomitant memory capacity are only realistic under the conditions of television-image storage, which are much less exigent in respect to the reliability of individual markings than what one can accept in the storage for a computer. In this latter case resolutions of one part in 20 to 100, i.e. memory capacities of 400 to 10,000, would seem to be more reasonable in terms of equipment built essentially along familiar lines.[11]

[10] For a fuller discussion of number systems, the interested reader may wish to consult D. E. Knuth, *The Art of Computer Programming*, vol. 2: *Seminumerical Algorithms* (Reading, Mass., 1969), pp. 161–180, 280–290.

[11] *Preliminary Discussion*, pp. 4, 5. It is interesting that the argument regarding the desirability of using microscopic storage elements as against macroscopic ones

The authors of the *Preliminary Discussion* decided not to use the serial mode of storage used in the EDVAC, since in a storage tube with an amplitude sensitive deflection system there was no requirement for such a mode. Instead, they were free to switch to any location on the face of a storage tube as rapidly as to any other. Thus it seemed natural — and virtually all modern computers do this today — to have the 40 storage devices arranged in parallel so that in corresponding places in each are located the 40 digits of a number. (It was decided that each number in the machine would consist of a sign plus 39 digits.) "Such a switching scheme seems to us to be simpler than the tecnique needed in the serial system and is, of course, 40 times faster. We accordingly adopt the parallel procedure and thus are led to consider a so-called *parallel machine,* as contrasted with the serial principles being considered for the EDVAC. . . . The essential difference between these two systems lies in the method of performing an addition; in a parallel machine all corresponding pairs of digits are added simultaneously, whereas in a serial one these pairs are added serially in time." [12]

Probably one of the principal reasons why Ordnance was able to underwrite the Institute project, at least in its early stages, was because it and the Moore School projects represented complementary schools of thought. It was therefore both courageous and correct for Gillon and Feltman to "hedge their bets" by encouraging both groups to proceed. Indeed, at that juncture in history it was certainly unknown which would succeed. As it happened, both were successful, but the Institute system, because of its parallel mode, was much faster: the Institute machine did a multiplication in about 600 microseconds, whereas the EDVAC took about 3 milliseconds; the former required about 25 microseconds to locate and read a word, whereas the latter took about 200 microseconds on the average; the former contained about 2,000 vacuum tubes, and the latter about 3,000 tubes and 8,000 crystal diodes.

Thus the parallel mode resulted in a system requiring less equipment and yet capable of operating at least five times and possibly even more faster. In the long run this inherent advantage of the parallel mode was to triumph, and the serial mode essentially disappeared. Equally, the simplicity of the binary mode prevailed after a number of years, and modern computers today are built to operate in the parallel mode on binary numbers.

---

was abandoned when the magnetic core was invented. To some extent it is being reconsidered in new devices using extremely small integrated circuits.

[12] *Ibid.,* p. 5.

How about the instructions or orders? As we mentioned earlier, they are the fundamental operations that the machine views as elementary. Let us consider one of them—addition. In principle at least, this operates on two different quantities and produces a third one. Generally there are three possible memory locations involved: one for each of the three quantities. Some designers even contemplated a fourth location, the place where the control was to go for its next instruction.

The Institute computer was designed so that each instruction contained only one memory location-number, or *address*, as it became known. Thus, addition was performed by at most three separate instructions: first, fetching from an address one quantity into the arithmetic organ; second, bringing a quantity from another address and adding it to what was already in that organ; third, storing the result at a third address. Under this dispensation of one address per order, the Institute computer was so designed that it permitted 10 binary digits to characterize an address and 10 to characterize the operation. Thus 1,024 or 2^{10} addresses and operations were possible. (In fact, there were less than 2^5 actual instructions.)

There were a number of sharp discussions with members of the staffs of the Ballistic Research Laboratory and of the Moore School regarding the question of how many addresses should be in each instruction, what number system should be used and, if binary, how to convert from binary to decimal and back again. The EDVAC adopted a four-address system against the advice of von Neumann and me. We felt that the last address, the location of the next instruction, was unnecessary in almost every case. Almost never does one need to assume a chaotic ordering of the instructions in the memory. With rare exceptions the instructions can be arranged linearly in the memory with one order succeeding the next by being at the next location-number. The control can be so designed that having executed the order at address n it then goes to $(n + 1)$ unless a transfer order is given sending it elsewhere for its next order. This system has now become quite generally the standard one.

The argument against a three-address code was that almost never did one actually need to add—or whatever other operation—two numbers in the memory and return the result there. The usual situation is that the machine has just produced a number which is in the arithmetic unit, and it is desired to operate on this quantity.

Also, often the result of the operation will be left in the arithmetic unit for further operations on it.[13]

Consideration of a two-address code proved it wasteful, and in the long run the one-address Institute system prevailed because of its economy and simplicity. Even in the early days such style-setting machines as EDSAC, Ferranti Mark I, Whirlwind I, UNIVAC I, IBM 701 and 704 used the Institute system. In fact, both the UNIVAC I and the IBM 701 also followed the Institute machine in storing two instructions in a single word.[14] This saved a little time in bringing the next instruction into the control and presaged what is now known as the "look-ahead" feature on some of the fastest and largest computers. In this system a block of orders is brought into the control at once to save the time of many memory references.

One of the most important reasons for storing instructions in the memory is the fact that one needs to modify them. The most obvious such modification is to change the address in an instruction. Only in this way can a sub-routine be useful in many different parts of a problem or in many different problems. To illustrate, suppose that, as was the case with the Institute machine, the operation of square root is not elementary but is performed by carrying out a small sequence of elementary operations. Suppose further that we will require the square roots of many numbers stored in very different locales in the memory. We may write a small program for effecting the square root of a number and leave open the address of the number. Then by an arithmetic operation on the instruction containing this number we can make the program effect the square root of any number we choose. This is how it was done in the Institute machine. A much more convenient way of changing addresses in instructions was invented for the Manchester machine; it was originally called a "B box" or "B register" but is now called an "index register." [15]

This ability to modify the addresses of instructions is not merely esthetically elegant, it is absolutely fundamental. Let us discuss this a little since it makes a fundamental distinction between ma-

[13] See C. V. L. Smith, *Electronic Digital Computing* (New York, 1959), pp. 35ff. – a good, clear discussion of the topic.

[14] By 1961 over half the systems in existence in the United States used one-address codes. M. H. Weik, *A Third Survey of Domestic Electronic Digital Computing Systems* (Aberdeen, Md., 1961), pp. 1028–1029.

[15] F. C. Williams, T. Kilburn, and G. C. Tootill, "Universal High-Speed Computers: A Small-Scale Experimental Machine," *Proc. IEE*, vol. 96 (1951), part 2, p. 13.

chines of the type of Babbage's Analytical Engine or the Harvard-IBM machine or the Bell Telephone Laboratories' relay computers and those of the type of the Institute machine. The earlier ones with paper tape memories could not modify memory locations or addresses, whereas the latter could. In 1964 Elgot and Robinson gave a rigorous proof of what the author and von Neumann had heuristically asserted in 1947. They showed that "self-modifying programs are more powerful (can compute more sequential functions) than nonself-modifying ones. . . ." [16] It is shown in their paper that "no finitely determined machine can compute all recursive sequential functions by fixed programs, i.e. by means of programs whose instructions are never altered or 'modified.'" [17] They go on to show that a machine in which instructions are alterable can so function. Thus, for example, a machine without the ability to alter instructions cannot form such simple expressions as $x_1 \cdot x_2 \cdot x_3 \cdots x_n$ for arbitrary values $n = 1, 2, \ldots$, whereas a very simple program in a machine of the Institute type can do so merely by altering the constant n, which is stored as an address in the program, as required.

In von Neumann's logical design paper, the *First Draft*, there was essentially nothing on programming or coding. He did, as mentioned earlier, program a sorting routine in 1945, but there had not as yet been any careful analysis of the problem of coding and programming. This took place more or less contemporaneously with the preparation of the *Preliminary Discussion* in 1946.

In the spring of that year von Neumann and I evolved an exceedingly crude sort of geometrical drawing to indicate in rough fashion the iterative nature of an induction. At first this was intended as a sort of tentative aid to us in programming. Then that summer I became convinced that this type of *flow diagram*, as we named it, could be used as a logically complete and precise notation for expressing a mathematical problem and that indeed this was essential to the task of programming. Accordingly, I developed a first, incomplete version and began work on the paper called *Planning and Coding* (cited above, n. 4). Von Neumann and I worked on this material with valuable help from Burks and my wife. Out of this was to grow not just a geometrical notation but a

---

[16] C. Elgot and A. Robinson, "Random-Access Stored-Program Machines, An Approach to Programming Languages," *Journal of the Association for Computing Machinery*, vol. 11 (1964), pp. 365–399.

[17] *Ibid.*, p. 397.

carefully thought out analysis of programming as a discipline. This was done in part by thinking things through logically, but also and perhaps more importantly by coding a large number of problems. Through this procedure real difficulties emerged and helped illustrate general problems that were then solved.

The purpose of the flow diagram is to give a picture of the motion of the control organ as it moves through the memory picking up and executing the instructions it finds there. The flow diagram also shows the states of the variables at various key points in the course of the computation. Further, it indicates the formulas being evaluated. As I pointed out once in a lecture on the subject, the flow diagram is really a Riemann surface with many sheets for the various inductions that enter. This notion has never been followed up in the detail it perhaps deserves.

Von Neumann was away frequently in Los Alamos and there are a few letters extant from that period which show the genesis of these flow diagrams from crude scrawls into a highly-sophisticated notation still in use today.[18] Finally by April 1947 the paper on *Planning and Coding* was completed and published. However, by the beginning of 1947 most of the notions of programming and coding were already well understood by von Neumann and me, and it was possible to give a public lecture on the subject at the 1947 Harvard Symposium on Large-Scale Digital Calculating Machinery.[19] The process of arriving at an analysis of what was happening and of evolving a notation that was both logically complete and also practically useful was itself an iterative one. Thus in the Fall of 1946 I sent von Neumann a draft of the planning and coding paper, but he found in it a gap: "It occurs to me now, that there is one contingency which was not described in your manuscript of the second part of our report." [20] Again in the spring of 1947 the manuscript was still not complete because of a missing notation. However, a last interchange of letters sufficed to straighten this out. It is amusing to quote a piece of von Neumann's letter since it illustrates so well his charm and his informal mode of expression.

I have to correct one item in the letter that I wrote to you yesterday—although this is probably late, because you will have noticed it.

[18] Letters, von Neumann to Goldstine, 16 September 1946, 2 March 1947, 9 July 1947, 19 July 1947.
[19] H. H. Goldstine, "Coding for Large-Scale Calculating Machinery," in *Proceedings, op. cit.*
[20] Letter, 16 September 1946.

This concerns the treatment of inductions. . . . You pointed out that the region at * . . . is confusing. . . . I suggested rewriting the * stretch like this: . . . I must have been feeble minded when I wrote this: . . .

Let us introduce a new type of box, called *assertion box*. . . .

I have carried out all corrections that you or I considered necessary, including the introduction of a new paragraph. . . ." [21]

I can perhaps describe best what was being done by quoting a little of the original text, since it serves to convey the spirit of the era and the general level of understanding as well. It also helps clarify the usage of the terms "programming" and "coding." (In the quotation the symbol C stands for the control organ.)

Before proceeding to the actual programming of such problems, we consider it desirable to discuss the nature of coding per se and in doing this to lay down a modus operandi for handling specific problems. We attempt therefore in this chapter to analyze the coding of a problem in a detailed fashion, to show where the difficulties lie, and how they are best resolved.

The actual code for a problem is that sequence of coded symbols . . . that has to be placed into the . . . memory in order to cause the machine to perform the desired and planned sequence of operations, which amounts to solving the problem in question. Or to be more precise: This sequence of codes will impose the desired sequence of actions on C by the following mechanism: C scans the sequence of codes, and effects the instructions, which they contain, one by one. If this were just a linear scanning of the coded sequence, the latter remaining throughout the procedure unchanged in form, then matters would be quite simple. Coding a problem for the machine would merely be what its name indicates: Translating a meaningful text . . . from one language (the language of mathematics, in which the planner will have conceived the problem, or rather the numerical procedure by which he has decided to solve the problem) into another language (that one of our code).

This, however, is not the case. We are convinced, both on general grounds and from our actual experience with the coding of specific numerical problems, that the main difficulty lies just at this point. . . .

To sum up: C will, in general, not scan the coded sequence of instructions linearly. It may jump occasionally forward or back-

[21] Letter, 2 March 1947.

ward, omitting (for the time being, but probably not permanently) some parts of the sequence, and going repeatedly through others. It may modify some parts of the sequence while obeying the instructions in another part of the sequence. Thus while it scans a part of the sequence several times, it may actually find a different set of instructions there at each passage. All these displacements and modifications may be conditional upon the nature of intermediate results obtained by the machine itself in the course of this procedure. Hence, it will not be possible in general to foresee in advance and completely the actual course of C, its character and the sequence of its omissions on one hand and of its multiple passages over the same place on the other as well as the actual instructions it finds along this course, and their changes through various successive occasions at the same place, if that place is multiply traversed by the course of C. These circumstances develop in their actually assumed forms only during the process (the calculation) itself, i.e. while C actually runs through its gradually unfolding course.

Thus the relation of the coded instruction sequence to the mathematically conceived procedure of (numerical) solution is not a statical one, that of a translation, but highly dynamical: A coded order stands not simply for its present contents at its present location, but more fully for any succession of passages of C through it, in connection with any succession of modified contents to be found by C there, all of this being determined by all other orders of the sequence (in conjunction with the one now under consideration). This entire, potentially very involved, interplay of interactions evolves successively while C runs through the operations controlled and directed by these continuously changing instructions. . . .

Our problem is then to find simple, step-by-step methods, by which these difficulties can be overcome. Since coding is not a static process of translation but rather the technique of providing a dynamic background to control the automatic evolution of meaning, it has to be viewed as a logical problem and one that represents a new branch of formal logics. We propose to show in the course of this report how this task is mastered.

The balance of this chapter gives a rigorous and complete description of our method of coding, and of the auxiliary concepts which we found convenient to introduce in order to expand this method. . . .[22]

---

[22] Goldstine and von Neumann, *Planning and Coding*, Part II, volume I, pp. 1–2.

Shortly after thinking through the broad outline of the first planning and coding paper von Neumann began to discuss with me the possibility of transforming the ENIAC into a stored program computer of a sort. He proposed using one or two of the ENIAC function tables as the place to store the orders describing a problem and to wire up the ENIAC once and for all to understand these orders. With a few other tricks, it was possible to use various devices in the ENIAC to control the execution of the orders.

The conversion of the ENIAC required a major programming chore that was to require a number of months. Inasmuch as von Neumann was very anxious to use the ENIAC in the new mode for Los Alamos, he had my wife made a consultant of that laboratory effective 7 June 1947.[23] By July 1947 von Neumann was writing: "I am much obliged to Adele for her letters. Nick and I are working with her new code, and it seems excellent." [24] This topic has been discussed (above, p. 233), so we need say no more at this time about it.

[23] Letter, A. E. Dyhre to Aydelotte, 16 May 1947; A. K. Goldstine to Dyhre, 28 May 1947; and Consultant Agreement between Regents, University of California and A. K. Goldstine.
[24] Letter, von Neumann to Goldstine, 19 July 1947.

Chapter 3

Automata Theory and
Logic Machines

More or less simultaneously with his interests in logical design, planning and coding, and solving hydrodynamical problems for Los Alamos, von Neumann had a profound concern for automata. In particular, he always had a deep interest in Turing's work and indeed offered Turing the position of his assistant at the Institute for Advanced Study in 1938.[1] Turing however decided to return to Kings College, Cambridge, where he was a Fellow.

Before describing the developments with automata, we should perhaps say a little about logic machines, since they touch at least tangentially on our story. The reader who wants a good account of the subject should however read Gardner's work on the subject.[2] Here we can touch on just a few names and facts.

Curiously, the first machine as such for solving problems in formal logics was made by a most remarkable British gentleman, Charles, Earl of Stanhope (1753–1816). He led a very full and colorful life both as a politician, first in the House of Commons and then in the Lords (1786–1816), and as a scientist. In addition to his work on logic machines, he designed and built a simple desk calculator in the tradition of Leibniz. He also invented a steamboat, a hand printing press, cylindrical bi-convex lenses whose ends have unequal curvatures, etc. He was the author of *On the Principles of Electricity, Containing Devices New Theories and Experiments, together with an Analysis of the Superior Advantage of High and Pointed Conductors* (1779). Stanhope's ability may be seen from the fact that at the age of nineteen he was elected to the Royal Society. He was however quite eccentric and without much influence in Parliament. In particular, he was a strong advocate of the French Revolution and debated unsuccessfully with Burke on this point. (Incidentally it was during this period that Burke said: "The age of chivalry is gone, and an age of economists and calculators has succeeded.") Stanhope's other claim to fame was the fact that he was William Pitt's brother-in-law by his first wife, and his eldest

[1] Sara Turing, *Alan M. Turing*, p. 55.
[2] Martin Gardner, *Logic Machines and Diagrams* (New York, 1958).

daughter acted as housekeeper for Pitt until Pitt's death, at which time she moved to Lebanon and set up there as a political power.[3]

Before abandoning the discussion of logic machines I would be remiss not to mention Lewis Carroll's contributions, which are set forth in his two works on the subject.[4] Charmingly, Carroll says his game "requires one player *at least*."[5] It is not too pertinent to our ends to describe his so-called Demonstrator beyond quoting Gardner's description: "Not only could it be used for solving traditional syllogisms by a method closely limited to the Venn circles; it also took care of numerical syllogisms (anticipating De Morgan's analysis of such form) as well as elementary problems of probability. In addition, it was based on a system of logical notation which clearly foreshadowed Hamilton's technique of reducing syllogisms to statements of identity by making use of negative terms and quantified predicates."[6]

The next big step was taken by another Englishman, William Stanley Jevons (1835–1882), who was primarily a political economist but also a significant figure in the field of logics. Indeed, he was the first person to construct a machine "with sufficient power to solve a complicated problem faster than the problem could be solved without the machines' aid."[7]

The next device worth calling attention to was designed by Allan Marquand (1853–1924), who began his career in logics and ethics but abandoned these to become professor of art and archaeology at Princeton University. He actually designed several devices, but the one of most interest to us is one built for him by a friend, Charles R. Rockwood, Jr., who was professor of mathematics at the university. This machine was described by the philosopher Charles Sanders Peirce as follows: "Mr. Marquand's machine is a vastly more clear-headed contrivance than that of Jevons. The nature of the problem has been grasped in a more masterly manner, and the directest possible means are chosen for the solution of it. . . . To work a simple syllogism, two pressures of the keys only are necessary, two keys being pressed each time."[8]

Gardner states somewhat curiously that after an extensive search Marquand's machine was found in the stacks of the university li-

[3] Gardner, p. 89.

[4] C. L. Dodgson, *The Game of Logic* (London, 1886) and *Symbolic Logic, Part I: Elementary*, 4th ed. (London, 1897).

[5] Gardner, p. 45.   [6] Gardner, pp. 80–90.   [7] *Ibid.*, p. 91.

[8] Peirce, "Logical Machines," *American Journal of Psychology*, vol. 1 (1887), p. 165. There is also an interesting letter extant from Peirce to Marquand about the use of electricity in machines, dated 30 December 1886.

brary. He also was given a copy of a circuit diagram prepared by Marquand for an electrified version of the machine.[9]

We need not go on with this brief enumeration of logic machines, since it carries us too far afield. However, we must mention another very different type of special purpose machine invented by Derrick Henry Lehmer, who has entered our tale earlier (above, pp. 202, 233). His father, Derrick Norman Lehmer, in 1929 brought out a number of card stencils which could be used to find rather easily so-called quadratic residues. These play a fundamental role in number theory. Using these stencils, it is possible to find whether a given number is prime or composite. In 1939 John Elder revised and extended Lehmer's stencils and put them on IBM punch cards.[10] By means of these cards any number less than about 3.3 billion could be examined to see if it was prime or not.

In 1933 the son, D. H. Lehmer, described an extremely fast photoelectric number sieve he had built for solving problems in number theory. What in effect he did was to mechanize further what his father had done with stencils. They created what is called a "Sieve of Eratosthenes" after its original inventor, who lived from about 275 to 195 B.C. and who made many great discoveries. Lehmer's machine consisted of a series of gears on a common shaft with holes near their peripheries which were rotated at considerable speeds. He arranged matters so that what was sought was a coincidence of holes in all gears. This was searched for by a photocell, and by this means 300,000 numbers could be processed in a minute. The machine was of great importance to number theory. During the EDVAC phase at the Moore School Lehmer was quite excited and sketched out the idea for a new machine using a bank of acoustic delay lines instead of gears. He succeeded in replacing his mechanical sieve by an electronic one. It is a modest solid-state device called the Delay Line Sieve (DLS-127). It was built as an unsponsored educational project of the departments of mathematics and electrical engineering at the Berkeley campus of the University of California.[11] This machine went into operation in 1965 and can process a million numbers per second.[12]

We must now return to the logic machines, an entirely different

[9] Gardner, p. 112.

[10] D. N. Lehmer, *Factor Stencils*, revised and extended by John D. Elder (Carnegie Institution of Washington, 1939).

[11] D. H. Lehmer, "Machines and Pure Mathematics," in *Computers in Mathematical Research*, edited by R. F. Churchhouse and J. C. Herz (Amsterdam, 1968).

[12] D. E. Knuth, *The Art of Computer Programming*, vol. 2: *Seminumerical Algorithms*, p. 347.

form of machine, and mention those of Post and Turing. They did not build actual devices but instead paper constructs. Their aim was to investigate a very deep problem in formal logics, and their automata were a way of describing how to cope with the basic problem.

What is an automaton in the Post-Turing sense? It is one of the proverbial "black boxes" which has a finite number of states which we may number 1, 2, . . . , n. The main point for us is to describe how the automaton can change its state from, say, i to j ($i, j = 1, 2, . . . , n$). The change takes place by an interaction with the world outside the automaton which is considered as being a tape. Let us suppose there is a long paper tape, divided into squares, associated with the automaton; further let us imagine we may (or not) write a 1 in a square or erase what is in one, i.e. write a 0. For simplicity imagine the automaton inspects the tape a square at a time; it can move the tape forward or backward one square at a time. While doing this it can read what is on the square being inspected, and it can also write or erase what is in that square. Thus if the automaton is in the state i and sees on a square the symbol $e = 0$, or 1, it can then move the tape p squares forward or backward—p may be 0 or $+1$, one forward, or -1, one backward—and write in the new field the symbol $f = 0$, 1. This causes the automaton to go into the state j and once k, p, f are given in terms of i and e the functioning of the automaton has been completely described.[13]

The fundamental point of Turing's analysis has to do with infinite sequences of binary digits. He asked which automata could construct given, preassigned sequences. Under his rules an automaton can construct such a sequence if it can be given as input or as its set of instructions a finite piece of tape, i.e. a tape with a finite number of squares, and can then write the infinite sequence, by running indefinitely. In other words, if the automaton can run indefinitely it will produce in a sufficiently long span of time any desired—finite—part of the infinite sequence.

Turing proved a most remarkable and unexpected result. There is a universal automaton in the sense that it can calculate any sequence that any special automaton can, provided only that it receives the appropriate set of input orders. This seems paradoxical, since in principle one can imagine automata of varying sizes and complexities. However, he showed it is true. In essence what he showed is that any particular automaton can be described by a

[13] For a detailed description see M. Davis, *Computability and Unsolvability* (New York, 1958).

finite set of instructions, and that when this is fed to his universal automaton it in turn imitates the special one.

Von Neumann was enormously intrigued with these ideas, and he started in 1947 working on two broad fronts: first, he wanted to find out how complex a device or construct needed to be in order to be self-reproductive; and, second, he wanted to investigate the problem of how to organize devices that must be made from parts that can malfunction.

These ideas are all closely related to Gödel's beautiful result whereby mathematical logics are reduced to a kind of computation theory (above, p. 173). Indeed, he showed that the basic concepts of logics are recursive, which is equivalent to saying they can be computed on a Turing machine.[14] It is thus not surprising to understand von Neumann's interest in automata theory; it combines his very early interest in logics with his newer interests in computers and neurophysiology. This latter one arose probably during his reading of the very important work by McCulloch and Pitts on neural networks.[15] Certainly he developed a great interest in neurophysiology around the time the paper appeared. In December 1944 Aiken, Wiener, and von Neumann proposed the formation of a small group to be called the Teleological Society to discuss "communication engineering, the engineering of computing machines, the engineering of control devices, the mathematics of time series in statistics, and the communication and control aspects of the nervous system." [16] The group was to consist of the founders plus S. S. Wilks of Princeton, W. H. Pitts of Kellex Corp., E. H. Vestine of Carnegie Institute, E. W. Deming of the U.S. Census, W. S. McCulloch of the University of Illinois Medical School, Lorente de No of the Rockefeller Institute, L. E. Cunningham of the Ballistic Research Laboratory, and myself.

Interestingly the McCulloch-Pitts work was originally intended as a mathematical model of the human nervous system. But it actually became much more: their model was equivalent to formal

[14] J. von Neumann, *Theory of Self-Reproducing Automata,* edited and completed by A. W. Burks (Urbana, 1966). Here Burks has in a very elegant way taken all von Neumann's work on the subject, including typescripts of a tape recording of five lectures he gave at the University of Illinois in December of 1949, and given them coherence and completion. He has also preserved very well indeed von Neumann's style.

[15] W. S. McCulloch and W. Pitts, "A Logical Calculus of the Ideas Immanent in Nervous Activity," *Bull. Math. Biophys.,* vol. 5 (1943), pp. 115–133. This paper was of great consequence to von Neumann, since he was to use their notations in the *First Draft* paper on the EDVAC and in several other works on automata as well.

[16] Letter, Wiener to Goldstine, 22 December 1944.

logics. Their result is that "anything that can be exhaustively and unambiguously described, anything that can be completely and unambiguously put into words, is ipso facto realizable by a suitable finite neural network . . . the converse statement is obvious. . . ." [17]

It had usually been assumed by thinkers on inanimate devices that their outputs must necessarily be less complex than themselves, and indeed this notion seems quite reasonable. Certainly all machines we can think of in the past had this quality. It was therefore not obvious that von Neumann's question as to whether there exist self-reproductive automata would have a positive answer. In fact it did.

In June of 1948 von Neumann gave a few lectures to a small group of us on the subject. These lectures were anticipatory to his paper for the Hixon Symposium, but they were quite detailed and indicated that he had progressed very far in his thinking. In fact there was no doubt that he already that spring saw in principle his way to the end. At that time he was thinking about a device which was in Burks' terms "a kinematic model" as contrasted with his later construct which was "a cellular model." All these things are beautifully presented by Burks in his completion of von Neumann's work. [18]

In his Hixon Symposium von Neumann presented his kinematic model and said that it was "feasible, and the principle on which it can be based is closely related to Turing's principle. . . ." He further discussed the fundamental notion of *complication* and surmised that automata whose complexity is below a certain level can only produce less complicated offspring, whereas those above a certain level can reproduce themselves "or even construct higher entities." It is interesting to conjecture what would have been the effect on von Neumann's work if he had known about DNA and RNA. He clearly stated in his writings and conversations on automata that the copying mechanism was performing "the fundamental act of reproduction, the duplication of the genetic material."

Somewhat after his Hixon lecture von Neumann shifted to a cellular model. According to Burks this was due to a conversation

---

[17] Von Neumann, "The General and Logical Theory of Automata," Collected Works, vol. v, pp. 288–328. This paper is a somewhat edited version of his Hixon Symposium lecture given at Pasadena on 20 September 1948.

[18] Von Neumann, *Theory of Self-Reproducing Automata*, ed. Burks. The kinematic model is discussed in Part I and the cellular one in Part II.

with Ulam, who persuaded him of the superiority of the cellular model.[19] He completed only a portion of that model and sent me a copy in 1952. In the covering letter he said:

This is the introduction – or "Chapter 1" – that I promised you. It is tentative and incomplete in the following respects particularly:

(1) It is mainly an introduction to "Chapter 2" which will deal with a model where every cell has about 30 states. It refers only very incompletely to "Chapter 3" in which a model with excitation – threshold – fatigue mechanisms alone will be discussed, and to "Chapter 4" where I hope to say something about a "continuous" rather than "crystalline" model. There, as far as I can now see, a system of non-linear partial differential equations, essentially of the diffusion type, will be used.[20]

In 1953 von Neumann delivered the Vanuxem Lectures at Princeton University – 2–5 March – in a series of four talks entitled "Machines and Organisms." At first he intended to have these lectures published by Princeton University Press, but later, under the pressure of his manifold activities and perhaps his failing health, he decided not to do this. Instead, John Kemeny of Dartmouth wrote an article for *Scientific American* more or less covering these lectures,[21] and a considerable portion of von Neumann's first three lectures appeared in another book by him.[22] Early in 1955 he was invited to deliver the Silliman Lectures at Yale University during the spring of 1956. However, on 15 March 1955 von Neumann was sworn in as one of the Atomic Energy Commissioners, and he and his wife moved to Georgetown in May. Then three months later – in August – catastrophe struck: "Johnny had developed severe pains in his left shoulder, and after surgery, bone cancer was diagnosed." [23]

He was unable to finish his manuscript. He had however asked Herbert S. Bailey, Jr., Director of Princeton University Press, to release him from his tentative obligation to write up his Vanuxem Lectures. This permission he of course received, and the resulting

[19] *Ibid.*, p. 94.

[20] Letter, von Neumann to Goldstine, 28 October 1952.

[21] Kemeny, "Man Viewed as a Machine," *Scientific American*, vol. 192 (1955), pp. 58–67.

[22] J. von Neumann, *The Computer and the Brain* (New Haven, 1958).

[23] *Ibid.*, Preface by Klara von Neumann.

unfinished manuscript represented what he wrote "in those days of uncertainty and waiting." [24]

As we have said, sometime in the early 1950s von Neumann was busy working on his cellular model.[25] At first he conceived of his device as three-dimensional and bought the largest box of "Tinker Toys" to be had. I recall with glee his putting together these pieces to build up his cells. He discussed this work with Bigelow and me, and we were able to indicate to him how the model could be achieved two-dimensionally. He thereupon gave his toys to Oskar Morgenstern's little boy Karl.

Von Neumann was somewhat later—I cannot date this very precisely, but it was before he became ill—rather diffident about his work on self-reproducing automata and talked to me at great length about its importance. His confidence in it had been shaken by a conversation, and strangely he needed reassurance. This was very atypical of him. But he had thought so much on the subject that he seemed to feel all his work was self-evident!

It is worth saying just a brief word about von Neumann's reference to the use of non-linear partial differential equations to study automata theory. He planned evidently to build a continuous model of a self-reproducing automaton and to study it by means of differential equation theory. This would have been a most remarkable achievement, since much of the difficulty was with mathematical logics. Von Neumann said on this point: "The reason for this is that it deals with rigid, all-or-none concepts, and has very little contact with the continuous concept of the real or of the complex number, that is, with mathematical analysis. Yet analysis is the technically most successful and best-elaborated part of mathematics. Thus formal logic is, by the nature of its approach, cut off from the best cultivated portions of mathematics, and forced onto the most difficult part of the mathematical terrain into combinatorics." [26] He went on in this 1948 Hixon lecture to say: "All of this re-emphasizes the conclusion . . . that a detailed, highly mathematical, and more specifically analytical, theory of automata and of information is needed." [27]

[24] I am indebted to Mr. Bailey for this information.

[25] See *Theory of Self-Reproducing Automata,* p. 94. Mrs. von Neumann dates this manuscript as being started in late September 1952. I think it must have been started earlier, since it was in typed form by the end of October and because in his letter of 28 October 1952 he said he was sending the chapter he had promised me.

[26] Von Neumann, "General and Logical Theory of Automata," Collected Works, vol. v, p. 303.

[27] *Ibid.,* p. 304.

Had he been able to bring the power of analysis to bear on formal logics and automata theory, von Neumann's results would certainly have been of greatest interest. In particular, Burks conjectured that he thought of differential equations in respect to his excitation –threshold–fatigue model. The problem, then, is largely concerned with the behavior of neurons when stimulated. This brings us very close indeed to the brilliant work of Alan L. Hodgkin and Andrew F. Huxley for which they received the Nobel Prize for medicine in 1963.[28] They described the behavior of nerve fibers by means of a non-linear partial differential equation.

Von Neumann's work on automata formed out of unreliable parts was an outgrowth, in part, of his interest in the Air Force's problem of reliability of its missiles, and in part of his observations on how the human nervous system allows us to work quite reliably even though individual neurons may be unreliable. He was on the Scientific Advisory Board of the Air Force from 1951 on and was struck with the need for highest reliability of functioning of missiles which had however lives of only a few minutes. They were, and perhaps still are, made of the highest precision parts obtainable and were assembled like fine Swiss watches; yet they need to function for only a very short time. This was not the sole reason for his interest, however, since already in his 1948 Hixon lecture he was mentioning the fact that the components out of which automata are built should be viewed as having a non-zero probability of failure. His thinking on the problem was then still very much in the formative stage, and possibly it was the Air Force problem which rekindled his interest.

In the discussion after his Hixon paper von Neumann illustrated a very simple way to guarantee the correctness of a calculation. He set the problem thus: Suppose one has a machine with a probability of 1 error in 10^{10} operations and that one wishes to solve a problem requiring 10^{12} operations. One would therefore expect about 100 errors in the course of the calculations. Suppose one connects together three such machines so that they compare results after each operation. If all three agree, they go ahead; if any two agree, the third is set to the agreed value and they go ahead; and if no two agree, they all stop.

Now the resultant system will yield the correct result unless at any point two of the three machines err simultaneously. The proba-

[28] A. L. Hodgkin, *The Conduction of the Nerve Impulse*, The Shennington Lectures VII (Springfield, Illinois, 1964).

bility of this is given by the sum of the probabilities that any two fail simultaneously. This is $3 \times 10^{-10} \times 10^{-10} = 3 \times 10^{-20}$ since there are three possible pairs of machines. Thus the probability that an error will occur in the given problem is $10^{12} \times 3 \times 10^{-20} = 3 \times 10^{-8}$, which is about one chance in 33 million! [29]

His thoughts on the organization of the central nervous system suggested to him that redundancy was the clue to the correct way to organize automata made with unreliable components, and the majority voting procedure just outlined is at least one way to consider. In a series of five lectures at the California Institute of Technology in 1952 he gave two models for organizing an arbitrarily reliable automaton out of unreliable components.[30] In the lectures he points out his inherent dissatisfaction with his procedure and his feeling that "error should be treated by thermodynamical methods, and be the subject of a thermodynamical theory, as information has been, by the work of L. Szilard and C. E. Shannon. . . . The present treatment falls far short of achieving this, but it assembles, it is hoped, some of the building materials, which will have to enter into the final structure." [31]

Perhaps we should spend a moment showing a few details of his scheme, since it has considerable interest even today. To make clear what he did, let us consider a basic organ as shown in Fig. 16.

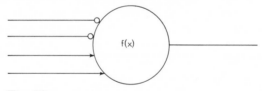

$f(x)$

Fig. 16

This device, a neuron-like one, has for illustrative purposes two inhibitory inputs and two excitatory ones. The former are shown with small circles at their ends and the latter with arrow heads. The device acts in this way: the output of the neuron is excited if

[29] Von Neumann, "General and Logical Theory," p. 323.

[30] "Probabilistic Logics and the Synthesis of Reliable Organisms from Unreliable Components," Collected Works, vol. v, pp. 329–378. In the introduction von Neumann acknowledged his indebtedness to K. A. Brueckner and M. Gell-Mann for conversations on this subject in 1951 at the University of Illinois, and to R. S. Pierce, who wrote up his lecture notes, which von Neumann then edited. (It should be clear to the reader that von Neumann often disliked writing up his work and tried artfully to avoid doing so.)

[31] *Ibid.*, p. 329.

and only if the number of stimulated excitatory inputs k satisfy the relation $k \geqslant f(l)$, where l is the number of stimulated inhibitory inputs and

$$f(x) = f_h(x) = x + h.$$

It is then a neuron which will be excited if and only if the excess of stimulations over inhibitions is at least h. While it is not too important for our exposition, we should mention the fact that there are time delays in neurons. The response of such an organ occurs one time unit after its stimulation. It is convenient to display only the number h inside the large circle. Thus in the three examples in Fig. 17 we have: a threshold 2 neuron with two excitatory inputs

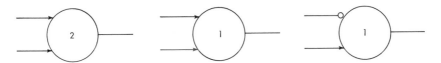

Fig. 17

and two threshold 1 neurons, one with two excitatory inputs, and one with one excitatory and one inhibitory input. According to our definition the first one is excited if and only if both inputs are; the second if either one is; and the third if the inhibitory one is not and the excitatory one is.

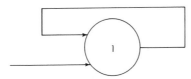

Fig. 18

Now consider the simple circuit in Fig. 18. It is stimulated if either input is. Thus if it receives a stimulus on its free input at time t it will give out a response at $t + 1$ and this will in turn stimulate its other input at $t + 1$. Clearly this will always stay stimulated. Such a device von Neumann thought of as a battery and symbolized it as $|\mathrm{i}||$.

He then introduced two organs: the well-known "Sheffer stroke" and another one he called a majority organ. The former is a threshold 1 neuron with two inhibitory and two excitatory inputs. The latter inputs are always stimulated, as we see from Fig. 19, and the former are labelled a and b. The device will be stimulated unless

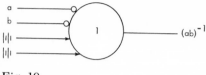

Fig. 19

both *a* and *b* are not. Now it is usual to write *a and b* in Boolean algebra as *ab* and *not-c* as c^{-1}. Thus the Sheffer stroke has as its output $(ab)^{-1}$ as indicated on the figure. It is perhaps of some interest to note that out of this kind of neuron one can achieve the three basic operations of Boolean algebra—*and, or,* and *not.* We indicate how in Figs. 20–22 below. They help to show how neural networks and automata can be achieved. To simplify the drawing let the Sheffer stroke be drawn as

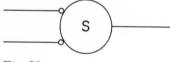

Fig. 20

Then the three operations are shown in the following simple neural networks:

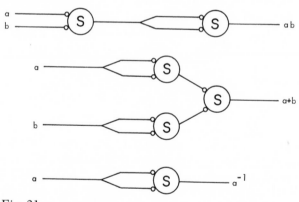

Fig. 21

The first of the drawings in Fig. 21 represents *and.* To see this, note that the output of the first Sheffer neuron in it is $(ab)^{-1}$ and this is now applied to both inputs of the second one. Thus the result is

$$[(ab)^{-1}(ab)^{-1}]^{-1} = [(ab)^{-1}]^{-1} = ab$$

since, as we recall from Boole's work, $x^2 = x$ and further that $(x^{-1})^{-1}$ $= x$. The second drawing represents *or*. Here the upper neuron's output is a^{-1} and the lower one's is b^{-1}; the final neuron then has as its inputs a^{-1} and b^{-1}. Its output is then $[(a^{-1})(b^{-1})]^{-1}$. A little reflection – perhaps the construction of a truth table – should satisfy the reader that this is $a + b$. (Indeed, $(xy)^{-1}$ is $x^{-1} + y^{-1}$.) The third drawing in Fig. 21 clearly depicts *not*.

Von Neumann's majority organ was a device for deciding whether at least two out of three inputs agree. He was to use it at many places in his automaton to decide what the majority had decided. Fig. 22 gives a schematic representation of a majority

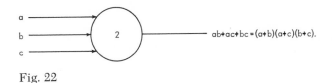

Fig. 22

organ. Von Neumann was concerned with the problem of how to design his automaton so that even if each neuron in it had a finite probability of error, nonetheless the response of the final outputs would be arbitrarily reliable. That is, could it be so designed that it would perform with a probability of error in its final result less than any preassigned number? He answered this affirmatively by two quite different constructs. In both cases his solutions require the use of redundancy. In this sense they mirror the central nervous system's design.

In one construct he triplicated his network O and then fed their outputs into two majority neurons. In Fig. 23 it is assumed for

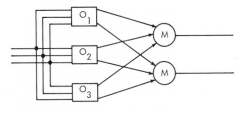

Fig. 23

simplicity that the network O has three inputs and two outputs. The networks O_1, O_2, O_3 are "carbon copies" of O, and the circles with Ms in them are majority organs.

Let us now make a heuristic evaluation of the resulting network. It can be shown that

$$\zeta \equiv \epsilon + (1 - 2\epsilon)(3\eta^2 - 2\eta^3)$$

is an upper bound for the probability of malfunctioning of the triplicated system O^* above where η is an upper bound for the probability of error of O itself.

Notice that the probabilities of errors on the input lines to either majority organ are independent, and that when the system is operating these inputs are all in the same excitation state—either all or none stimulated. Then an upper bound for at least two of the input lines carrying the wrong information is

$$\xi = 3\eta^2 - 2\eta^3$$

and therefore

$$(1 - \epsilon)\xi + \epsilon(1 - \xi) = \epsilon + (1 - 2\epsilon)(3\eta^2 - 2\eta^3)$$

is a safe estimate for the probability of failure in an output line. Now it is not hard to show that with $\epsilon < \frac{1}{6}$ and $\eta \sim \epsilon + 3\epsilon^2$ the expression ζ above can be made smaller than η. This means that the probability of the system O^* malfunctioning is smaller than that of O. Thus things become more reliable, and iterations of the procedure will result in increased reliability.

Von Neumann's other trick was to replace single lines, either input or output, by bundles containing large numbers N of lines. If $\Delta < \frac{1}{2}$, then he decided that the stimulation of at least $(1 - \Delta)N$ lines meant the bundle was stimulated whereas the stimulation of at most ΔN meant it was not. He then created a majority organ for bundles and also a restoring one. This latter device restored the stimulation level of a bundle and therefore prevented successive degeneration of information.

We need not go into more details beyond noting that von Neumann was able to show that both his schemes were successful. Up till now at least, neither system has seen practical usage since the amount of extra equipment required is very great.

By 1956 the subject of automata theory had become quite popular and a number of excellent papers had appeared on the subject.[32]

[32] See P. Elias, "Computation in the Presence of Noise," *IBM Journal of Research and Development*, vol. 2 (1958), pp. 346–353. In particular note the references, including one to a paper by E. F. Moore and C. Shannon on reliable circuits made from less reliable relays. See also, *Automata Studies*, edited by J. McCarthy and C. Shannon, Annals of Math. Studies, No. 34 (Princeton, 1956).

The paper by Elias was a tentative first step in the application of information theory, or rather thermodynamical ideas, to the problem of analyzing computation with an unreliable computer. His work was followed by a profound analysis carried out by Winograd and Cowan, which surely would have been viewed by von Neumann as carrying out his program to the end.[33] In effect what they did was to adapt Shannon's ideas on information to computation.[34] To do this they succeeded in defining the notion of computational capacity for unreliable modules. Moreover, they showed that arbitrarily reliable networks may be built out of such modules using a module redundancy that "need only be greater than a certain minimum whose value is determined by the computation capacity of modules comprising these networks."[35] Their result depends upon certain, not too unreasonable, assumptions concerning the complexity of the modules and the independence of module malfunction on its complexity. The price they paid was that the complexity of the modules had to increase. They also extended their work to show that reliable machines can also be built, even though there are errors of interconnections between modules.

In thinking about von Neumann's contributions, I am of the opinion that he perhaps viewed his work on automata as his most important one, at least the most important one in his later life. It not only linked up his early interest in logics with his later work on neurophysiology and on computers but it had the potentiality for allowing him to make really profound contributions to all three fields through one apparatus. It will always be a fundamental loss to science that he could not have completed his program in automata theory, or at least pushed it far enough to make clear, for example, what his ideas were on a continuous model. He never was given to bragging or to staking out a claim where he didn't deserve to be; I am therefore confident he had at least a heuristic insight into this model and perhaps some idea as to how it would interact on logics and neurophysiology.

Finally, it is interesting to note that von Neumann worked on his theory of automata alone. This was in rather sharp distinction to most of his later work where his practice was almost always to work with a colleague. Very possibly he wanted his automata work to stand as a monument to himself—as indeed it does.

[33] S. Winograd and J. D. Cowan, *Reliable Computation in the Presence of Noise* (Cambridge, Mass., 1963).

[34] C. E. Shannon and W. Weaver, *The Mathematical Theory of Communication* (Urbana, 1949).

[35] Winograd and Cowan, p. 3.

The branch of mathematics concerned with numerical calculation is very old, going back at least to the day of Archimedes who had, along with so much else, a procedure for finding upper and lower bounds for square roots of both small and large numbers.[1] Many of the great names of mathematics, such as Newton and Euler, are attached to procedures for carrying out various numerical tasks. One of the greatest of these is the Prince of Mathematicians, Carl Friedrich Gauss (1777–1855). Bell says of him: "Archimedes, Newton, and Gauss, these three, are in a class by themselves among the great mathematicians, and it is not for ordinary mortals to attempt to range them in order of merit." [2]

This is not the place to describe even a few of Gauss's fabulous accomplishments beyond noting that he was one of the greatest calculators of all time. For nearly twenty years he occupied much of his time with astronomical calculations. One of the very big tasks he undertook was to calculate the orbit of the minor planet Ceres and to work out a general procedure for determining the orbit of any celestial body given three observations of its position. Ceres was first discovered in 1801 but was soon lost. Gauss, then a young man, was attracted by the problem and developed his method, which he applied to Ceres. The lost planet was then rediscovered where he predicted it would be.[3] Partly as a result of this brilliant work he was appointed director of the Göttingen observatory in 1807. Bell says rather grudgingly that as a result of Gauss's method what would have taken Euler three days to calculate could then be done in a few hours.

Gauss made many great discoveries in mathematics, electricity — he invented the electric telegraph in 1833 — physics, and astronomy.

[1] D. C. Gazis and R. Herman, "Square Roots Geometry and Archimedes," *Scripta Mathematica*, vol. 25 (1960), pp. 229–241.

[2] Bell, *Men of Mathematics*, pp. 218–269.

[3] See F. R. Moulton, *An Introduction to Celestial Mechanics* (London, 1931), pp. 191–260. Moulton says of Gauss's contribution: "This work, written by a man at once a master of mathematics and highly skilled as a computer, is so filled with valuable ideas and is so exhaustive that it remains a classic treatise on the subject to this day."

In the calculational realm he discovered the method of least squares for finding the most probable value from a set of independent measurements of a quantity,[4] of Gaussian elimination for solving systems of linear equations (above, p. 125), of numerical integration for finding an approximation to the value of an integral, and many others as well. As Bell says, "He lives everywhere in mathematics."

The state of numerical mathematics stayed pretty much the same as Gauss left it until World War II, with one fundamental exception. In 1928–29 a paper appeared by Courant, Friedrichs, and Lewy that was an essential step forward.[5] Prior to their work it was really not possible to solve numerically the partial differential equations that enter into heat or fluid flow problems except fortuitously.

When a differential equation is "solved numerically" what actually happens is that a related approximate problem is solved. It is in general not possible to solve numerically most problems in mathematical analysis with exactitude. This is so because the operations of differentiation and its inverse, integration, are transcendental for digital machines. Thus, given any problem involving such operations, one must replace them by approximations in terms of operations that are basic to digital devices.

These things have been known since at least Newton's time. Fortunately, the types of problems met with in pre-World War II applied mathematics were usually of a sort for which this need to approximate caused little difficulty. This was due to two things: on the one hand, most problems involved so-called total differential equations, and, on the other, not many multiplications could be performed by human computers—we shall see why both points are relevant in a few paragraphs.

With the sole exception of the Courant, Friedrichs, and Lewy paper the pre-World War II literature on approximation mathematics is devoted to discussions and analyses of the errors of approximation or truncation. That is, this literature is concerned with the accuracy—the size of the error—of the various approximation formulas for evaluating definite integrals, step-wise methods for integrating total differential equations, etc.[6]

[4] Legendre also discovered this method and felt Gauss had appropriated his ideas (Bell, p. 259).

[5] R. Courant, K. Friedrichs, and H. Lewy, "Über die partiellen Differenzengeichungen der Matematischen Physik," *Math. Ann.*, vol. 100 (1928–29), pp. 32–74.

[6] Whittaker and Robinson, *The Calculus of Observations* (London, 1937), gives a good picture of the prewar situation.

While all this suffices for total differential equations and many other types of problems, the *numerical* solution of certain types of partial differential equations gives rise to a totally new phenomenon discovered by Courant and his colleagues. It is precisely these types of equations which are all-important in our present era. Typically, for example, the problem of the numerical forecasting of the weather depends on such equations as do studies of supersonic flows of fluids. In the famous 1928–29 paper cited above the authors showed that even though one replaced the original differential equation by a set of numerical approximants that were arbitrarily close (in a well understood sense) to the given equation, the resulting solution might not have any relationship to the true one.

This result was totally astonishing since the notion of numerical — as contrasted to physical — stability was not at all understood. Courant, Friedrichs, and Lewy analyzed a step-wise finite approximant to the partial differential equations expressing compressible, non-viscous flow in which the continuous time variable was replaced by a succession of discrete time intervals, Δt and the space variable by a similar succession of intervals, Δx. What they showed was that the solution of the finitistic problem would be relevant to the true one provided that a sound wave could not travel more than a distance Δx in Δt seconds.

This condition has become known as the "Courant condition" and is central to a study of numerical stability. It made possible modern numerical analysis. The importance of the concept to this field was understood by von Neumann at once, and the ideas were somehow filed away in his memory. Then in the 1940s, when he began to push his Los Alamos colleagues into numerical calculations, he formulated what he liked to call the Courant "criterium" in a new way that was highly useful for practical calculation. He did this in a series of unpublished lectures given at Los Alamos in February 1947. His procedure was heuristic as he himself knew but very powerful. It generalized the Courant-Friedrichs-Lewy work to more complicated situations, and it provided an excellent machinery for testing of numerical stability.[7] It has since been rigorously justified.

Early in World War II von Neumann started a small project at the Institute to explore various topics in hydrodynamics from a numerical point of view. In this he had as most able collaborators Pro-

[7] Von Neumann, Collected Works, vol. v, p. 664.

fessors Valentine Bargmann, now of Princeton University, and Deane Montgomery of the Institute for Advanced Study. This collaboration resulted in a number of reports on shock waves issued by Warren Weaver's Applied Mathematics Panel of the National Defense Research Committee and by the Bureau of Ordnance of the U.S. Navy Department in the period from 1942 through 1945.[8] Then in the fall of 1946, Bargmann, Montgomery, and von Neumann wrote a very significant paper in which they started to grapple with an absolutely fundamental problem in numerical analysis: how best to solve a large system of linear equations. This problem for which Gauss devised a method plays a key role in virtually all phases of numerical work.[9]

In this paper, perhaps for the first time, we find a preliminary analysis and some discussion of numerical stability. The authors say that a "method of computation is called stable if the rounding errors, which are inevitable in any one stage, do not tend to accumulate in a serious way." The point about the accumulation of errors may be explained as follows.

In a numerical calculation each multiplication or division is inexact inasmuch as the product of two n digit numbers is a $2n$ digit number, and the quotient of such numbers is in general an infinitely many digit number. Now calculations are normally done with a certain fixed number n of digits, so either humans or machines normally "round off" their products or quotients to this many digits. In a sense these are errors not malfunctions, since they create results which differ from the correct ones. After a lengthy calculation involving many multiplications and divisions, it is clear many "errors" have been committed, and it is pertinent to ask whether they have accumulated in such a way that the result of the calculation is in doubt.

This phenomenon occurs in all manner of numerical work even when there are no approximate formulas used. An excellent example is given by the problem of solving a large system of linear equations. This problem involves only a finite number of additions, subtractions, multiplications, and divisions. Thus every operation is elementary to a digital computer, and yet the practical solution of such a problem is not at all easy. Indeed, Hotelling, a first-class statistician undertook a heuristic analysis that seemed to show

[8] Von Neumann, Collected Works, vol. VI, pp. 178–379.

[9] V. Bargmann, D. Montgomery, and J. von Neumann, "Solution of Linear Systems of High Order," 25 October 1946. This was a report prepared for the Bureau of Ordnance, Navy Department and appears in von Neumann's Collected Works, vol. V, pp. 421–477.

Gauss's method for solving systems of linear equations by elimination was unstable. If n is the number of equations, he estimated the
error in the final result may be as large as 4^n times the error in a
single "round off." [10] If this result were true, it would mean that to
achieve five-digit accuracy in solving a linear system of 100 equations in as many unknowns, one would have to carry about $100 \times$
$\log_{10} 4 + 5$ or approximately 65 digits throughout the calculation!

Unfortunately Bargmann, Montgomery, and von Neumann accepted tentatively this very pessimistic conjecture of Hotelling
and turned away from the Gaussian procedure to iterative ones.
Von Neumann was however not at all happy about the situation
regarding Gauss's procedure. He and I discussed this many times,
and we did not feel it reasonable that so skilled a computer as
Gauss would have fallen into the trap that Hotelling thought he
had noted.

By the late fall of 1946 or very early in 1947 von Neumann and I
finally understood how matters really stood. Von Neumann remarked one day that even though errors may build up during one
part of the computation, it was only relevant to ask how effective
is the numerically obtained solution, not how close were some of
the auxiliary numbers, calculated on the way to their correct
counterparts. We sensed that at least for positive definite matrices
the Gaussian procedure could be shown to be quite stable.

With this as our goal, we rapidly satisfied ourselves in a heuristic
fashion that for such matrices this was true. It remained then to
produce a completely rigorous analysis and discussion of the
procedure. It seemed to me very important at this time to start
modern numerical analysis off on the right foot, so I drafted a paper
which went in depth into errors, numerical stability, and pseudo-
operations; and then I embarked on the main problem. In mid-
1947 I was informing von Neumann, "I am writing the final sections
of the matrix inversion paper and rewriting others I wrote last
week." [11] Von Neumann agreed with what I had written, and after
several iterations a paper eventuated in September 1947.[12] Wilkinson has said of this paper that it "may be said to have laid the foun-

[10] H. Hotelling, "Some New Methods in Matrix Calculation," *Ann. Math. Statist.*,
vol. 14 (1943), pp. 1–34.

[11] Letter, Goldstine to von Neumann, 19 July 1947. In this letter I gave the final
error estimates as well as some other analysis which we did not use. There is also
an undated letter from me to him in regard to statistical estimates and some analysis
of how we might proceed.

[12] J. von Neumann and H. H. Goldstine, "Numerical Inverting of Matrices of
High Order," *Bull. Amer. Math. Soc.*, vol. 53 (1947), pp. 1021–1099. See also, von
Neumann, Collected Works, vol. V, pp. 479–572.

dation of modern error analysis in its rigorous treatment of the inversion of a positive definite matrix by elimination. It is however a fairly difficult paper to read. . . ." [13]

Around the same time Turing noted the practical successes of his colleagues Fox, Huskey, and Wilkinson in inverting large matrices.[14] This inspired him to investigate why this was so. When he visited us in January 1947, we discussed our ideas with him. He persevered with his ideas and produced results in accord with ours.[15]

The significant fact shown by Turing and by us was that the total error in finding the inverse of a positive definite matrix A was proportional to the ratio of the largest to the smallest eigenvalue and to the square of the order n of A. This made clear where the difficulty lay since the ratio of those two eigenvalues in effect measured how nearly parallel the hyperplanes represented by the linear equations were. Thus Gauss's method was shown to be very good indeed provided the original problem was not "ill-conditioned"; in other words, the procedure was stable.

The error estimates we obtained were however upper bounds, assuming in every case that the worst possible difficulties occurred. It seemed desirable to ask what the situation would be on the average. That is, in a busy computation center where large numbers of matrices arising in many different situations were processed, would the error bounds then be lower? Accordingly, I started such an analysis shortly after the previously cited work was completed. It turned out however to be very hard for us, and both von Neumann and I had difficulties in carrying through the analysis. We finally succeeded late in 1949.[16] To our chagrin, the year after our paper was published an improved and easier analysis was published by Mulholland.[17] However the essence of both papers was that under reasonable probabilistic assumptions the error estimates of the previous paper could be reduced from a proportionality of n^2 to n, the order of the matrix. (Parenthetically, it is amusing to

[13] J. H. Wilkinson, *Rounding Errors in Algebraic Processes* (Englewood Cliffs, N.J., 1963).

[14] L. Fox, H. D. Huskey, and J. H. Wilkinson, "Notes on the Solution of Algebraic Linear Simultaneous Equations," *Quart. Jour. Mech. and Appl. Math.*, vol. 1 (1948), pp. 149–173.

[15] A. M. Turing, "Rounding-Off Errors in Matrix Processes," *Quart. Jour. Mech. and Appl. Math.*, vol. 1 (1948), pp. 287–308.

[16] H. H. Goldstine and J. von Neumann, "Numerical Inverting of Matrices of High Order, II," *Proc. Amer. Math. Soc.*, vol. 2 (1951), pp. 188–202.

[17] H. P. Mulholland, "On the Distribution of a Convex Even Function of Several Independent Rounding-Off Errors," *Proc. Amer. Math. Soc.*, vol. 3 (1952), pp. 310–321.

remark that we were so involved with matrix inversion that we probably talked of nothing else for months. Just in this period Mrs. von Neumann acquired a big, rather wild but gentle Irish Setter puppy, which she called Inverse in honor of our work!)

The extensive activity in numerical analysis stimulated considerable interest among many mathematicians, and the list of those who were either continuously or occasionally connected with the mathematical group of the Institute's computer project is large. It included: Valentine and Sonia Bargmann, Nils A. Barricelli, James H. Bartlett, Eduard Batschelet, Salomon Bochner, Andrew and Kathleen Britten Booth, Arthur W. Burks, Walter Elsasser, Foster and Cerda Evans, Carl-Erik Fröberg, Werner Gautschi, Donald Gillies, Joseph Gillis, Gilbert A. Hunt, Toichiro Kinoshita, Manfred Kochen, Margaret Lambe, Richard A. Leibler, Edward J. McShane, Hans Maehly, Benoit Mandelbrot, John Mayberry, Nicholas Metropolis, Deane Montgomery, Francis J. Murray, Maxwell H. A. Newman, Leslie G. Peck, Chaim L. Pekeris, Robert D. Richtmyer, J. Paul Roth, Hedi Selberg, Ernest S. Selmer, Seymour Sherman, Daniel Slotnick, Charles V. L. Smith, Jean Spitzer, Abraham H. Taub, John and Olga Taussky Todd, Elenore Trefftz, Bryant Tuckerman, Max Woodbury, and William Woodbury.[18] There were probably others as well, but I have forgotten their names and duly apologize for my oversights. Many publications ensued from all these people's efforts, which served to show the academic, governmental, and industrial worlds the power of the electronic computer. Their work was sponsored by the Office of Naval Research, and for a long time was under the supervision of Mina Rees (above, p. 211) and later Joachim Weyl and, under their direction, a number of very fine people including C. V. L. Smith, who spent a sabbatical year at the Institute writing an excellent book on computers.[19]

This very energetic mathematical program at the Institute for Advanced Study played an extremely important role in the establishment of a fine and vigorous intellectual tone for the subject of numerical analysis. The groups in engineering, logical design, and meteorology also served to lead the world in these fields. The Institute fulfilled in its Electronic Computer Project its mission

[18] *The Institute for Advanced Study, Publications of Members, 1930–1954* (Princeton, 1955).

[19] C. V. L. Smith, *Electronic Digital Computers* (New York, 1959).

for encouraging, supporting, and patronizing learning – "of science in the old, broad, undifferentiated sense of the word." [20]

A few at least of the topics pursued should be mentioned in some detail. Von Neumann had been interested, probably after a very fruitful discussion with Ulam, in the statistical properties of sequences of numbers. He therefore persuaded George Reit-wiesner of the Ballistic Research Laboratory to undertake what was in those days an extensive calculation of the digits of e and π on the ENIAC. [21] Later he, Metropolis, and Reitwiesner did an interesting statistical analysis to see if there were any significant deviations from randomness in π and e. They detected none for the former but serious ones for the latter. Von Neumann also collaborated with Tuckerman on a similar study of the digits of $2^{1/3}$ on the Institute computer. They carried out a continued fraction expansion of this number and formed over 200 partial quotients. [22]

The mathematical group attempted – I believe quite successfully – to preach the doctrine that the classical methods of numerical analysis were inadequate for the new machines, and that there was need to develop new ones suited to their inner economies. To understand this, we should realize that in the days before the electronic calculators the economy of computation was characterized by cheapness of storage – pads of paper or decks of punch cards – and expensiveness of multiplication (and division). That is, the human computer or the human using punch card equipment usually was happy to avoid multiplying and preferred to record voluminous amounts of intermediate data. For this reason most books on numerical analysis of the pre-World War II era contain large numbers of algorithms in which differencing techniques replace multiplying. This was in the spirit of Babbage's Difference Engine.

On the other hand, the electronic computer had a quite different economy. Storage was expensive – 20 words in the ENIAC and 1,000 in the Institute machine – but multiplication was cheap – 300 per second in the ENIAC and around 1,300 for the Institute machine. This reversal of the computing economy of machines meant that

[20] *The Institute for Advanced Study, Publications,* Preface by Robert Oppenheimer.

[21] G. Reitwiesner, "An ENIAC Determination of π and e to more than 2000 Decimal Places," *Math. Tables and Other Aids to Computation,* vol. 4 (1950), pp. 11–15.

[22] J. von Neumann and B. Tuckerman, "Continued Fraction Expansion of $2^{1/3}$," *ibid.,* vol. 9 (1955), pp. 23–24.

new algorithms were needed which emphasized multiplication and played down large-scale storage of intermediate data.

One other point was crucial and needs to be pointed out once again. The early pre-war computations involved so few multiplications that numerical instability was generally unnoticed. The enormous speeds of the post-war machines meant that in a typical problem many millions of multiplications would be performed. This in turn meant that rounding errors could potentially accumulate in a disastrous way.

It was thus the goal of the mathematical group to nurture new computational techniques well suited to the new machines. As an example of this, when we completed the papers on matrix inversion we began a collaboration with Professor Francis J. Murray, then of Columbia University and now of Duke University, to discover a stable method for finding the fundamental invariants of conic sections in n-dimensional space when they are subjected to rotations of the coordinate axes. We discovered such a procedure and worked out in detail an error analysis for it, which I presented in August 1951 at a symposium on simultaneous linear equations and the determination of eigenvalues. This meeting was sponsored by the National Bureau of Standards at the University of California at Los Angeles. After the paper had been presented, Professor Alexander M. Ostrowski of Basel arose and in a kindly way pointed out the fact that we had rediscovered a hundred-year-old algorithm of Jacobi.[23]

This procedure became the accepted procedure for finding eigenvalues of symmetric matrices until Givens in 1954 found a simpler method and Householder in 1958 found a still better one.[24]

All through this period von Neumann's interest in hydrodynamics remained very high, both on the theoretical and the calculational levels. He continued to lead the Los Alamos people into ever more computations on a variety of machines, including several on the ENIAC, one on the IBM Selective Sequence Electronic Calculator (SSEC, below, p. 327), and several on the Institute's computer. Almost all the time there were members of the Los Alamos Laboratory

---

[23] C. G. J. Jacobi, "Über ein leichtes Verfahren, die in der Theorie der Säcular-storungen vorkommenden Gleichungen numerisch aufzulösen," *J. reine und angewandte Math.*, vol. 30 (1846), pp. 51–95.

[24] See J. H. Wilkinson, *The Algebraic Eigenvalue Problem* (Oxford, 1965). He says in this excellent book, on p. 343, that "The present interest in Jacobi's method dates mainly from its rediscovery in 1949 by Goldstine, Murray, and von Neumann, though it was used on desk computers at the National Physical Laboratory in 1947." Our paper did not appear until 1959 under the title "The Jacobi Method for Real Symmetric Matrices," *J. Assoc. Comp. Mach.*, vol. 6 (1959), pp. 59–96. It is in volume V of von Neumann's Collected Works, pp. 573–610.

working at the Institute under von Neumann's aegis. In volume VI of his Collected Works there are well over 250 pages constituting 14 papers devoted to work on hydrodynamics. From our parochial point of view two are relevant. In calculations involving the flow of compressible fluids shock waves generally arise. These are surfaces across which such quantities as pressure, density, temperature, velocity, etc., exhibit discontinuities, and they present great complications numerically. Von Neumann and Richtmyer devised an ingenious scheme for automatically treating such discontinuities by artificially introducing a dissipative mechanism into the fluid flow equations.[25] In the other paper von Neumann and I undertook the numerical analysis and calculation of the behavior of a blast wave proceeding spherically from a very powerful point-source explosion in an ideal gas. Our aim was to develop an explicit treatment for shocks; in this sense it was antithetical to the previously cited work.[26]

The calculations done by Los Alamos were all inspired by suggestions of von Neumann's and in all cases with large measures of his superabundant wisdom and detailed help. In several of these calculations Klara—or Klari as we all called her—von Neumann, his second wife, was the programmer who prepared substantial parts of the codes. This work seemed to give her considerable satisfaction and made her summers at Los Alamos with him very satisfying and fruitful for her. The physicists working on these problems were Metropolis and Foster and Cerda Evans.

The calculation on the SSEC was undertaken by Richtmyer and my wife, who also helped on at least one of the ENIAC calculations for the AEC. The one done on the SSEC was an extremely large one. The planning started in October of 1949, and the actual running was not completed until the end of January 1951, or possibly later.[27]

Perhaps it is worth mentioning a few other pieces of work to indicate the breadth of the numerical analysis that was undertaken at the Institute. One of the most important projects was the development of the now well-known Monte Carlo method, which was apparently first suggested by Ulam to von Neumann.[28] The essence

[25] J. von Neumann and R. D. Richtmyer, "A Method for the Numerical Calculation of Hydrodynamic Shocks," *Journal of Applied Physics,* vol. 21 (1950), pp. 232–237; Collected Works, vol. VI, pp. 380–385.

[26] H. H. Goldstine and J. von Neumann, "Blast Wave Calculation," *Communications of Pure and Appl. Math.,* vol. 8 (1955), pp. 327–353; Collected Works, vol. VI, pp. 386–412.

[27] File, A. K. Goldstine, Los Alamos bills.

[28] Letter, von Neumann to Richtmyer, 11 March 1947; Collected Works, vol. V, pp. 751–762.

of the idea is quite elegant, and it became possible only with the advent of the electronic computer.

What is this Monte Carlo method? Very roughly, the idea is to replace a given precise mathematical procedure by one involving random processes. There are two broad classes of Monte Carlo problems: probabilistic situations and deterministic ones. In the former case a Monte Carlo approach would be to find a set of random numbers distributed according to the probabilistic laws of the given problem and then to infer the solution of that problem from the behavior of the random numbers.

A nice illustration of how this is done is given by a problem worked out by the late Professor Sam Wilks of Princeton University for the Applied Mathematics Panel of NDRC. He and his group were confronted with the problem of determining how many bombs would need to be dropped to blast a safe corridor through an enemy mine field. Wilks solved the problem by assuming certain random behavior to the bombing. He then put random numbers in two hats, corresponding to the behavior of the bombing across and along a strip representing the mine field. One number was drawn from each hat, and the pair represented the coordinates of a hypothetical crater. He then assumed all craters were ellipses of fixed size and orientation. He played this game many times, keeping track, of course, of how many bombs needed to be dropped in each game to make a connected path through the field.

Another example was the one used by von Neumann for studying neutron diffusion and multiplication. For such a study he chose random numbers to represent the random actions of the neutrons. (If a neutron is produced isotropically, then its direction of motion at birth is equidistributed; further the fate of the neutron has to be decided by other random processes to account for phenomena such as fission, absorption, or scattering.) This, like the bombing example, is a simulation of a highly complex probabilistic device by a simpler probabilistic model. The model can then, of course, be experimented with at will by the designer without the need to build real devices and test them at great costs.

Some deterministic processes also lend themselves to the Monte Carlo method. Suppose we desire to find the volume of a portion of n-dimensional space whose boundary consists of various surfaces. One way to do this is to write down, according to the usual rules of the calculus, a multiple integral and then evaluate this either analytically or numerically. Normally the latter procedure requires systematical divisions of the volume into small cells. The Monte

Carlo approach is instead to select from uniformly distributed random numbers the coordinates of points in the total space. Then the proportion of those inside the volume to the total number measures approximately the volume. By doing this probabilistic experiment enough times we can approach the correct volume with arbitrarily high probability. This Monte Carlo method has been widely used in many fields with considerable success, and there are many papers and books on the subject today.[29]

We shall close our brief discussion of the work of Ulam and von Neumann on the Monte Carlo method by quoting a lovely remark von Neumann made when speaking at a symposium on the subject. He said on that occasion: "Any one who considers arithmetical methods of producing random digits is, of course, in a state of sin. For, as has been pointed out several times, there is no such thing as a random number—there are only methods to produce random numbers, and a strict arithmetic procedure of course is not such a method."[30]

We had a conviction back in the 1940s that the computer would play a fundamental role in gaining heuristic insights into areas of both pure and applied mathematics which were stalemated from the lack of hints how to proceed. "We can now make computing so much more efficient, fast, and flexible that it should be possible to use the new computers to supply the needed heuristic hints. This should ultimately lead to important analytical insights."[31]

An instance of this sort of thing happened when Fermi reviewed the calculation by Richtmyer and my wife Adele. He noticed that the results were curiously insensitive to variations in one parameter of the problem. Accordingly, he fixed it at zero and saw that the resulting equations were so simplified he could solve them analytically. This obviated the need for continuing huge calculations and gave the Los Alamos physicists a much clearer insight into what was happening.

At this time there was a very great number theorist, Emil Artin, on the faculty of Princeton University. He became interested in a conjecture by a well-known nineteenth-century mathematician, Ernest E. Kummer, who played an important role in the develop-

[29] See, e.g., J. M. Hammersley and D. C. Handscomb, *Monte Carlo Methods* (London, 1964).

[30] J. von Neumann, "Various Techniques Used in Connection with Random Digits," *Journal of Research of the National Bureau of Standards*, Appl. Math. Series, vol. 3 (1951), pp. 36–38; Collected Works, vol. v, pp. 768–770.

[31] Goldstine and von Neumann in von Neumann's Collected Works, vol. v, p. 5.

ment of what are called ideals. If the conjecture could be proved true, it would have some important implications, and Artin wanted to find out if it was at least a plausible one. Kummer had himself done a numerical study which involved testing his conjecture on 45 primes. Artin persuaded us to examine the problem more extensively for him. We carried out the analysis for primes less than 10,000 and concluded that the conjecture was probably false.[32] This possibly saved Artin from the futile and back-breaking task of trying to prove a dubious theorem. Unfortunately, to date not very much has been done of this sort except in number theory. That is, not much exploration of special cases has been undertaken by mathematicians seeking hints as to theorems. Turing did a calculation at about the time we did the Kummer calculation on the Manchester computer and verified the famous Riemann conjecture in a certain region.[33] Emma and Derrick H. Lehmer also have done a number of very nice explorations of number theoretical conjectures.[34]

While it is not our intention to enumerate in detail the work of each member of the mathematics group, it is interesting to run rapidly through the calculations and analyses to get a flavor for what went on. To this end I have somewhat arbitrarily decided to mention the topics contained in a report to the Office of Naval Research in 1954.[35]

A large study of the difficulties inherent in the numerical integration of systems of differential equations was undertaken. This work centered around several systems: one was a study with Professor Martin Schwarzschild on the internal structure of stars, mainly red giants; another was a set of quite extensive tabulations of certain Bessel and Cylinder functions for Professor S. Chandrasekhar of Yerkes Observatory; a third was a study with Doctor

[32] J. von Neumann and H. H. Goldstine, "A Numerical Study of a Conjecture of Kummer," *Math. Tables and Other Aids to Computation,* vol. 7 (1953), pp. 133–134.

[33] A. M. Turing, "Some Calculations on the Riemann Zeta-function," *Proc. London Math. Soc.,* vol. 3 (1953), pp. 97–117.

[34] E.g., D. H. and E. Lehmer and H. S. Vandiver, "An Application of High-Speed Computing to Fermat's Last Theorem," *Proc. Nat. Acad. Science, USA,* vol. 40 (1954), pp. 25–33. This work was done on the National Bureau of Standards Computer SWAC in the Institute for Numerical Analysis at the University of California at Los Angeles. This Institute was founded at the Bureau by John H. Curtiss, now professor of mathematics at the University of Miami, who was for many years a leader in the Washington community in computers and computing.

[35] Report on Contract No. N-7-ONR-388, Task Order I, Institute for Advanced Study, April 1954. For more details the interested reader may wish to consult the previously cited book, *Institute for Advanced Study, Publication of Members, 1930–1954.*

Walter R. Beam on a travelling wave amplifier; and lastly, a study for Professor Milton White of the stability of particle orbits in what was to become the Princeton-Pennsylvania accelerator.

In each of these studies important numerical analyses were undertaken. Indeed, the problems were chosen to combine significant numerical studies with important physical ones. Thus in the first one various methods of numerical integration were evaluated; in the second these investigations were extended to equations containing singularities; and in the third the use of redundancies such as the constancy of total energy was exploited.

All this work helped to show many people in a variety of scientific disciplines the tremendous breadth of applicability of the electronic computer. It certainly was seminal and, I believe, it was vital in conditioning scientists to accept and to welcome the computer as a basic new tool both for the experimentalist and the theoretician. Thus in a very real sense the activities of the mathematical group at the Institute prepared a market for the great industrial concerns which were soon to enter the picture.

As we mentioned earlier, the aim of the Institute Electronic Computer Project was a threefold thrust into numerical mathematics, some important and large-scale application, and engineering. For the second effort von Neumann chose numerical meteorology. This choice was probably dictated by his profound understanding of hydrodynamics and by his desire to show the fundamental importance of the modern computer to our society. Moreover, he knew of the pioneering work during World War I in this field by Lewis F. Richardson, whom we shall discuss presently. Richardson failed largely because the Courant condition had not yet been discovered, and because high-speed computers did not then exist. But von Neumann knew of both.

A very substantial effort was mounted in this area, and a great field opened up. Today the subject is routinely taught worldwide in universities with meteorology departments; and the weather bureaus of virtually all great countries use techniques for their daily forecast inspired by what was done at that time. This accomplishment is another reason why von Neumann's name will long endure in science and why the Institute for Advanced Study is one of the great intellectual centers of the world.

Originally, von Neumann tried to bring Carl-Gustav Rossby (d. 1957) to the Institute on a long-term basis but only succeeded in bringing him as a visitor during the second term of the academic year 1950–51. Rossby was one of the leaders in meteorology as early as 1931 when he became professor of that subject at MIT, an assistant chief of the United States Weather Bureau, and later a distinguished service professor at the University of Chicago. Then he went back to Sweden as professor in Stockholm more or less until his death. During the war he was a consultant on meteorology to the Secretary of War and to General H. H. Arnold, the commanding general of the U.S. Air Force.

The leader, both intellectual and titular, of the meteorology project was Jule G. Charney, who is now a professor at MIT. He was at the Institute from 1948 to 1956 and was of the greatest importance in bringing about the accomplishments mentioned above. He formed a group of first-rate colleagues all of whom deserve mention: Arthur Bedient, Roy Bergren, Jacob Blackburn, Bert R. J. Bolin, James Cooley, George Cressman, E. T. Eady, Arnt Eliassen, Ragnar Fjörtoft, John Freeman, K. Gambo, Bruce Gilchrist, Ernst Hovmöller, Glenn Lewis, Adolph Nussbaum, Hans Panofsky, Norman A. Phillips, George W. Platzman, Paul Queney, Irving Rabinowitz, Frederick G. Shuman, Joseph Smagorinsky, A. L. Stickles, Philip D. Thompson, and George Veronis. Some of these men have gone on to other distinguished careers, but many are still leaders in numerical meteorology in the United States, Denmark, France, Great Britain, Norway, and Sweden, both in weather bureaus and in universities. The work carried on by this group was sponsored by the geophysics branch of the Office of Naval Research under Gordon Lill. Much of this work has appeared in numerous journal articles and some books.[1]

The fruits of this activity at the Institute were first harvested in July 1954 when the Joint Numerical Weather Prediction Unit came into being in Washington, D.C. under the joint staffing and support of the U.S. Weather Bureau, the Air Weather Service of the U.S. Air Force, and the Navy Weather Service. By May 1955 this unit was forecasting by numerical means on an operational, daily basis. Bedient, Blackburn, and Thompson were at the Institute under the sponsorships of the Air Force; Cressman, Shuman, and Smagorinsky, of the Weather Bureau; and Stickles, of the Navy. We should also mention Harry Wexler (d. 1962) who was head of meteorological research for the Weather Bureau for many years, and who played a key role in helping form this unit. He was also very helpful and influential during the formative days of the meteorology group.

There was close cooperation among the three groups making up the Electronic Computer Project, and this was quite significant. There were numerous mathematical problems underlying the equations of meteorology, and these were handled by close cooperation among the relevant groups with everyone "pitching in" who could.

Thompson's book is dedicated to Rossby and von Neumann. Of them he says: "Rossby, with rare insight into the essential dynamical processes of the atmosphere, provided a workable formulation of the problem, and von Neumann, with incomparable grasp of the theory of computing, developed the means for solving it."[2]

[1] E.g., P. D. Thompson, *Numerical Weather Analysis and Prediction* (New York, 1961).
[2] *Ibid.*, Preface.

The early beginnings of numerical meteorology can be traced quite directly to a remarkable Englishman, Lewis Fry Richardson (1881–1953). He was a Quaker who had studied under J. J. Thomson at Cambridge. After leaving there he worked in the government, university, and industrial research worlds of Great Britain. In his work he early came upon difficulties in solving differential equations analytically and was driven into numerical analysis. He then learned about finite difference techniques, and his great work on the subject "grew out of a study of finite differences and first took shape in 1911 as the fantasy which is now relegated to Ch. 11/2. Serious attention to the problem was begun in 1913 at Eskdalemuir Observatory with the permission and encouragement of Sir Napier Shaw, the Director of the Meteorological Office. . . ." [3]

Richardson joined the Friends' Ambulance Unit during World War I and continued his work while in France attached to a motor ambulance convoy. He says in his preface: "The manuscript was revised and the detailed example of Ch. IX was worked out in France in the intervals of transporting wounded in 1916–1918. During the battle of Champagne in April 1917 the working copy was sent to the rear, where it became lost, to be re-discovered some months later under a heap of coal."

In an obituary notice it is said that he worked out a six-hour forecast for 20 May 1910.[4] To do this he used a horizontal space grid about 200 km on a side and four vertical layers. This calculation was "strikingly at variance with the observed facts." Richardson realized that his time had not come. He wrote: "Perhaps some day in the dim future it will be possible to advance the computations faster than the weather advances and at a cost less than the saving to mankind due to the information gained. But this is a dream." [5]

Richardson of course could not know of the as yet undiscovered Courant condition (above, p. 288), and he had grossly violated it. Chapman, in his very informative introduction to Richardson's work, quotes Charney as follows: ". . . to the extent that my work in weather prediction has been of value, it has been a vindication of the vision of my distinguished predecessor Lewis F. Richardson, a former Secretary of this Society." [6]

[3] L. F. Richardson, *Weather Prediction by Numerical Process*, with a new Introduction by Sydney Chapman (Dover Edition, 1966), p. xii. The original edition appeared in 1922.

[4] *Ibid.*, Introduction, p. vii.

[5] *Ibid.*, p. xi.

[6] *Ibid.*, p. ix. These remarks were made upon the award of the Symons Gold Medal of the Royal Meteorological Society to Charney.

In 1950 Charney, Fjörtoft, and von Neumann published the first paper on their work.[7] This historic paper contains both an analysis of the equations used and a summary of the 24-hour forecasts made for four days in early 1949. These days were chosen since then the weather systems were so extensive that linearized techniques could not have predicted their true behavior in time. The forecasts were not too bad considering the necessity for doing the work on the ENIAC — the Institute computer was not yet completed — using a very coarse mesh. The grid used was in fact much coarser than Richardson's: it was 736 km on a side with 15 by 18 such cells — the choice was dictated by the small memory capacity of the ENIAC. At first, a time interval of one hour was used, but this was increased to three hours. They felt at that time that using the Institute machine with a finer grid would make it possible to realize Richardson's dream "of advancing the computation faster than the weather . . . at least for a two-dimensional model."

While it is not within our scope to discuss the details of the forecast work, it is perhaps instructive to say a little on the subject.

First of all what is being predicted? It is change in the atmospheric flow. Why this? The knowledge of this motion is necessary — unfortunately it is not always sufficient — to predict the phenomena we customarily think of such as cloudiness, humidity, precipitation, temperature, etc. To understand this motion is not at all easy since the physical processes involved are not well understood, and the differential equations which derive from a knowledge of these processes present considerable difficulties.

To describe a weather prediction model one customarily writes down equations which guarantee that the so-called laws of the conservation of mass, momentum, and energy are observed. This results in a set of highly complex equations. Not only are these equations complex but Thompson, who has analyzed the situation, said that for a space interval of about 167 km the time steps need to be around 10 minutes so that the Courant condition will not be violated.[8]

Of course, unless a forecast for the next twenty-four hours can be made in a short time — say about an hour — it is not very useful. Therefore, there is an enormous premium on speed of computation.

In 1948 Charney made a very important contribution when he

---

[7] J. G. Charney, R. Fjörtoft and J. von Neumann, "Numerical Integration of the Barotropic Vorticity Equation," *Tellus*, vol. 2 (1950), pp. 237–254; Collected Works, vol. VI, pp. 413–430.

[8] Thompson, *Numerical Weather Analysis and Prediction*, p. 17.

discovered that by making certain approximations of a not unreasonable nature he could significantly alter the equations of motion
and permit a much larger time step.[9] The 1950 calculations based
on these assumptions gave very encouraging results. The forecasts
obtained were comparable to what could be done by a skilled meteorological forecaster using the traditional technique of synoptic
forecasting. This old technique consisted essentially of an extrapolation forward in time of the weather conditions using the insight
and training of highly-skilled humans.

The first model produced was two-dimensional. There was only
one layer in the vertical dimension. (Actually the level chosen was
the 500 millibar one, but this is not important from our point of
view.) When this model was run on the Institute computer and certain mathematical refinements introduced, 24-hour forecasts were
produced in 6 minutes. These forecasts predicted atmospheric
flows over an area of about 5,400 by 5,400 kms with a space mesh of
300 km. Thus by 30 June 1953 the time for a 24-hour forecast by
machine was reduced from 24 hours on the ENIAC to 6 minutes – a
reduction factor of 240.[10]

Next the meteorologists turned to multi-layer models to achieve
greater accuracy and higher predictability. The first such model was
a two-level one and the next a three-level one. Most effort went into
the latter. In particular, a great deal of the work centered around
the period 23–26 November 1950, because on 24 November a great
storm hit Princeton and its environs causing considerable damage.
The three-level model "gave a good prediction for the central period when the storm was developing." [11] The three-level forecasts,
when first started, took 48 minutes to run, which was still very acceptable. The forecasts themselves occasioned considerable excitement, since they exhibited cyclogenesis, the development of
cyclonic motion.

Since those early, pioneering days the daily production of forecasts has gone forward pretty much keeping pace with new computer developments, and much research continues to be done on
ever more accurate and longer range forecasts. This work in the
United States is part of the overall responsibility of the Weather
Bureau under George Cressman, its Director. The daily work and

[9] J. Charney, "On the Scale of Atmospheric Motions," *Geofysiske Publikasjoner*
(Oslo), vol. 17, 2 (1948).

[10] Final Report on Contract No. DA-36-034-ORD-1023, Staff of the Electronic
Computer Project, Institute for Advanced Study, April 1954, pp. II-94 through II-
137.

[11] *Ibid.*, p. II-134.

the developmental problems are the concern of the National Meteorological Center of the Bureau under its Director, Frederick Shuman; and the research work of a more speculative nature is undertaken at the Bureau's Geophysical Research Laboratory in Princeton, which is headed by Joseph Smagorinsky. In addition, of course, much work goes on at numerous universities around the world, at many weather bureaus, and also at the National Center for Atmospheric Research in Boulder where Philip Thompson is Associate Director.

The great increases both in speed of computation and in memory capacity have naturally changed the ways of handling the physical numerical aspects of the prediction problem in the direction of ever greater accuracy. However, it seems very fair to trace all this worldwide activity to the meteorology group at the Institute for Advanced Study.

The engineering activity at the Institute was the keystone in the arch built there. Without it, all the rest would have been pointless and in a very real sense incomplete. With it, the total contribution to society by the relevant Institute members has been of very large proportions.

We were very fortunate indeed in having Julian H. Bigelow as our chief engineer during the formative stages of the project. He had visionary ideas on circuitry and the intellectual toughness to force these ideas to fruition. Without his leadership it is doubtful that the computer would have been a reality. All the staff felt great respect and admiration for him both as an engineer and as a human being. Bigelow was responsible for the overall design concept as well as for the details of the arithmetic and control organs. He was ably seconded by James H. Pomerene, who, as we mentioned earlier, succeeded him at a later stage, and who saw the machine to completion when Bigelow went on leave to study on a Guggenheim Fellowship. Not only did Pomerene carry the managerial responsibility, he also brilliantly carried out the development of several important parts of the device. The other engineers were all excellent and all deserve mention. They were not all at the Institute during the same period, and some were visitors from other institutions, but all made important contributions and should be remembered. They were: Heinz Billing, Ames Bliss, Hewitt Crane, Norman Emslie, Gerald Estrin, Ephraim Frei, Leon Harmon, Theodore W. Hildebrandt, William Keefe, Gordon Kent, Richard W. Melville, Peter Panagos, Jack Rosenberg, Morris Rubinoff, Robert Shaw, Ralph J. Slutz, Charles V. L. Smith, Richard L. Snyder, Erik Stemme, Willis H. Ware, and S. Y. Wong.

Not only did the group develop and build a pioneering and successfully operating machine, they also founded a line of machines both in direct and collateral descent. Those in direct descent were the ORDVAC and the first ILLIAC—both built by the Digital Computer Laboratory at the University of Illinois under the joint leadership of Professors Ralph Meagher and Abraham H. Taub (ORDVAC for the Aberdeen Proving Ground and ILLIAC for the University of Illinois). This activity at Illinois is especially important to the computer field not only because it is still pioneering in the development of new systems but also because of the important educational role that Laboratory — now the Department of Computer Science — has always played.[1] This project at Illinois owed its existence to the late Louis Ridenour, who as Dean of the Graduate School was instrumental in moving the university into the computing field — in fact Illinois was one of the first American universities to have an electronic computer. In his endeavors Ridenour had strong backing from Professor Nathan Newmark of the Civil Engineering Department. He brought Taub and Meagher to the university in 1948 and persuaded the Ballistic Research Laboratory to contract for a sister machine to the one being developed at the Institute for Advanced Study. He also persuaded the university to make a copy for itself. The laboratory had an excellent staff which had the distinction of getting the ORDVAC through its acceptance tests on 10 March 1952 and the ILLIAC on 1 September 1952.[2] Since then it has turned out a number of other advanced machines under varying leaderships.[3]

Before the Institute computer was finished, Willis Ware left to become head of a group at the Rand Corporation in Santa Monica, California. He and his colleagues built the JOHNNIAC (named for Johnny von Neumann), which passed its acceptance test in March 1954. Meanwhile, the Atomic Energy Commission's laboratories, Los Alamos, Argonne, and Oak Ridge also built copies.[4]

Other machines which derived fairly closely from the Institute one, but which may be considered as being of collateral descent, were built in the United States, Europe, Russia, and Israel.[5]

[1] The interested reader may wish to consult *Computers and Their Role in the Physical Sciences*, edited by S. Fernbach and A. H. Taub, Chapter 3 by H. H. Goldstine, pp. 51–102.

[2] M. H. Weik, *A Survey of Domestic Electronic Digital Computing Systems* (1955).

[3] Of these, the ILLIAC II was also derived from the Institute design. C. V. L. Smith in his book, *Electronic Digital Computers*, p. 421, says, "This design is a natural development of the single-address asynchronous logic due to von Neumann, Goldstine, and their associates of the Institute for Advanced Study. . . . It is interesting to note the vitality of this approach as shown by this latest extension of it."

[4] The Los Alamos group was under Metropolis' direction with J. Richardson as chief engineer; its machine was named MANIAC, and it ran successfully in March 1952. The Argonne group was headed by J. C. Chu, whom we first met as an engineer on the ENIAC. They designed a total of three machines of Institute type: first the AVIDAC for Argonne; then the ORACLE for the Oak Ridge National Laboratory (with consultative assistance from Burks); and finally GEORGE for Argonne.

[5] Some of these are: the Stockholm machine BESK, the Lund University SMIL, the Munich Technical Institute PERM, the Moscow BESM, the Israeli WEIZAC, the University of Sidney SILLIAC, and the Michigan State University MSUDC.

Perhaps, however, the most important lines of machines whose organization stems at least collaterally from the Institute for Advanced Study work were the famous 700 and 7000 series of the International Business Machines Corporation and the UNIVAC 1100 series of the Sperry Rand Corporation.[6] More on these later.

We now see how the Institute for Advanced Study ideas on machine organization were, and essentially still are, the basic ones on which most modern systems rest. Of course these ideas have evolved and been modified, but in essence they rest on the pioneering efforts at the Institute. We must also mention again the very significant modification instituted by Kilburn of the Manchester group (above, p. 265).

But let us return to our main theme: the accomplishments of the Institute engineering groups. In a certain real sense this is made much easier and perhaps even superfluous by C. V. L. Smith's excellent account of the engineering work.[7] However, for completeness of this account, I shall single out some of the advances due to the engineering group: first, they pioneered the asynchronous mode of operation and "the notion that all transfers of information must be done by what are now known as positive latches, that is, at no time is information stored even for extremely short intervals of time in capacitors or inductors but must always be in some such device as a flip-flop or toggle. . . ."[8]; second, they exploited the potentialities of F. C. Williams' ideas on storage tubes (above, p. 248) and made great contributions to making these ideas practically workable; third, they did a great deal to lead the computing field away from conventional chassis designs.

Let me say a little on these advances since they merit some consideration. Bigelow had a strong conviction that a computer should operate asynchronously, i.e. it should not be under the control of a clock which allows a fixed number of beats for each operation. Instead he felt it would result in a faster machine if each operation signalled when it was completed and thereby caused the next one to start. Now in such a regimen it is dangerous to transfer information from place to place unless it is positively stored in devices that cannot lose their information by the passage of time. Therefore, Bigelow designed a means for transferring information by a very reliable means of so-called positive latches that ensured no loss of

[6] Smith, *Electronic Digital Computers*, p. 37.

[7] In his *Electronic Digital Computers*.

[8] *Computers and Their Role in the Physical Sciences*, ed. Fernbach and Taub, p. 95.

information. When this was done it was considered quite radical.

Jack Rosenberg, under Bigelow's direction, designed the adder for the arithmetic unit using these ideas. Gerald Estrin wrote what may well be the first engineering type of diagnostic code. It was so designed that every toggle and gate in the machine could be examined for correctness of operation. This was most important to us and perhaps pointed the way for present-day diagnostic codes.[9]

The work on storage tubes was in some real sense perhaps the most critical thing done by the engineers. At the beginning of the project we were closely allied to the RCA group (above, p. 242) and hoped to receive from it a memory. The RCA group under Zworykin and Rajchman opted to build a storage tube that became known as the Selectron, and which was in fact used in the Rand JOHNNIAC. It came out in a 256-binary-digit-per-tube version and was a work of great engineering virtuosity. Unfortunately, it was very complex in its structure and required technologies that were perhaps ahead of their time. The engineers at RCA elected to build into their tube two grids of equally spaced bars at right angles to each other and forming many "windows." By properly applying voltages in a very elegant way exactly one window could be opened at a time, and an electron beam was allowed to flow through to a screen where information could be read or written. Otherwise, all windows were opened and thereby whatever was stored on the screen would be maintained. The complications resulting from the need for the windows, plus certain other difficulties, made the development of the Selectron a major undertaking.[10]

The RCA engineers decided upon their window means at a great cost in effort to ensure fidelity in switching to a given location on the storage screen. Unfortunately for them Williams came along just when he did and showed that the normal techniques used for switching the beam of a standard cathode ray tube were accurate enough. We need not here enter into the intricacies of the matter. It suffices to remark that the event proved Williams correct. It is also important to note again that a group under Forrester at MIT was also busy developing a storage tube which was somewhere

---

[9] See, for example, *Symposium – Diagnostic Programs and Marginal Checking for Large-Scale Digital Computers, IRE Convention Record* (1953). At this meeting a number of interesting papers on the subject were presented by J. P. Eckert, G. Estrin, L. R. Walters, and M. Wilkes and S. Bartin.

[10] Jan A. Rajchman, "The Selectron – A Tube for Selective Electrostatic Storage," *Proceedings of a Symposium on Large-Scale Digital Calculating Machinery* (Cambridge, Mass., 1948), pp. 133–135; a follow-on to this was presented by Rajchman in September 1949 at a later Symposium at Harvard.

between those of Rajchman and Williams in principle. It was brought into an operating state, but it and all other electrostatic storage tubes were supplanted by Forrester's magnetic cores which were to become much superior to the tubes.[11]

The development of the magnetic core was one of the most important steps in making computers reliable and capable of having very large memories. They are still in widespread use — albeit enormously improved.[12] They were first suggested in 1947 independently both by Forrester and Andrew D. Booth of the University of London, who lectured on the idea there. The major part of the development work was done — again independently — by Forrester in his Digital Computer Laboratory and by Rajchman at RCA. I well recall Rajchman's effort. He had bought a simple hand-operated press for squeezing tablets, such as aspirin, from powder. By a small conversion he made the press squeeze out small "doughnuts" of a ferrite material.

The beauty of Williams' system was that standard cathode ray tubes could be used with no internal modifications, although most cathode ray tubes had defects in their phosphor screens which rendered them unusable for our purposes but still entirely usable in oscilloscopes. Both the Allen B. DuMont Laboratories and the Radio Corporation of America were very helpful to us in making it possible to get suitable numbers of tubes for testing. We actually used about 19% of the tubes we tested.[13]

Even with this rejection rate the Williams' tube was very cheap compared to all others and quite satisfactory. The first use of the tube in a true random-access or, as it is often called, parallel memory was in 1951 at the Institute for Advanced Study and at the University of Illinois. Originally, Williams inserted and recovered information from his tubes in a serial fashion because this imposed the least exigent requirements on the tubes. Bigelow, when he heard of Williams' work, realized at once that it could be used exactly as we wished. We sent him over to Manchester in late June or early July 1948, since I wrote Williams on 21 June introducing Bigelow.[14] Williams also visited us in September 1949.

[11] J. W. Forrester, "Data Storage in Three Dimensions," Project Whirlwind Report M-70, Massachusetts Institute of Technology, Cambridge, Mass., April 1947; and "Digital Information Storage in Three Dimensions Using Magnetic Cores," *Jour. Appl. Phys.*, vol. 22 (1951), pp. 44–48.

[12] Forrester had an operable and full-scale core memory on Whirlwind in 1953.

[13] Final Report on Contract No. DA-36-034-ORD-1023, April 1954.

[14] Letters, Goldstine to Williams 21 June 1948; Williams to Goldstine, 30 July 1948. The latter letter noted Bigelow had by that date left for home.

My records indicate that on 16 June 1948 Pomerene started doing experimental work on the Williams tube and entered his results in his laboratory notebook. On 21 June I wrote Williams acknowledging his report, which was brought to me by Hartree on a visit. I also noted in 1952 that I might have heard of this work before Hartree's visit from Uttley, on an earlier visit.[15] (All this was important to Williams in connection with a patent application in the United States covering his ideas.)

Perhaps we should now ask how the Williams tube was used. In the Institute machine and its "copies" or derivative machines there were exactly as many tubes as binary digits in a word. All deflection systems of the tubes were connected in parallel, so that the electron beams in all were pointing at corresponding locations in all tubes. By this means the digits of a word were stored in corresponding positions in the memory. A single electrical command sufficed to switch all beams to the same location in each tube for reading, writing, or regenerating. What does that mean?

When information is stored in the form of an electrical charge in a phosphor, it gradually leaks away—about $1/10$ second—unless something is done about it. If the information is reread and restored about every $1/30$ second, it will essentially stay there forever. Both Pomerene and Meagher arranged matters so that whenever no commands to read or write were being obeyed the contents of the Williams tube memory were being systematically—one location after the next—reread and rewritten. This regeneration phase could be interrupted at will to do "useful" work. It required about 25 microseconds to read or write a number, about 30 to add, around 20 times as long to multiply, and 30 times as long to divide.

There were several very considerable engineering obstacles that had to be got over before the Williams tube memory could be used reliably. One was the great sensitivity of the tubes to extraneous, outside electromagnetic "noise" of all sorts. Without something drastic being done about such things as electrical disturbances, such as the energy being given out by fluorescent bulbs, etc., the system would have been inoperative. In the summer of 1949 Pomerene made a series of tests on the effects of various metal shields around the tubes on cutting out these extraneous disturbances. These shields worked very well. Pomerene next had to go to very delicate designs for his circuits, but in the end he made a 34-hour error-free run on 28–29 July 1949 using two of these tubes. This

[15] Letter, Goldstine to Williams, 10 July 1952, also Goldstine to Williams, 8 November 1948.

showed that the system would have the requisite stability.[16] Thus 1949 was in many ways the watershed for the Institute engineering project. Although very much had yet to be done after that date, there was little doubt that it would or could be. Prior to then there was considerable doubt. Much of the correspondence between von Neumann and me during this period reflects this, as did many of our conversations.

The fact that a word could be reached anywhere in the Williams memory of the Institute machine in about 30 millionths of a second, in contrast to the average time of about 0.5 thousandths of a second for the EDVAC type, meant that storage access time for the former was about 60 times that for the latter. This gave the Institute type computer a great inherent advantage over the EDVAC type and meant in the long run the ascendancy of the former. James Pomerene has kindly told me of the long discussions he and his colleagues had with M. V. Wilkes regarding the relative virtues of serial versus parallel organizations. We had, of course, made the decision to "go" parallel against the accepted views of the engineering world, a matter discussed in our *Preliminary Discussion* paper.[17] We did this because we knew we could gain a speed advantage of 40 (for a 40-bit word) simply by this option. Wilkes, whose EDSAC was a serial machine, apparently argued strongly for his organization and Pomerene for ours. As it turned out, our system had fewer tubes in it than his and was much faster. (The EDSAC contained about 4,000 tubes and the Institute machine about 2,300.)

At the time the decision in favor of parallel over serial was made (1946) it was not at all obvious that the number of tubes would be smaller for the former than for the latter. Indeed, the *a priori* view was to the contrary, since a serial arithmetic unit contains much less equipment than a parallel one. The real saving occurred in the control, which was much more complex in the serial machine. This more than outweighed the savings in the arithmetical parts. As Pomerene has said in the event our decision was the correct one, since with no more equipment we had a speed advantage of about 40 just from the fact of being parallel. Time has more than justified this decision.

Let us return to the memory discussion. While the electrostatic storage tube was faster than the supersonic delay line, it had two major problems: the electromagnetic noise we mentioned above and another. If a particular spot was read or written on too many

[16] J. H. Pomerene, Electronic Computer Project, Engineering Notebook I.
[17] Burks, Goldstine, von Neumann, *Preliminary Discussion*, p. 5.

times without the whole vicinity being regenerated, it was possible for stray electrons to "leak" into neighboring spots and change them from one binary state to another erroneously. We see this today on television when we see a very white shirt, for example, against a very black suit. Under these conditions the boundary between the two is frequently uneven and eroded in appearance.

While roughness in appearance may be perfectly tolerable in a television program, this phenomenon may spell catastrophe in a calculation, since erroneous changes may materially alter numbers or perhaps even worse the very instructions defining a given problem. It was soon determined that there was a given "read-around" ratio for the system beyond which one could not go. To avoid it, we placed a burden on our programmers. They were forced to resort to a variety of tricks to avoid too high a read-around ratio. Later it was learned how to increase markedly the stability of the memory, so this burden was eased. Also, in the beginning the read-around ratio would fluctuate daily and perhaps even hourly. Here again Pomerene at the Institute and Meagher at Illinois learned how to design their circuits so that frequent "tune-ups" of the memory were not needed.

Much has been said or written about how to test for errors and how to diagnose them when they do occur. This was always in the forefront for us because the read-around problem was always with us. We devised all manner of specialized and artificial arithmetical tests for parts of the machine. However, we could run these on a given morning with complete success, then put in a real problem and have a catastrophic failure in seconds. After much travail we decided that the only real diagnostic test of an arithmetical type we could devise was to make the machine do a very large and "real" problem—that is, a problem that filled the memory and used all parts in some statistically random fashion. If such a problem ran reproducibly, we were able to use the machine successfully. Accordingly, we took such a problem and broke it up into segments of about a half-dozen arithmetic operations each. Each such segment—typically it might be $[(ab + c)d + e]/f$—was then coded in two very different ways. For example, another form of the expression above would be $a[(bd)/f] + c(d/f) + e/f$. The machine would calculate each of the pair, store all partial results of the calculation in the segment, and not proceed until the end result of the segmental calculations agreed. If they did not, the machine came to a halt, indicated which segment it was in, and we could look at the evidence of the "crime." In this manner we finally achieved a diagnostic test that satisfied us. The use of this, together with Gerald Estrin's elegant engineering diagnostic procedure (above, p. 309), made our lives quite bearable.

Another minor but very useful diagnostic tool we developed was an instruction built into the computer at a small cost that summed the entire contents of the memory. We were able with certain known test problems to find this sum at a number of known times, and then we could check if this quantity was correct periodically.

We had originally planned for a memory of $4,096 = 2^{12}$ binary digits, but the practicalities of life were otherwise. The Selectron as finally built had only a $256 = 2^8$ binary-digit storage capacity, and the Williams tube had $1,024 = 2^{10}$ bits. The latter was achieved by a raster of 32×32 points on the screen of a five-inch cathode ray tube used to store one of two possible patterns at a point. This should be contrasted with the performance of the same tube in a television set, where it stores information at approximately $500 \times 500 = 250,000$ points and at each of them a continuum of tones from black to white!

It is worth again emphasizing the crucial importance of the MIT Radiation Laboratory radar work on the whole computer field. Out of that Laboratory and that school were to come all memories for post-ENIAC machines. Recall that the preliminary experiments both on supersonic delay lines and electrostatic storage tubes were done in the laboratory and that the magnetic core was developed at the school. This highlights the premier role of MIT in the technological world.

However, to return to the Institute. As I said, the 1,024-word memory for the computer was too small to hold both the program and digital information for a problem. Therefore a study of magnetic drums was initiated by Bigelow, and the work was done by him in concert with Peter Panagos, Morris Rubinoff, and Willis Ware under a small contract with the Office of Naval Research. This work terminated 31 June 1948.[18] Rubinoff continued these studies which resulted in a small—8 inch—magnetic drum containing 2,048 additional words of memory made up of two sections each of 1,024 words. This was done by using 80 tracks on the drum and 1,024 locations on each track. This drum was added to the Institute computer and placed on a limited test basis in May and June

[18] Bigelow, et al., "Fifth Interim Progress Report on the Physical Realization of an Electronic Computing Instrument," Institute for Advanced Study, Princeton, 1 January 1949.

1953.[19] It proved very useful, and early in 1955 the need for still more memory caused us to procure a 16,384-word drum from Engineering Research Associates, now merged into Sperry Rand, Inc.[20] The integration of this to the system was done in a very elegant fashion by Harmon under Pomerene's direction. It embodied a procedure that is not used today but could well be. It was possible to read or write an arbitrary length block of words starting at any designated word on the drum. This was achieved by giving the address of the starting point and the number of words in the block, and allowed us great flexibility of operation.

The first input-output organ for the Institute computer was developed for us by the National Bureau of Standards, which was anxious to achieve its proper role in the computer field. Accordingly, the Bureau modified some teletype equipment for our use in late 1948 and early 1949.[21] This equipment was then importantly modified by Richard Melville. This change substantially speeded up its operating rates. The modified teletype allowed us to load the memory in about 8 minutes and to print it out in 16. The former speed was bearable, but the latter was quite out of the question. We had to have a print-out of the memory everytime we tried to diagnose a malfunction of the machine. To spend a quarter of an hour on this was intolerable and resulted in having our machine inoperable a great deal. This was no reflection on the Bureau's work but rather indicated what could be achieved with a serial paper tape.

In fact, the Bureau became very much involved in the field under the joint influences of the late Samuel N. Alexander, who headed the electronics work on computers at the Bureau, and John Curtiss,

[19] H. H. Goldstine, J. H. Pomerene, and C. V. L. Smith, "Final Progress Report on the Physical Realization of an Electronic Computing Instrument" (Part I, Text), Institute for Advanced Study, January 1954; and Staff of the Electronic Computer Project, "Final Report on Contract No. DA-36-034-ORD-1023," Institute for Advanced Study, Princeton, April 1954, p. I-2.

[20] Engineering Research Associates was formed at the war's end by a group which played a very active role during the war and developed great expertise in computing devices in general and in particular in relation to magnetic drums. A summary of their work may be found in their book, Staff of Engineering Research Associates, Inc., *High-Speed Computing Devices* (New York, 1950). On p. 219 there is a list of some of their reports for ONR and for the National Bureau of Standards on drum technology. There are excellent bibliographies in the book for the student of this era. The book itself is a careful analysis of the electronic field as of 1950 and was in very large measure written by the late Professor C. B. Tompkins (he was Gamow's Tommy) of UCLA, with help from J. H. Wakelin and W. W. Stifler, Jr.

[21] Bigelow et al., "Third Interim Progress Report on the Physical Realization of an Electronic Computing Instrument," Institute for Advanced Study, 1 January 1949, p. 25.

who headed the mathematical section.[22] They built several pioneering machines called the SEAC, Standards Electronic Automatic
Computer, which passed its acceptance test in May 1950 and operated in the Bureau's Washington facility; the SWAC, Standards
Western Automatic Computer, operated in the Bureau's Institute
for Numerical Analysis on the UCLA campus; and the DYSEAC,
Second Standards Electronic Automatic Computer. The second
machine was financially supported by the Wright Air Development
Center and later by ONR, while the third was built in a truck for
the U.S. Army Signal Corps. It went into operation in April 1954.
The group at the Bureau also made significant engineering developments, including a so-called carry look-ahead adder which is
still widely used in computers.[23]

These efforts on the part of the National Bureau represented a
real effort on its part to assume leadership for the various agencies
of the Federal Government in the computer field. This was not to
be, however, since the great industrial concerns were to supplant
the Bureau in this task. The effort was nonetheless very worthwhile
and produced many fine people in the field.[24]

In any case, we found the paper tape medium impossible for our
purposes and were fortunately able to prevail upon IBM to help
us.[25] The electronic work needed to adapt the new equipment — an
IBM 514 reproducing punch — to our computer was carried out by
Hewitt Crane, who had shortly before this left IBM to join us. This
was most fortunate since he understood the inner workings of the
equipment and could adapt it to our purposes. This equipment
made all the difference. It changed the speeds for input-output
operations from minutes to seconds. The speed-up was about ten-
fold for reading and twenty-fold for punching out. These increases

[22] Under them were a number of able people including F. Alt, E. W. Cannon, R.
Elbourn, A. Leiner, W. Notz, R. Slutz, J. Smith, A. Weinberger, and perhaps others I
have forgotten.

[23] The interested reader will find references to their papers in the name index in
C. V. L. Smith, *Electronic Digital Computers.*

[24] In particular, the Institute at UCLA was very influential in making that school a
center for numerical analysis. There is still an excellent tradition for this work there,
not only in the mathematics department but also in the business and medical
schools, and Gerald Estrin heads a far-ranging group interested in computer developments. (He also took leave from us while he was still at the Institute for Advanced
Study and went to the Weitzmann Institute in Rehovoth, Israel, to help build their
first computer.)

[25] Memorandum, C. C. Hurd to J. C. McPherson, 8 February 1952. Memorandum,
D. R. Piatt to W. W. McDowell, Report on Trip to Institute for Advanced Study . . . ,
9 April 1952. (This material has kindly been made available to me through the
courtesy of J. C. McPherson.)

made decent operation possible, and the change-over took place in
November 1952. The Institute computer was publicly announced
on 10 June 1952, and the problem used to illustrate its operation
was that of Kummer (above, p. 297). I think the problem may not
yet have been completed on that date, but we must have felt very
sure of ourselves to have picked that date for announcement of the
machine. The machine was certainly operating prior to this date.
Smith says: "This development was successful, and before the end
of the year 1951 the IAS machine and a version of it built at the University of Illinois were successfully operating with random-access
or 'parallel' memories, and the Ferranti Mk I, based upon Williams' original work at the University of Manchester, was successfully using a serial memory." [26]

There were two other devices designed by Pomerene for the Institute system that are worth mentioning because they were both
important. One was a Williams tube that was controlled by a 40-
position switch and located by the operating console. With it we
could view the contents of any tube or the difference between the
contents of any two. The programmers made frequent and ingenious use of the device as did the engineers. Glenn Lewis and
Norman Phillips, for example, arranged an iteration routine in
the meteorological calculation so that they could see on the view-
ing tube how the iteration was proceeding. We were thus able to
watch the progress of an iteration without need for any print-outs
and could try out various alternatives for speeding it up with great
ease.

The other device was very pioneering. It was a programmed
graphical display on the face of an oscilloscope and was tied to
the magnetic drum. This was very useful then just as its modern
counterpart is today.[27]

There are pictures extant from those days showing among others J.
Robert Oppenheimer, who was Director of the Institute from 1947
on, and von Neumann. Oppenheimer was always solidly behind
the computer project and I feel sure from frequent personal conversations with him that he fully appreciated its importance. His
relations with von Neumann, though, were complex, and he therefore stayed well out of things until von Neumann left to sit on the
Atomic Energy Commission, at which point he became accessible
and was very helpful to me. He and von Neumann were never

[26] *Electronic Digital Computers*, p. 278.
[27] Burks, Goldstine, and von Neumann, *Preliminary Discussion*, p. 40.

intimate friends, but they deeply appreciated and respected each other.

Indeed, von Neumann once remarked to me that the entire Los Alamos venture would in his opinion have been quite impossible without Oppenheimer. He also felt very strongly that Oppenheimer was treated terribly by the United States Government. He felt that had Oppenheimer been in England he would have been created an earl and placed above censure. In fact, he felt quite bitter about the desire of democracies to cut down the figures that stood out. (These are nearly his words.) The two men of course held quite opposite views on the question of the super or thermonuclear bomb, but von Neumann never questioned for a moment Oppenheimer's integrity or his loyalty. He testified in this sense at the hearings on Oppenheimer's clearance and said, as to his loyalty and integrity "I have no doubts . . . whatever." As to his "discretion in the handling of classified material and classified information," he said, "I have personally every confidence." [28]

It is not without interest today to reread von Neumann's words with respect to the use of the Institute computer for the Atomic Energy Commission:

Q. Would you say anything about the role done at the Institute with respect to the development of computers?

A. We did plan and develop and build and get in operation and subsequently operate a very fast computer which during the period of its development was in the very fast class.

Q. Did Dr. Oppenheimer have anything to do with that?

A. Yes. The decision to build it was made 1 year before Dr. Oppenheimer came, but the operation of building it and getting it into running took approximately 6 years. During 5 of these 6 years, Dr. Oppenheimer was the Director of the Institute.

Q. When was it finally built?

A. It was built between 1946 and 1952.

Q. When was it complete and ready for use?

A. It was complete in 1951, and it was in a condition where you could really get production out of it in 1952.

Q. And was it used in the hydrogen bomb program?

A. Yes. As far as the Institute is concerned, . . . this computer came into operation in 1952 after which the first large problem that was done on it, and which was quite large and took even un-

[28] *In the Matter of J. Robert Oppenheimer, Transcript of Hearing before Personnel Security Board, Washington, D.C. April 12, 1954 through May 6, 1954,* United States Atomic Energy Commission, Washington, 1954, pp. 643–656.

der these conditions half a year, was for the thermonuclear program. Previous to that I had spent a lot of time on calculations on
other computers for the thermonuclear program.[29]

There is much else that one could write about the details of the
work at the Institute, but that is not the aim of this account. We
have said enough to give the flavor of the period. The Institute for
Advanced Study continued the computer project until about 1958.
After von Neumann left, I headed the project, with Charney,
Pomerene, and Maehly leading meteorological, engineering, and
mathematical groups; and work continued very actively with excellent results. But the sands of time were beginning to run out on
the project. The Institute as an institution and we as individuals
had accomplished our original purpose. Oswald Veblen felt this
very strongly. He saw in his mind – by this time he was technically
blind, but he did have some peripheral vision – that the era of the
academic institution making its own computers was drawing to a
close. He did not see, unfortunately, the present era of these institutions as great users and innovators of ever-new applications,
nor did he realize the impact they have on industry by articulating
their needs and requirements. Neither did he appreciate the impact
the computer would have on research workers in universities nor
how it would change the working habits of students the world over.
Veblen felt that the Institute had done what it set out to do and that
it should get out of the field. Oppenheimer disagreed with this
opinion strongly and felt there was a continuing role for the computer at the Institute. However, his views did not prevail, possibly
because I felt that the next thrust in the field would come from industry and accordingly left in 1958 to join IBM.

In looking back I am convinced that neither the University of
Pennsylvania nor the Institute for Advanced Study received the
credits which were their due. We have discussed the university
and why it did not. The Institute did not, I believe, for several
reasons:

First, it was probably unwise from the Institute's point of view
for us to put our basic ideas into the public domain instead of
obtaining patents covering them. What is free is perhaps only very
nominally valued by the world. The ideas were picked up and
utilized by everyone, exactly as we wanted them to be, but perhaps
without proper intellectual credits always being given. Our mailing
lists included the large corporations, so that they had fullest access

[29] *Ibid.*

to all our reports, both logical and engineering as they were written.

Second, the fact that the Institute dropped out of the computer field when it did tended to obscure its accomplishments in the eyes of the world which see only what is above ground and not the roots. Moreover, industrial concerns do not have as their prime responsibility the attribution of credits in the way that scholars and scientists do. They therefore do not customarily credit others; instead, they naturally point at their own accomplishments.

Third, the problem of what to do with the computer project at the Institute caused some sharp discussions and antedated by perhaps five years a schism that has taken place in the structure of mathematics: on the one side are the more traditional areas of mathematics and on the other those which are closely related to computers and computing. While this may be an oversimplification, it is not too far off the mark. The split has occurred in virtually every great university of the world, but like all great events it has not been without trauma. When the computer project was closed up at the Institute, very likely some of this was reflected, and the School of Mathematics returned to its highly distinguished path through the areas called "pure" mathematics. This is not written in a denunciatory sense but rather to indicate why the Institute did not more aggressively claim its just credit. Today, it has had a chance to see the fullness of its accomplishment, and time has dulled the differences that arose. It is therefore once again becoming conscious of the greatness of its attainments and proud to have been the home of one of the great accomplishments of man. Moreover, under the leadership of its new Director, Carl Kaysen, a School of Social Science is emerging that has considerable use for computation. Kaysen, while still in the federal government, was quite active in urging even greater use of computers to provide statistical bases on which to make governmental decisions.

Of course while all this was going on the rest of the world was not sitting quiescent. On the contrary, the computer revolution swept very rapidly throughout the United States, Europe, Israel, and Japan. Most of the work abroad was done at universities or research institutions and resulted in a large number of one-of-a-kind computers. None of these was in itself particularly important except that each served to condition some country to the need for electronic computers. Thus the integrated effect of all this activity was to create a demand for the electronic computer, first in the scientific and then a little later the commercial communities of the world. Since the economic picture abroad was in general not good after the war there was a few years' time lag in foreign computer developments as contrasted to those here. This lag was most noticeable in the "hardware" side and least on the "software" one.

It is not surprising that those countries hardest hit by the war required more time to recover from the effects of that cataclysm before embarking on the construction of computers. It is however remarkable that Great Britain had such vitality that it could immediately after the war embark on so many well-conceived and well-executed projects in the computer field. These include the construction by Wilkes of the EDSAC at Cambridge, by Williams of MADM and its successors at Manchester, by Turing (and his successors) at Teddington, and by Booth at Birkbeck College, London.[1] Out of these developments, which essentially started with the visits of Hartree and of Womersley, grew a considerable industry in Britain. Similarly the visit by Professor Stig Ekelöf (above, p. 249), started an extensive interest in computation and computer science in Sweden which has resulted in that country becoming a leader in a number of fields related to computers—numerical analysis and process control, to mention just two.

A detailed enumeration of worldwide computer accomplishments would add little to our general aim of sketching the main intellec-

---

[1] The work of Wilkes, Williams, and Turing has been discussed previously (pp. 217–219, 247–248); for Andrew Booth, see Appendix.

tual thrusts in the field. We have however provided a brief sketch of these developments in the Appendix and now resume our account of general developments.

The rapid construction of many different types of computers both in the United States and abroad resulted in the erection of a veritable tower of Babel. Each computer had its own language, and there was not too much incentive to create standardized languages. A substantial contribution industry—particularly IBM, since its computers became so widely adopted—made to this field was to establish quite a large community with the same machine or with a very few different types of machines. This made it natural for the members of this community and for the nascent industry to work towards the goal of a few common languages, so that meaningful communication could occur among computer users. Thus, the next big steps to be described are the industrialization of the computer field and the beginnings of scientific languages for computers. Of course, the paths we need to trace are by no means direct ones but are like comparable ones in the field of evolution: we must expect to find paths that start out most promisingly and then disappear, as well as others that start out modestly and then suddenly become main highways.

A significant example of the former type of path was that started by a UNESCO General Conference in the latter part of 1948. At that time it was realized that electronic computers were likely to be of inestimable importance to the world at large, and it was believed that for a considerable time they would be in very short supply. It was therefore felt that a consortium of nations should form an international computation laboratory to serve its members much as CERN has done so well in the high-energy physics world. The history of this undertaking is of interest since it illustrates a project that started out with an excellent rationale but whose importance and necessity diminished unexpectedly with the rather sudden availability of good computers on the open market. On 16 June 1949 the Director-General of the United Nations Educational, Scientific, and Cultural Organization asked the member states to comment on a report prepared by a U.S. National Research Council Committee on UNESCO. This report grew out of an instruction to the Director-General by the Third General Conference on UNESCO (Beirut, November-December, 1948) "to consider the possibility of an International Computation Centre and plans for its establishment. . . ." [2] The United States report grew out of a meeting of the

[2] *Third Session of the General Conference of UNESCO*, Resolution 3.81.

National Research Council Committee on UNESCO on 14 January 1949 under the chairmanship of Professor Bart Bok. Bok is professor of astronomy at the University of Arizona and vice-president of the International Astronomical Union. The committee set up a sub-committee consisting of C. I. Bliss of the Connecticut Agricultural Experiment Station, John Curtiss, W. J. Eckert, Philip M. Morse of the Massachusetts Institute of Technology, and Harlow Shapley of Harvard College Observatory.[3]

This resulted in responses to the Director-General from a number of nations including Italy, the Netherlands, Switzerland, and Denmark. The Danish Academy of Technical Sciences Committee was expanded into a Danish National Commission for UNESCO by the addition of several senior scientists, including Harald Bohr. With these additions the Commission embraced the disciplines of astronomy, biology, the engineering sciences, geodesy, mathematics, physics, and statistics. This group under the chairmanship of Professor Richard Petersen (see Appendix below) then prepared a reply to UNESCO. They said: "The plan originally put forward—as early as at the meeting of the Preparatory Commission of UNESCO in June 1946—was to establish a United Nations Computation Center equipped with one of the modern calculating machines in some East Asian country, preferably China. . . ."[4] The Commission appreciated the desire to help with the scientific development of the Far East but nonetheless strongly urged the establishment of the Centre in Europe and specifically in Denmark.

In the spring of 1951 a Committee of Experts was formed by UNESCO, and it met in Paris in May. Out of this meeting came a report which led to a Conference for the establishment of an International Computation Centre in the fall of that year.[5] (It was at the spring meeting I first had the pleasure of meeting Petersen and of forming what became, I believe, a warm friendship with him and his wife.) The expert committee had referred to it proposals from the Italian, Dutch, and Swiss governments seeking to be the site for the

[3] *Preliminary Memorandum on the Establishment of an International Computation Center under the General Sponsorship of UNESCO and ECOSOC*, Harlow Shapley, Chairman of the Sub-Committee, undated. The Sub-Committee met however on 17 March 1949 at the Watson Scientific Computing Laboratory, New York City.

[4] *Report on the Establishment of an International Computation Centre Submitted to UNESCO by the Danish National Commission for UNESCO*, Paris, 16 December 1949 (UNESCO/NS/ICC/3).

[5] UNESCO/NS/ICC/8, Paris, 21 May 1951; 6C/PRG/7, Paris, 18 June 1951.

proposed center. (I am no longer clear why Denmark withdrew.) In any event, the Director-General, Dr. Jaime Torres Bodet, and the expert committee asked me if I would evaluate the various proposals and advise the Director-General on their respective merits. This I did.[6] By this time Switzerland had withdrawn and I was forced to choose between the quite excellent computing facilities in Amsterdam and those in Rome. After much internal debating and reading of papers by members of the two centers, I chose Rome. This choice was upheld in November by a Conference held in Paris at which the United States did not send a delegate, only an observer. I was this observer.[7] In the event the Centre had great difficulties in getting started because the big countries such as the United States and Great Britain did not subscribe to the Convention. Then, as it gradually got started its *raison d'être* lessened. In spite of these difficulties, it did a very good job under the direction of Stig Comét of Sweden (see Appendix).

It is worth remarking on the worldwide interest in the Centre. Delegates attended from the following places: Central and South America – Brazil, Mexico, and Peru; Europe – Belgium, Denmark, France, Italy, Netherlands, Norway, Sweden, and Switzerland; Asia – Israel, Japan, Syria, and Turkey; Africa – Egypt, Liberia. In addition, there were official observers from the following countries: Austria, Great Britain, India, Iran, Uruguay, Union of South Africa, and the United States; also representatives from the International Union of Telecommunications, the International Council of Scientific Unions, the International Union of Pure and Applied Physics, the International Union of Theoretical and Applied Mechanics, the International Mathematical Union, the Union of International Technical Associations, the French National Centre for the Study of Telecommunication (CNET), Battelle Memorial Institute, Bull Machines Co., IBM France, Ferranti Ltd., SAMAS and SEA (Société d'Electronique & d'Automatisme). Thus, by the time the first modern computers were in their first days of running, we see twenty-five countries already concerning themselves with these machines!

[6] UNESCO/NS/ICC/14, Paris, 26 November 1951. My covering letter was dated 12 November 1951 and stated my authority as 6C/PRG/7 and that it was with the approval of a commission of the General Conference held 18 June to 11 July.

[7] UNESCO/NS/92, Paris, 8 April 1952.

Now that we have at least in broad brush painted the worldwide computer scene in the academic and governmental communities as of 1957 or thereabouts, we must examine the industrial responses to the challenges of these communities. The academic and governmental activities both in the United States and throughout the world undoubtedly developed a substantial market-place for computers, not only for scientific purposes but also for commercial ones. None of these groups was however equipped to carry the field forward. Each made its contribution by its training of students, engineers, and scientists to appreciate and to use these new tools. What was needed now was a mechanism to mass-produce machines.

The American Institute of Electrical Engineers and the Institute of Radio Engineers held a computer conference at the end of 1951. An examination of the conference committee shows that, of 16 members, 2 were from government, 2 from universities, 2 from technical journals, and 10 from industry. The companies involved are also of interest: Bell System (3), Burroughs Adding Machine (2), Engineering Research Associates (1), General Electric (1), International Business Machines (2), and Technitrol Engineering (1). In spite of this strong industrial representation, only 6 out of 19 papers were concerned with industrially-produced machines. The six were on the Burroughs Laboratory Computer; the ERA 1101; the IBM Card-Programmed Calculator; the seven electromechanical systems, Models I–VI (there were two Model V's) produced by the Bell Telephone Laboratories; and two on the Census UNIVAC system of the Eckert-Mauchly Computer Company. The other papers concerned university and governmental systems or were general discussions of a more philosophical nature.[1]

Thus, by the start of 1952 the scales were just beginning to tip slightly away from the government-university community to the in-

---

[1] See *Review of Electronic Digital Computers,* Joint AIEE-IRE Computer Conference (New York, 1952), pp. 1–114. There is one paper which is anomalous in that it fits both categories: it is one on a University of Manchester machine built jointly with Ferranti Ltd.

dustrial one. By the start of 1956 they had already overbalanced. Weik notes that out of 44 manufacturers of electronic systems in use in the United States only 17 were from the former and the balance of 27 from the latter.[2]

Eckert and Mauchly had developed the BINAC in August 1950 and the UNIVAC in March 1951. When their business was acquired by Remington Rand in March 1950 (95% then and the remaining 5% in 1951), that company had a significant technological lead in the field of electronic computers. This was further augmented when the Engineering Research Associates were acquired in April 1952 by Remington Rand, which itself then merged with Sperry Gyroscope to become Sperry Rand in mid-1955.

The ERA 1101 was delivered in December 1950 to the Georgia Institute of Technology; this was one of the first magnetic drum machines to be built. It was an asynchronous, binary machine using a one-address code, and its multiplying time was 260 microseconds.[3] A synchronous system, the 1102, with a drum half the capacity of the 1101, was built for the United States Air Force. Three of these were made for the Arnold Engineering Development Center of the USAF at Tullahoma, Tennessee. ERA also built some special-purpose systems, including the Logistics Computer for the Office of Naval Research located at George Washington University, completed in 1953. Later the ERA 1103 and a modification, the 1103A, were built. The first one passed its acceptance test in August 1953. Originally the 1103 had a Williams tube memory of 1,024 words, a 16,384 word ERA drum, and either Raytheon or Potter magnetic tape units. The 1103A had a 4,096-word core memory and UNIVAC type tape units. By the end of 1955 Weik reports that 10 units had been produced and installed.[4]

The UNIVAC also sold well, and by the time of Weik's first survey (1955) fifteen had been installed. Moreover a magnetic core memory version called UNIVAC II had been built; it had a memory with a range of from 2,000 to 10,000 words. (Some of the UNIVAC machines just mentioned were probably UNIVAC II's; Weik has an inconsistency in his text: he says in one place that 15 had been installed and in another that 22 were in current operation.)

During this period IBM was not idle. Soon after the Harvard-IBM machine was installed, Mr. Watson suggested to his engineer-

[2] M. H. Weik, *A Survey of Domestic Electronic Digital Computing Systems* (1955), pp. 204–207.

[3] F. C. Mullaney, "Design Features of the ERA 1101 Computer," *Review of Electronic Digital Computers*, Joint AIEE-IRE Computer Conference, pp. 43–49.

[4] Weik, *op. cit.*, pp. 189–191.

ing staff that a "follow-on" be built. This was done, and the resulting machine, the Selective Sequence Electronic Calculator, the
SSEC, was installed at the IBM World Headquarters, 590 Madison
Avenue, New York City, in January 1948. It ran until August 1952
when it was dismantled. Bowden has described the machine in this
way: "It could be seen from the street by passing pedestrians who
affectionately christened it 'Poppa.' It was a very large machine
and contained 23,000 relays and 13,000 valves [tubes]. All arithmetic operations were carried out by the valves and so it was more
than 100 times as fast as the Harvard Mark I machine."[5] The SSEC
had a hierarchy of memories: a small high-speed store in vacuum
tubes, a larger one in relays, and a very large one in 80-column
paper tapes. The instructions were also stored on this kind of tape.
There was provision for transferring the control automatically. The
machine multiplied 14-decimal digit numbers in about 20 milliseconds. It was very useful to a number of groups, including the
AEC, as we have mentioned above. Indeed the first calculation
done on the SSEC was of great importance to society and recalls for
us the nineteenth century interest in computation.

> The computation of improved positions of the moon, based on
> Brown's theory, was set as the demonstration problem of the
> Selective Sequence Electronic Calculator of International Busi-
> ness Machines at its dedication in January 1945. An exhaustive
> comparison of the new positions with Brown's *Tables* was made
> for selected dates. . . . All discrepancies were accounted for and
> the superiority of the new positions was established. . . .
> The longitude, latitude and horizontal parallax of the Moon, at
> intervals of one-half day from 1952 to 1971 inclusive, have been
> calculated under the immediate supervision of W. J. Eckert, Wat-
> son Scientific Computing Laboratory.[6]

Thus we find Wallace Eckert (above, p. 108) using the newest
tools for his life-long work of improving Brown's tables.

The great importance of the SSEC perhaps lay in the fact that by
this mechanism IBM's management became interested in the possibilities of electronic computers as contrasted to its calculators.
These latter were electronic punch card machines. They seem to
have evolved from the pair of relay calculators made by IBM for

[5] Bowden, *Faster than Thought*, pp. 174–175.
[6] *Improved Lunar Ephemeris 1952–1959, A Joint Supplement to the American
Ephemeris and the (British) Nautical Almanac* (United States Government Printing Office, 1954).

the Ballistic Research Laboratory at Aberdeen and installed there in December 1944; in addition to these two, IBM made one for the Naval Proving Ground at Dahlgren and two for the Watson Scientific Computing Laboratory at Columbia University.[7] Out of these emerged a line of calculators, the first made with vacuum tubes being the 603 Electronic Calculator, which appeared in the spring of 1946. Two years later the 604 was introduced. This was an extremely popular machine, and Weik reports that over 2,500 were produced by the end of 1955.[8] These machines could be controlled as to what operation to perform by information on a punch card. They were therefore card programmed.

This line of calculators reached its peak in 1949 in the Card-Programmed Calculator, which represented an ingenious lashing together of several existent IBM machines, the 402-417 Accounting Machine with the 604 and the 941 Auxiliary Storage Unit.[9] This system was also quite successful and Weik reports in his survey that over 200 were installed.[10]

However, the day of the computer was fast arriving at IBM. On 7 April 1953 the now famous 701 was formally dedicated at a luncheon at which Oppenheimer was the principal speaker. The machine used electrostatic storage tubes, a magnetic drum, and magnetic tapes. It was essentially competitive with the ERA 1103. It was developed in two years to meet the challenge of competition. The project was directed jointly by Jerrier A. Haddad and Nathaniel Rochester. The former was charged "with the complete technical and executive responsibility for the development and design program" and the latter "with the executive responsibility for the complete systems planning operation." In all, 19 of these systems were built and installed, and IBM was launched into the new world of electronic computers.[11]

[7] J. Lynch and C. E. Johnson, "Programming Principles for the IBM Relay Calculators," Ballistic Research Laboratory Report No. 705, October 1949; W. J. Eckert, "The IBM Pluggable Sequence Relay Calculator," *Math. Tables and Other Aids to Computation*, vol. III (1948), pp. 149–161.

[8] Weik, *op. cit.*, pp. 57–58. For a description of the machine see P. T. Nims, "The IBM Type 604 Electronic Calculating Punch as a Miniature Card-Programmed Electronic Calculator," *Proc. Computation Seminar*, August 1951 (IBM, 1951), pp. 37–47.

[9] C. C. Hurd, "The IBM Card-Programmed Electronic Calculator," Seminar on Scientific Computation, November 1949 (IBM, 1949), pp. 37–51; J. W. Sheldon and L. Tatum, "The IBM Card-Programmed Electronic Calculator," *Review of Electronic Digital Computers, op. cit.*, pp. 30–36; W. W. Woodbury, "The 603-405 Computer," *Proc. of a Second Symposium, op. cit.*, pp. 316–320.

[10] Weik, pp. 55–56.

[11] M. M. Astrahan and N. Rochester, "The Logical Organization of the New IBM

When the Joint Meteorological Committee of the Joint Chiefs of Staff decided to establish the Joint Numerical Weather Prediction Unit on or about 1 July 1954, it first established an Ad Hoc Group with Cdr. Daniel F. Rex as chairman. This group decided "to conduct a series of competitive test calculations to determine the suitability of the IBM Type 701 and ERA Model 1103 computers for the numerical weather prediction problem." [12] Joseph Smagorinsky of the U.S. Weather Bureau and I were asked to supervise these tests and to make a recommendation.

At this point in time — January 1954 — there was one 1103 available; it and the 701 installed at 590 Madison Avenue in New York were used in a test of a meteorological calculation. The report issued by us stated that the "speeds of the two machines are comparable with a slight advantage in favor of the 701 taking into account so-called optimal programming for the 1103." [13] However the input-output equipment for the 701 was significantly faster. A Technical Advisory Group — with von Neumann as chairman and Rex as secretary — to the Ad Hoc Group met on 28 January 1954 to consider our report. The report was accepted and "it was the unanimous recommendation of those present that the IBM type 701 computer be selected for use in the Joint Numerical Weather Prediction Unit."

In my opinion it was Thomas Watson, Jr. who played the key role in moving IBM into the electronic computer field. When he came out of the Air Force in 1945 his experience as a pilot had apparently convinced him of the fundamental importance of electronics as a new and prime technology for our society. He therefore exerted considerable pressure on IBM to move it in the direction of adapting electronic techniques to the business machine field. He early sensed the importance scientific computing would have to the entire field, and he brought to IBM in 1949 a mathematician, Cuthbert C. Hurd, who had been at the Oak Ridge National Laboratory since 1947.

At Oak Ridge, Hurd was exposed to a strong tradition for computation engendered there by Alston S. Householder. He also came to

---

Scientific Calculator," (*Proc. ACM* (1952), pp. 79–83; W. W. Buchholz, "The System Design of the IBM 701 Computer," *Proc. IRE*, vol. 41 (1953), pp. 1262–1275; C. E. Frizzell, "Engineering Description of the IBM Type 701 Computer," *Proc. IRE*, vol. 41 (1953), pp. 1275–1287.

[12] Letter, Rex to Goldstine, 16 December 1953.

[13] Report to the Ad Hoc Group for Establishment of a Joint Numerical Weather Prediction Unit: The ERA 1103 and the IBM 701, 27 January 1954.

know von Neumann and to have a deep admiration for his genius. He also knew in detail about our work at the Institute for Advanced Study. It was therefore quite natural for him to turn to von Neumann for help in bringing IBM into the "brave new world" of the modern electronic computer.

By October 1951, through Hurd's interest, IBM offered von Neumann a consulting contract. This relationship was a very important one for IBM since it made available to the technical people in that company virtually all that was known about the scientific application of the computer. Moreover, it ensured the highest quality scientific advice and, I believe, went a long way towards setting an ultimate standard of excellence.

It is quite clear in retrospect how profoundly and directly the university community affected industry: the ideas that were developed during the height of the Moore School era influenced Remington-Rand quite directly, just as those from the Institute for Advanced Study influenced IBM. Interestingly enough, both these industrial directions owed, and indeed still owe, extraordinarily much to von Neumann. With all due respect for the brilliance of his accomplishments in more orthodox mathematical areas, perhaps his greatest influence on our culture will be judged to be what he did for the computer field.

But let us now return to our discussion of early machines. We should mention the IBM Type 702, a machine using binary coded decimal and alphabetic symbols intended more for commercial than scientific work.[14] The first of these was accepted on 1 February 1955, and by the end of 1955 fourteen had been produced; it was then replaced by the Type 705. This latter was just in production when Weik issued his first report in 1955, but he notes that over 100 were on order at that date. The 701 was followed by the 704, a magnetic core memory machine, and as of Weik's report one had been built by the end of 1955, with 35 others then on order. The first one was to be delivered to the Radiation Laboratory, Livermore, at the University of California.

We must also mention the IBM Type 650, a magnetic drum machine put out with a 1,000- or 2,000-word drum, 60 words of magnetic core memory, and magnetic tapes. It did a multiplication in 2 milliseconds, as compared to 444 microseconds for the 701. But

---

[14] C. J. Bashe, P. W. Jackson, H. A. Mussell, W. D. Winger, "The Design of the IBM 702 System," *Trans. AIEE*, vol. 74 (1956), pp. 695–704; C. J. Bashe, W. Buchholz, and N. Rochester, "The IBM Type 702, An Electronic Data Processing Machine for Business," *Jour. ACM*, vol. 1 (1954), pp. 149–169.

the computer scene in American universities changed with its appearance. Prior to this system, universities built their own machines, either as copies of someone else's or as novel devices. After the 650, this was no longer true. By December 1955, Weik reports, 120 were in operation, and 750 were on order. For the first time, a large group of machine users had more or less identical systems. This had a most profound effect on programming and programmers. The existence of a very large community now made it possible, and indeed desirable, to have common programs, programming techniques, etc. We shall discuss later how this started off a field that today employs hundreds of thousands of people.

The next machine of IBM we shall mention was called the Naval Ordnance Research Calculator, NORC. It was a very large and very fast one-of-a-kind, electrostatic storage tube machine using the decimal system. (It did a multiplication in about 30 microseconds.) On the occasion of its first public showing on 2 December 1954 von Neumann made a speech which concluded as follows:

> The last thing that I want to mention can be said in a few words, but it is nonetheless very important. It is this: in planning new computing machines, in fact, in planning anything new, in trying to enlarge the number of parameters with which one can work, it is customary and very proper to consider what the demand is, what the price is, whether it will be more profitable to do it in a bold way or in a cautious way, and so on. This type of consideration is certainly necessary. Things would very quickly go to pieces if these rules were not observed in ninety-nine cases out of a hundred.
>
> It is very important, however, that there should be one case in a hundred where it is done differently and where one uses the definition of terms Mr. Havens quoted a little while ago. That is, to do sometimes what the United States Navy did in this case, and what IBM did in this case: to write specifications simply calling for the most advanced machine which is possible in the present state of the art. I hope that this will be done again soon and that it will never be forgotten.[15]

Von Neumann truly meant these words, as I was to learn some time after this talk. He and I were discussing ways to encourage industry to make bold new strides forward when suddenly he asked me whether it would not be a good idea to give contracts to both IBM and Sperry Rand to build the most advanced type computers

[15] Von Neumann, Collected Works, vol. V, pp. 238–247.

that they would be willing to undertake. Out of this conversation grew a quite formal set of arrangements between the Atomic Energy Commission and each of these companies. The results were the LARC of Sperry Rand and the STRETCH of IBM. Both projects strained each company's engineering staffs virtually to the breaking point but resulted in major advances in computer technology that were rapidly reflected in the product lines of these companies.

As I suggested earlier, one of the great contributions industry made to the computer field was the development of common lines of machines, so that users could potentially communicate with one another. Undoubtedly, this was a great impetus to the development of scientific languages such as FORTRAN. In the succeeding pages I shall discuss the programming developments of the period 1946–1957, since they foreshadow many of the modern ones.[1]

While the electronic computer produced a revolution by increasing incredibly the speed of processing data, it still left a large task for the human: the task of programming the problems to be run. It is therefore not at all surprising that from the start great emphasis was put upon methods for alleviating the burden of the programmer. Thus, what we need to discuss is the genesis of automatic programming.

To do this it is perhaps useful to look in a little depth into a computer of the 1946–1950 era. To simplify things, let us consider the Institute for Advanced Study one with its one-address code and ask what instructions or orders are really like and what a program looks like.

It is the totality of orders that makes up the *language* a machine understands; it is usually referred to as *machine language*. This is in modern parlance the most primitive or lowest level language of machines. Let us therefore ask a little about the structure of this language. By 1 July 1952 the Institute computer had a basic vocabulary of 29 instructions.[2] Each such order consisted, in general, of ten binary digits to express a memory location $-2^{10} = 1024-$ and 10 additional ones to express the specific operation. Thus the operation of clearing the contents of the accumulator and adding whatever is stored in, say, memory location 1000, into it is then

---

[1] For a detailed account of the later period see J. Sammet, *Programming Languages: History and Fundamentals* (New York, 1969) or S. Rosen, ed., *Programming Systems and Languages* (New York, 1967).

[2] H. H. Goldstine, J. H. Pomerene, and C. V. L. Smith, *Final Progress Report on the Physical Realization of an Electronic Computing Instrument* (Princeton, 1954), pp. 22–41.

1111101000   1111001010

Memory Location   Operation

Thus an order is a 20-binary-digit expression. What then does a simple program look like? Consider the following trivial one: the parameters a and b are stored in locations 10 and 11; we wish to form the quantity $c = a + b + ab$ and store it in location 12. Further, let us put the instructions in consecutive locations starting at 0. The code, then, is this:

    0.   00000010101111001010,
    1.   00000010111111001000,
    2.   00000011001110101000,
    3.   00000010101111001100,
    4.   00000010111111000110,
    5.   00000011001111001000,
    6.   00000011001110101000.

While this enumeration is virtually unintelligible to a human, it is exactly what the machine *understands*. What does it mean in a language we can understand? This is set forth below.

   0.   Clear the accumulator and add number stored in location 10 into it. S(10) → Ac+.
   1.   Add number stored in location 11 to contents of accumulator. S(11) → Ah+.
   2.   Transfer the number in the accumulator to location 12. At → S(12).
   3.   Clear arithmetic register and transfer into it number stored in location 10. S(10) → R.
   4.   Clear accumulator and multiply number stored at location 11 by the number in the arithmetic register placing the 39 most significant digits in the accumulator. S(11) × R → A.
   5.   Add number stored in location 12 to contents of accumulator. S(12) → Ah+.
   6.   Transfer the number in the accumulator to location 12. At → S(12).

This is a translation into English of the machine language statements. It is however quite verbose without being very informative. We therefore adopted another language for writing out our programs. This was a symbolic or mnemonic one based on the one above. We would have written the seven orders above as this:

	Coding Column		Explanatory Column	
0.	10	c	A	a
1.	11	h	A	$a + b$
2.	12	S	12	$a + b$
3.	10	R	R	a
4.	11	×	A	ab
5.	12	h	A	$ab + a + b = c$
6.	12	S	12	c

This was very close in form to what we customarily wrote as programs. It was then necessary for someone to translate this *symbolic language* into *machine language*.[3] In our case this was done by humans.

One of the first developments in automatic programming was introduced in the fall of 1949 on the EDSAC, where the conversion from the symbolic form to the machine one was done by the computer itself. "It is highly desirable that the machine itself should carry out as much of this work as possible; the chance of error is then reduced and the programmer is left free to concentrate his attention on the more essential aspects of the problem."[4]

While the symbolic language we proposed in 1947 was not very good in the sense that it was not close to a standard mathematical one, it or variants on it were used for a number of years on most machines. We did not work on what are now called higher-level languages. Attention instead was focussed on developing of libraries of programs that could be used repeatedly to save the labor of rewriting them many times. The third volume of our planning and coding series was intended for precisely this purpose. As we wrote:

> We wish to develop here methods that will permit us to use the coded sequence of a problem, when that problem occurs as part of a more complicated one, as a single entity, as a whole, and avoid the need for recoding it each time when it occurs as a part in a new context, i.e. in a new problem.

The importance of being able to do this is very great. It is likely

---

[3] For complete details, see H. H. Goldstine and J. von Neumann, *Planning and Coding of Problems for an Electronic Computing Instrument, Report on the Mathematical and Logical Aspects of an Electronic Computing Instrument*, Part II, vol. I (1947), vol. II (1948), and vol. III (1948).

[4] M. V. Wilkes, D. J. Wheeler, and S. Gill, *The Preparation of Programs for an Electronic Digital Computer* (New York, 1951), p. 15.

to have a decisive influence on the case and the efficiency with which a computing automat of the type that we contemplate will be operable. This possibility should, more than anything else, remove a bottleneck at the preparing, setting up, and coding of problems, which might otherwise be quite dangerous.[5]

We further said there that "the possibility of substituting the coded sequence of a simple (partial) problem as a single entity, a whole, into that one of a more complicated (more complete) problem, is of basic importance for the ease and efficiency of running an automatic, high speed computing establishment in the way that seems reasonable to us. . . . We call the coded sequence of a problem a *routine* and one which if formed with the purpose of possible substitution into other routines, a *subroutine*. As mentioned above, we envisage that a properly organized automatic, high speed establishment will include an extensive collection of such subroutines, . . . i.e., a 'library' of records in the form of the external memory medium, presumably magnetic wire or tape." [6]

In using a subroutine it is evident that it may be used in different parts of given routines, and hence its memory locations and memory references need adjustment. Moreover, various parameters and free variables of the subroutine will in general change from one usage to the next. We said about these things: "It seems, however, much preferable to let the machine itself do it by means of an extra routine, which we call the *preparatory routine*. We will speak accordingly of an *internal preparation* of subroutines, in contradistinction to the first mentioned process, the *external preparation* of subroutines. We have no doubt that the internal preparation is preferable to the external one in all but the very simplest cases." [7]

In *Planning and Coding* we then discussed and coded in detail such preparatory routines, which was the point of the volume. Our effort was followed in 1951 by a most interesting book on *The Preparation of Programs for an Electronic Digital Computer* by Wilkes and the Cambridge (England) group. They said there that the "book contains a detailed description of the library of subroutines used in the Mathematical Laboratory of the University of Cambridge . . . and of the way in which programs can be constructed with its aid." [8]

Now all this work on subroutines and preparatory routines was

[5] *Op. cit.,* vol. III, p. 2.
[6] *Ibid.,* p. 2.
[7] *Ibid.,* p. 4.
[8] Wilkes, Wheeler, and Gill, *op. cit.,* p. 1.

essential at that time in history, but it did only a little to ameliorate the condition of the programmer, who was still forced to write simple mathematical expressions in languages which were far from being natural to the problem at hand.[9] Efforts were however being directed towards higher-level languages, i.e. ones that come closer to the language of a scientific, engineering, or commercial discipline than does machine language or any of the mnemonic ones. Some preliminary work in this direction had been done by R. Logan, W. Schmitt, and A. Tonik in October 1952. They prepared a UNIVAC short code which in a tentative way expressed mathematical formulas in a system that was a first step on the way towards introducing formulas into programming languages. The idea for this code was due to Mauchly.[10]

Around the same period the late Heinz Rutishauser of the ETH in Zurich was thinking quite systematically about automatic programming and about ways to express mathematical concepts in a reasonable way on a computer.[11] He also conceived of a practical procedure for translating such expressions into machine language. Thus he probably invented the first *compiler,* and his is also the first *problem-oriented* language.[12] Rutishauser's work was based in part on some earlier work of Zuse.[13] (It should be remarked at this point that Rutishauser's work both in programming and numerical analysis was first-class, and it is sad that he died so young.) He was followed by one of his students, C. Böhm, who carried his ideas fur-

[9] In 1952 the topic was still much discussed as is indicated by *Proceedings of the Association for Computing Machinery, Jointly Sponsored by the Association for Computing Machinery and the Mellon Institute, Pittsburgh, Pa., May 2 and 3, 1952* (1952). In these proceedings there were seven papers on programming, and six had to do mainly with subroutines. J. Alexander, H. H. Goldstine, J. H. Levin, H. Rubinstein, and J. D. Rutledge, "Discussion on the Use and Construction of Subroutines"; J. H. Levin, "Construction and Use of Subroutines for the SEAC"; R. Lipkis, "The Use of Subroutines on SWAC"; D. J. Wheeler, "The Use of Subroutines in Programs"; J. W. Carr, "Progress of the Whirlwind Computer towards an Automatic Programming Procedure"; G. M. Hopper, "The Education of a Computer"; C. W. Adams, "Small Problems on Large Computers."

[10] Sammet, *Programming Languages,* pp. 129–130.

[11] H. Rutishauser, *Automatische Rechenplanfertigung bei programmgesteuerten Rechenmaschinen,* Mitt. No. 3, Inst. für Angewandte Math. der ETH Zürich (Basel, 1952); Rutishauser, "Bemerkungen zum programmgesteuerten Rechen," *Vorträge über Rechenlagen, Göttingen, March 1953,* pp. 34–37; and Rutishauser, "Massnahmen zur vereinfachung des Programmierens (Bericht über die in 5-jähriger Programmierungsarbeit mit der Z4 gewonnen Erfahrungen)," *Nachrichten-technische Fachberichte* (Braunschweig), vol. 4 (1956), pp. 60–61.

[12] K. Samelson and F. L. Bauer, "Sequential Formula Translation," pp. 206–220 in S. Rosen, ed., *Programming Systems and Languages.*

[13] K. Zuse, "Über den allgemeinen Plankalkül als Mittel zur Formulierung schematisch-kombinativer Aufgaben," *Arch. Math.,* vol. 1 (1948/49), pp. 441–449.

ther.[14] Rutishauser's ideas apparently were ahead of their time, and neither his nor Böhm's work evoked much response until a few years later when they were taken up by the Munich group under Bauer and Samelson.[15] This work prospered and ultimately led to a committee from the (American) Association for Computing Machinery and the (German) Association for Applied Mathematics and Mechanics (GAMM) being formed to write the algorithmic language ALGOL.[16] The very capable chairmen were Alan J. Perlis of Carnegie-Mellon University and Samelson of the Munich Institute of Technology. The group first met in 1957, and according to Rosen "the most active member of the (American) committee was John Backus of IBM. He was probably the only member of the committee whose position permitted him to spend full time on the language design project, and a good part of the 'American proposal' was based on his work." [17]

At the 1952 Association for Computing Machinery meeting Charles Adams expressed the aim of programmers — at least of that period. He said: "Ideally, one would like a procedure in which the mathematical formulation together with the initial conditions can simply be set down in words and symbols and then solved directly by a computer without further programming." [18]

At the same meeting John Carr described the programming work at the Massachusetts Institute of Technology. It is clear that the Whirlwind group there was very alive to the needs of the programmer. Adams and J. T. Gilmore extended the ideas of Wilkes, Wheeler, and Gill, and there evolved from this a symbolic address procedure, an idea that seems to have been independently created by Rochester and his colleagues at IBM.[19] The Whirlwind group

---

[14] C. Böhm, "Calculatrices digitales du dechiffrage des formules logico-mathématiques par la machine même dans la conception du programme," *Annali di mat. pura. appl.*, series IV, vol. 37 (1954), pp. 5–47.

[15] K. Samelson, "Probleme der Programmerungstechnik," *Inter. Kolloquium über Probleme der Rechentechnik* (Dresden, 1955), pp. 61–68.

[16] "ACM Committee on Programming Languages and GAMM Committee on Programming, Report on the Algorithmic Language ALGOL," edited by A. J. Perlis and K. Samelson, *Num. Math.* (1959), pp. 41–60; H. Zemanek, "Die algorithmische Formelsprache ALGOL," *Elektronische Rechnanlagen*, vol. 1 (1959), pp. 72–79 and 140–143; S. Rosen, *op. cit.*, pp. 48–78, 79–117.

[17] Rosen, *op. cit.*, p. 9. The authors of the report on ALGOL were P. Naur, J. W. Backus, F. L. Bauer, J. Green, C. Katz, J. McCarthy, A. J. Perlis, H. Rutishauser, K. Samelson, B. Vauquois, J. H. Wegstein, A. van Wijngaarden, and M. Woodger.

[18] C. W. Adams, "Small Problems on Large Computers," *Proceedings of the ACM* (1952), p. 101.

[19] J. W. Carr, *ibid.*, pp. 237–241; N. Rochester, "Symbolic Programming," *Trans. IRE, Prof. Group on Electronic Computers*, vol. EC-2 (1953), pp. 10–15.

also pioneered in the development of a so-called interpretive algebraic coding system. Laning and Zierler wrote a programming language in 1952–53 that permitted the programmer to write mathematical expressions in a symbolism that was close to the normal mathematical one.[20]

Perhaps we should pause to say a few words about the meaning of the concepts interpreter and compiler. Both are programs written in languages which are not understood by a computer. Both must therefore be *translated* by the computer into a machine-language program before they can be carried out. In the case of the interpreter, the translation or decoding of each statement is done every time that statement is read; with the compiler the decoding of each statement is done *a priori*, and from there on the computer deals only with the machine-language program. The Laning-Zierler program was an example of an interpreter program and Rutishauser's one of a compiler program. The disadvantage with both techniques is that they entail a loss of operating speed. In the case of the compiler, the act of compiling-translating takes time, and the machine code produced may not be as good as could be produced by a human programmer. In the case of an interpretive system, the translation is being done over and over with a consequent loss of time. The compiler is, *coeteris paribus*, clearly faster than the interpreter, but it forces the programmer to make many decisions sooner than he would like to. In a paper written in 1954 Goldfinger in discussing interpretive routines said of them that they "would increase running time by a factor of ten or worse." [21]

In spite of these speed disadvantages automatic coding won out over hand coding because it is so very much easier. It makes possible the use of computers by students in elementary schools as well as in universities and colleges. In effect, at the cost of running more slowly computers have assumed the burden that would otherwise fall on the human.

The Whirlwind group early recognized the difficulty of asking students to use machine language, and Adams and Laning accordingly wrote an interpretive language for the use of students on Whirlwind.[22] The entire Whirlwind programming activity was

[20] J. H. Laning, Jr. and W. Zierler, *A Program for Translation of Mathematical Equations for Whirlwind I*, M.I.T., Engineering Memorandum E-364, Instrumentation Lab., Cambridge, Mass. (January 1954).

[21] R. Goldfinger, "New York University Compiler System," *Symposium on Automatic Programming for Digital Computers, Navy Mathematical Computing Advisory Panel, 13–14 May 1954* (ONR, Washington, D.C., 1954), pp. 30–33.

[22] C. W. Adams and J. H. Laning, Jr., "The M.I.T. System of Automatic Coding,"

excellent, and Sammet says it was "probably the most significant of all the early work." [23]

The UNIVAC group was fortunate in having Grace Hopper as the head of its programming effort. (Previously she had been a colleague of Howard Aiken at Harvard.) She was untiring in her efforts to further the use of compilers.[24] As a result, the UNIVAC had perhaps the first automatic coding systems in the United States. They were called A-0 and A-1. However, it was not until A-2 was developed in 1955 that one of their compilers received extensive usage.[25] Next the A-3 was developed, but it was produced concurrently with an algebraic translator AT-3, later called MATH-MATIC.[26] This code was competitive with one developed at IBM which largely supplanted its competitors.

IBM produced its 701 in 1952 as we saw above. The machine had a one-address code like the Institute computer. At the beginning of 1953 it set up a project under John Backus to develop an automatic code to simplify programming. This resulted in the so-called Speedcoding System.[27] Backus is one of the outstanding figures in this whole field. His contributions have been absolutely fundamental to programming. He played a key role in developing Speedcoding, FORTRAN, ALGOL, and the so-called Backus normal form.

The development of FORTRAN, Formula Translating System, began in the summer of 1954. By 10 November 1954 a preliminary report was issued internally at IBM. In this unsigned document the credo of the designers, Backus and Ziller, is expressed. This was to enable the 704 "to accept a concise formulation of a problem in terms of a mathematical notation and to produce automatically a high-speed 704 program for the solution of the problem." The system itself came out in 1957, and it soon became the most widely used higher level programming language.[28]

Symposium on Automatic Programming for Digital Computers (ONR, Washington, D.C., 1954), pp. 40–68.

[23] Sammet, *Programming Languages*, p. 132.

[24] G. M. Hopper, "The Education of a Computer," *Proceedings of the ACM* (1952), pp. 243–249.

[25] *The A-2 Compiler System*, Remington Rand Inc., 1955.

[26] R. Ash et al., *Preliminary Manual for* MATH-MATIC *and* ARITH-MATIC *Systems*, Philadelphia, 1957.

[27] *Speedcoding System for the Type 701 Electronic Data Processing Machines*, IBM Corp., 24-6059-0 (Sept., 1953); J. W. Backus and H. Herrick, "IBM 701 Speedcoding and Other Automatic Programming Systems," *Symposium on Automatic Programming for Digital Computers* (1954), pp. 106–113; also J. W. Backus, "The IBM 701 Speedcoding System," *Jour. ACM*, vol. 1 (1954), pp. 4–6.

[28] Many papers have appeared on the subject; see, e.g., J. W. Backus et al., "The

In the paper by Backus and his colleagues an example is given of the power of this language. A program of 47 FORTRAN statements was written in four hours. When it was compiled, it produced about 1,000 machine instructions, and it took about six minutes to do this compilation. Thus we see here the saving attained by this language: a programmer needed to write only 5% of the total machine-language instructions to achieve the same result. It is of some interest to note that FORTRAN did not easily become accepted largely because users felt they could write better codes than could a machine. Ultimately this fear—if it was that—was stilled and the language became very widely accepted. Further acceptance was achieved with modified version, FORTRAN II, which had certain superior features.

Many other systems were created, but we have space only to mention two others. A Datatron computer was installed at Purdue University's Statistical Laboratory. Fortunately, Alan J. Perlis, whom we have just met as chairman of the ALGOL committee, was there. He designed one of the first algebraic compilers for that machine, and when he went to what was then Carnegie Institute of Technology he redesigned it for the IBM 650. This program was called IT (Internal Translator).[29]

Finally, Remington Rand developed a system called UNICODE in 1957–58 for the 1103A and 1105 machines. Sammet says of it: "UNICODE cannot be said to have contributed anything significant to the improvement of scientific languages since it introduced no new concepts of its own." [30]

Later—after 1958—other commercial languages were of course developed. They are outside our time span, however, and will not be discussed here.

Fortran Automatic Coding System," *Proc. WJCC*, vol. 11 (1957), pp. 188–198. Backus' colleagues on the project were R. J. Beeber, S. Best (MIT), R. Goldberg, L. M. Haibt, H. L. Herrick, R. A. Nelson, D. Sayre, P. B. Sheridan, H. Stern, I. Ziller, R. A. Hughes (Livermore Radiation Laboratory), and R. Nutt (United Aircraft Corporation). A list of other papers on the subject may be found in Sammet, pp. 302–304.

[29] A. J. Perlis and J. N. Smith, *A Mathematical Language Compiler*, Automatic Coding, Monograph No. 3, Franklin Institute (Philadelphia 1957), pp. 87–102.

[30] Sammet, p. 138.

Now that we have traced some of the ideas and met some of the people behind computer development in the period from 1623 through 1957, it is time we reflect a little on what we have learned in the way of underlying themes and trends.

There are several such themes that suggest themselves at once: the fact that mankind chose to automate computing rather than some other phase of the human condition; that this area of human occupation should prove to be so extraordinarily productive and useful to man; that the automation of computation should alter irreversibly the *Weltanschauung* of mankind in civilized nations; and, finally, that the computer attained its true importance only with the advent of electronics. In the closing pages I shall discuss these topics briefly and reflect on their significance.

Since Galileo's time and perhaps since even much earlier, man has striven hard to understand and to control the mechanical and physical world around him, just as in an earlier era he devoted great energy to understanding and coming to terms with the spiritual one. The tools provided by Galileo, Kepler, and Newton were sufficiently powerful to enable man gradually to understand the observable phenomena of nature and finally to control some of them. This trend has been a long one; it still goes on today and is usually described as the Industrial Revolution. It is not our aim here to discuss this evolution of society from one in which human or animal muscles provided the energy needed for all tasks to one in which fossil fuels give up their stored energy at man's command. Much has been written on the subject, and it is well understood today how there has been and still is a complex interplay between scientific, engineering, and socio-economic factors. It is this interplay that has characterized our society for better or worse.

The early work on computers really culminated in Babbage's efforts. In a sense, he came along at precisely the correct moment in time. The great inventive thrust in Great Britain was at a high level when he began his efforts, and it was an era of unbounded enthusiasm for invention. During his life many great inventions were made, and very likely he was swept along in his ideas by the enthusiasms of his times. In any case, he, together with his predecessors, started a new trend—the Computer Revolution. It got off to a slow start, and it has only been in the last quarter century that this revolution has become of importance to society; and, in a relative sense, it is still in its earliest infancy, even though it has moved at a prodigious rate in the years since the end of World War II.

All this work started because of man's perception in some highly intuitive way that, in the words of James Clerk Maxwell, "the human mind is seldom satisfied, and is certainly never exercising its highest functions, when it is doing the work of a calculating machine." This realization was crucial. Just as the use of man's muscles for pumping water was inefficient, so was the use of his mind for carrying out calculations. We have seen that early electronic computers such as the one at the Institute for Advanced Study had total vocabularies of only about two dozen words, and they could have done with perhaps as few as half a dozen. Computing is thus subhuman in that it calls on very few of man's manifold abilities and yet is fundamental to many of his other activities, as Leibniz so clearly perceived. This then is basically why computing was chosen as a human task to be mechanized. Yet even though the work is subhuman in character we may still ask, quite properly, what made it possible to automate the processes involved. After all there are other subhuman tasks that have not been automated. Why this one?

Consider mathematical programs as we knew them in the 1950s. Even for an extremely large scientific problem its program was an essay of perhaps 750 words written in pidgin English with a total repertoire of only about two dozen words. Contrast this with almost any other form of human activity! There is no other that I know of where it is possible to express an entire highly complex situation in so short an essay in such a childish language. This is what made it possible to mechanize or automate computation. Recall that each word in the vocabulary of a computer must in some— usually quite real—sense be built out of components that are expensive or rare. The great power of the computer lies in its ability to iterate repeatedly the same short description of a basic mathematical process. Thus, to calculate the trajectory of a satellite on a flight to Mars it suffices for the mathematician to describe how the satellite moves forward only an extremely small distance. The computer then iterates this process extraordinarily many times but at such an enormous rate of speed that the total solution is rapidly achieved.

If man had attempted to automate almost any other intellectual pursuit, he would have been blocked by the impossibly large amount of equipment needed merely to distinguish the diverse concepts involved. It is only mathematics which is today known to be so simple. To see this in another way, consider that mathematics is essentially unique in being able to express ideas of greatest complexity by means of a very few equations which somehow contain in themselves the essence of the whole situation. Contrast this for a moment with sociology or modern physics, where it is not even certain if all the concepts are yet known and where their mutual interrelations are often at best obscure. These are the reasons man chose to act as he did and also why he succeeded.

Why is computing so important to mankind? We might have felt at one time that calculating would make up only a tiny part of human activity. Computing would seem, in the post-Galilean era, to have one foot in the physical sciences and the other in the accounting world. For this reason the average person might expect that little of his intellectual activity need be spent in computing. To him, this subject would seem to be mainly the concern of physical scientists and accountants, with perhaps some minor excursions into neighboring sciences and arts. Indeed, von Neumann and I in writing about the subject in the 1940s were primarily concerned with the applications of the computer to those areas of the sciences where mathematization had penetrated most deeply, namely, the physical sciences and parts of economics.

All this sounds fine but still leaves the average man largely unmoved, since computing appears to be relevant only to the most technical questions and therefore is largely irrelevant in the daily world. Perhaps the first great example showing that this is not so is that of our daily weather forecast. While we can address many other examples, most of them seem to affect people in their daily lives somewhat obliquely or indirectly. Not so the weather, as Mark Twain pointed out. Here is the case where daily life is directly impacted by the electronic computer.

But this example is still in the realm of the physical sciences, and the computer has influenced our society just as much in other areas. While it is not my purpose to catalog areas where the computer has profoundly influenced our lives, I would like to discuss the point a little.

Although it is certainly true that the computer can solve very many problems in areas that can be rendered into a mathematical form, this is a rather sterile and not very useful definition since it suggests largely scientific and engineering applications far removed from the man in the street. It is therefore better to recognize that what a computer really deals with is not just numbers alone but rather with *information* broadly. It does not just operate on numbers; rather, it transforms information and communicates it. Perhaps the greatest importance of the stored-program concept lies precisely at this point. As we have seen, information — the instructions characterizing a problem — can be coded into numerical form and then altered at will at the computer as the computation proceeds. This may well be the genesis of the idea of encoding information into digital form and then transforming it as desired.

Herein lies the key to the importance of electronic computers. Their universality makes them as useful for sorting information as for multiplying numbers. It is indeed prophetic that the first modern stored program was one for sorting and merging data (above, p. 209). Now all Social Security records, Internal Revenue and Census data, corporate stockholder listings, inventories of all sorts, etc. are grist to the computer's mill. (Recall that Babbage called his arithmetic organ a mill.) These huge files cannot only be sorted, they can be updated and decisions made based on the signs of certain numbers. Thus they can issue checks, determine the amounts to be paid, etc.

Still another example of a different sort is provided by the Bell System. In an earlier day all communication via telephone was routed by humans — usually young women — who sat at switchboards and were in plain fact the switches. Today computers and computer-like devices have taken over this function, which was mainly a subhuman one. Indeed, it is said that there are not enough young women available today to handle the present-day communications load if it had not been automated. Exactly what was the task performed by the switchboard operator? She was told the number to be called by a subscriber, found it, and manually made the connection. This function is another sorting problem and is clearly better done by computers.

In sum, the importance of the computer to society lies not only in its superb ability to do very complex tasks of an abstruse mathematical nature but also in its ability to alter profoundly the communication and transformation of all sorts of information. It is the latter capacity that has been so useful to the humanist and the sociologist as well as to the businessman. Since it is not my aim here to write at length on the implications of the computer to man and society, I leave this tempting topic without going into further depth beyond mentioning one illustration. In the preparation of manuscripts today it is perfectly possible to utilize computer programs for expediting the work. One can store the text in a computer's memory, bring out portions for emendations, and call for printings of these revised versions with self-justified margins if one wishes. Or the text can be typed into a computer that controls a photo-composition machine whose output is a set of photographic images of the final composed work in suitable form for printing with minimal human intervention.

Instead of pursuing this topic further let us switch to the related topic of mankind's *Weltanschauung* and how the computer has affected it. Just a decade ago the computer was an exotic, arcane instrument which in some mysterious way was imagined to be some sort of replica of a human brain. Today, even the sixth-grade children in the little suburb where I live are using terminals connected with computers and are quite conversant with the concepts of computers and programming languages. It is now a part of their normal life. In the world today the computer has ceased being an object of awe and wonder and has taken its place as a comprehensible, useful tool in our technologically-oriented society.

When the Institute for Advanced Study closed down its Computer Project in 1958, it gave its machine to Princeton University, where it was for a while the university's prime computing instrument. Today that university has a very large computer, makes it available to all who need it, and possibly spends for computing services an amount comparable to that for library services. Thus in about a decade the computer has emerged from the role of an anomaly to a necessity for many and at least a commonplace thing for others.

The average man today is constantly made aware of the computer. He meets it in connection with his pay check, his income tax, his charge accounts, his bank account, his credit cards, and in many other of his daily activities. Moreover, these usages are not just conveniences; they are essentially necessities. The real increases in our gross national product have been accomplished in an important part by increases in workers' productivity made possible by computers. Many workers are today providing services not goods. Their productivity has been essentially altered, increased markedly, by the computer. An obvious example of this is the ticket agent in an airlines office. Her ability to find space and issue a

ticket very quickly rests entirely on computerized airline reserva-
tion systems.

Thus the world-view of mankind has been irreversibly altered;
man's way of life is changed and will continue to change in response to the challenges and problems raised by the computer in
society.

Finally, we may well ask why the computer only came into its
own with the advent of electronics. The answer of course is
because of speed. But why is speed so important? Clearly it was
not merely to do quicker those problems already being done by
more primitive means. That is to say, the whole point of great
increases in speed is to handle totally novel situations not possible before. Once one can predict the weather 24 hours in advance
in a quarter of an hour, it is not worth a great deal to be able to cut
this to one second; but it would be worth enormous amounts if one
could predict the weather for 30 days in advance in a quarter of an
hour.

We may summarize the pre-electronic days of digital computing
by recognizing that in that era electromechanical digital computers
eased man's burden in a significant but modest number of ways
without however creating a new way of life for him. It was not
until the electronic era that totally new burdens could be undertaken by machines to better mankind's way of life.

Of course, all these changes have not been without social consequence, and it is therefore not surprising that some very thoughtful people have been concerned about them. We are probably now
at about the comparable point in time in the Computer Revolution
to that in the Industrial Revolution when the Luddites were vainly
smashing equipment. Our society is being asked to make many
changes in our day, some of which are of a painful sort. These
changes have induced some unrest and concern in areas such as
data banks; jurists, legislators, and sociologists will have to come to
terms with these changes and their consequences in constructive
ways so that our way of life can avoid potential disadvantages and
take full advantage of the benefits of computerization.

We have in the body of the text discussed the world-wide developments that directly affected progress in the computer field. There were however many others that had extremely important local effects, and we discuss or at least sketch them here.

GREAT BRITAIN. As we saw, the British were very active from the beginning and produced a variety of excellent machines (above, pp. 217–219, 246–249). We need say no more about their activities, except to say a few words more about Andrew D. Booth. He was in those days — 1947 — at Birkbeck College, London University, working on relay calculators. His ARC, Automatic Relay Computer, was his first machine and was completed in cooperation with the present Mrs. Booth, Kathleen H. V. Britten. This was followed by a very small machine SEC, Simple Electronic Computer, and then by a series of APEC's, All Purpose Electronic Computers, built for Birkbeck for x-ray crystallography and for a number of other users. These were built in the 1949–1951 era and were at least partly the results of Booth's and Miss Britten's year at the Institute for Advanced Study in 1946–47, a visit made possible by a grant from the Rockefeller Foundation. The Booths were clearly somewhat ahead of their time in some respects; as mentioned earlier, he conceived of the magnetic core and lectured on it at about the same time as did Forrester.

SWEDEN. The Swedish Government, as we mentioned (above, pp. 249ff), sent Stig Ekelöf to the United States in mid-1946 to look into the computing field. In November or December 1948 Sweden had appointed a five-man board headed for a short while by Rear-Admiral Stig Hanson-Ericson and then in April 1950 by the Under-Secretary of Defense, Gustav Adolf Widell, as chairman, with Gunnar Berggren, Ekelöf, Carl-Erik Fröberg, and Commodore Sigurd Lagerman as advisors. In December 1948 the Board authorized a highly talented man, Conny Palm, to head "a Working Group" to do research and development work on machines. The first accomplishment of the group was a relay machine.[1] Palm and Gösta Neovius built the BARK, Binär Automatic Relä-Kalkylator, at the Royal Institute of Technology in Stockholm with funding by the Swedish Telegraph Administration. They were supported by Fröberg, G. Kjellberg, and a number of other talented people. The design of the machine was due to Neovius and Harry Freese. It was completed in February 1950.[2]

Fröberg, Kjellberg, and Neovius were members of the group who spent the academic year 1947–48 in the United States. The group consisted of four first-rate men, both physicists and engineers: Carl-Erik Fröberg, now Professor at Lund University and head of its Institute of Computer Sciences, spent his time at the Institute for Advanced Study, where we became good friends. He is both a distinguished and delightful person, with a fine sense of humor. Göran Kjellberg spent the year with Howard Aiken's people at Harvard; Gösta Neovius was at the Massachusetts Institute of Technology; and Erik Stemme, now Professor at the Chalmers Institute of Technology in Gothenberg, visited at RCA, Princeton and then with us at the Institute for Advanced Study.

Stemme became chief engineer of a project to build an electronic computer. The Working Group under Stemme then built the BESK, Binär Electronisk Sekvens Kalkylator, at the Royal Institute along the lines of the Institute for Advanced Study machine. It was completed in November 1953, and then in September 1956 its Williams tube memory was replaced by a 1,024-word magnetic core memory.

Fröberg built his SMIL, Siffermaskinen I Lund, as a magnetic drum machine in organization like BESK, and hence like the Institute for Advanced Study machine. It was put into operation in June 1956 and decommissioned on February 3, 1970.[3] This machine, as well as BESK, played an important role in training of Swedish students in computer engineering and science.

Since those times the computing people in Sweden have turned their attention with great success to numerical analysis and the computer sciences away from machine construction.[4]

---

[1] Much of the material here is contained in a few mimeographed bulletins issued by the Board in its early days.

[2] G. Kjellberg and G. Neovius, "The BARK, A Swedish General Purpose Relay Computer," *MTAC*, vol. 5 (1951), pp. 29–34.

[3] SMIL means "smile" and BESK is the slang word for "beer" and means "bitter."

[4] Among the distinguished numerical analysts who were involved from the earliest days with the Working Group, under the direction first of Palm (who died prematurely in December 1951), Neovius, and Stig Comét, and then Gunnar Havermark,

DENMARK. In Denmark there was also considerable interest in computing due largely to one man, Professor Richard Petersen of the Technical University in Copenhagen. He was a student, close personal friend, and mathematical collaborator of Harald Bohr from 1933 until Bohr's untimely death. Petersen in addition to being a fine mathematician was a truly fine person of tremendous integrity, goodwill and sincerity.[5]

As early as 1946 the Danish Academy of Technical Sciences formed a committee chaired by Petersen and containing a number of leading figures in Denmark with an interest in computation. Among others, the committee contained Niels E. Nørlund and Bengt G. D. Strömgren. The former was Niels Bohr's brother-in-law and is the author of many fundamental papers on finite difference methods and on geodesy.[6] I first met him at Petersen's home about fifteen years ago and had the pleasure of seeing the beautiful maps he had made in his laboratory, the Danish Geodetic Institute.[7] Bengt Strömgren is now professor of astrophysics at the University of Copenhagen, president of the Royal Danish Academy of Sciences and Letters and of the International Astronomical Union. He has had a most distinguished career both in Europe and in the United States, where he was a distinguished service professor at the University of Chicago and later a professor at the Institute for Advanced Study.

In any case the Danish committee started construction of both a differential analyzer and an electronic equation solver. Then came the UNESCO proposal, and Richard Petersen had occasion to meet and talk with American and European colleagues in Paris. These talks helped convince him of the desirability of electronic digital

---

are Carl-Erik Fröberg, O. Karlqvist, and Germund Dahlquist, now Professor at the Royal Institute of Technology in Stockholm, who was head of the mathematical staff in the Working Group from 1 May 1956. He today is a very important numerical analyst with a worldwide reputation. It is indeed interesting how all these people in Sweden are carrying on a century-old tradition dating from Scheutz's time.

[5] In a touching tribute to him his biographer has truly written: "His friends, his colleagues, and the thousands of students whom he had the opportunity of teaching, retain innumerable valuable and merry recollections of a man in perfect mental equilibrium, always on the move towards new goals, always ready to help, and thoroughly enjoying every task which he took upon himself." Erik Hansen, Obituary of Professor Dr. Richard Petersen, translated by James Steffensen and privately distributed by Mrs. Karen Richard Petersen.

[6] E.g., N. E. Nørlund, *Vorlesungen Über Differenzenrechnung* (Berlin, 1924).

[7] During World War II the Nazis directed this institute to work for them. Nørlund managed to persuade them that he be allowed to prepare maps of medieval Danish ports and harbours! These maps were of course no use whatsoever to the Nazis but are of great importance to scholars.

machines. He thereupon applied to the Carlsberg Foundation for financial support. Fortunately the Board of Directors of this very enlightened foundation, which included Prof. Børge Jessen, the distinguished mathematician, was very helpful, and Petersen set up the Regnecentralen—Danish Institute of Computing Machinery—as a division of the Danish Academy of Technical Sciences. He also secured the blessing of the Swedish Board to make a modified copy of the BESK. In particular, the intention was to use a core memory. This machine was dedicated in the summer of 1957. Fortuitously, my family and I arrived in Copenhagen in time to take some small part in the dedication festivities of the DASK, the first machine produced by the Regnecentralen.[8]

NORWAY. The Norwegian government was much less active than were the Swedish or Danish ones. But it did send Dr. Ernest S. Selmer of the University of Oslo to the Institute for Advanced Study for the second terms of 1950–51 and 1951–52. His university then procured APE(X)C from Booth. It also built a small machine called NUSSE that was described by the Nobel Laureate in economics, Ragnar Frisch.[9]

THE NETHERLANDS. The computing activity in the Netherlands started on 11 February 1946 with the establishment of the Mathematisch Centrum. When started it consisted of four departments headed by distinguished scientists: pure mathematics under J. G. van der Corput and F. J. Koksma; mathematical statistics under D. van Dantzig; applied mathematics under B. L. van der Waerden; and computation under A. van Wijngaarden. The latter is one of its most vigorous members and has become a leader in the ALGOL community. He first built a relay machine in 1948–1951 called ARRA. Then a vacuum tube version with a magnetic drum was built and finished in 1954 by van Wijngaarden. This was in turn followed by ARMAC, Automatische Rekenmachine Mathematisch Centrum, in June 1956.[10]

[8] This organization subsequently became a commercial venture. Petersen's biographer has noted that: "He doubtless considered his work in connection with the institute to be an extension of his work at the Technical University, and he did not conceal his regret that the hopes of a closer connection between these two institutions were not fulfilled."

[9] R. Frisch, *Notes on the Main Organs and Operation Technique of the Oslo Electronic Computer* NUSSE, Memorandum of the Socio-Economic Institute of the University of Oslo, October 1956.

[10] A brief résumé of the machine's characteristics may be found in *Journal of the ACM*, vol. 4 (1957), pp. 106–108. This is a portion of an Office of Naval Research

It would be very wrong not to mention van Wijngaarden's colleagues: G. Blaauw, B. J. Loopstra and C. S. Scholten. Another Hollander of this period is W. L. van der Poel who independently built a small machine at the Postal Telephone and Telegraph Laboratory in the Hague called PTERA; he has long played a significant role in the programming field.[11]

FRANCE. We have already mentioned Louis Couffignal's activities at the Institute Blaise Pascal. In addition the Compagnie des Machines BULL brought out in 1956 its GAMMA 3 ET, a small electronic machine.[12] Quite soon after the war an organization known as SEA, Société d'Electronique et d'Automatisme, was formed by F. H. Raymond, an exceedingly good engineer. This organization built a series of electronic computers under the generic name of CAB, Calculatrice Arithmetique Binaire.

SWITZERLAND. We have already mentioned Konrad Zuse and his key position in the German computing field. His Z4 was the starting point for the Swiss at the ETH in the year 1950. There Edward L. Stiefel founded an Applied Mathematics Institute and gave life to a whole school of excellent numerical analysts who have made Zurich prominent in the field.[13]

Another man from this Institute is a first-class engineer, Ambros Speiser, who received his engineering degrees from the ETH in 1948 and 1950. He spent the year 1948–49 working in Aiken's laboratory at Harvard preparing himself to lead the engineering staff at the ETH in building the ERMETH, Electronische Rechenmaschine der Eidgenössischen Technischen Hochschule. The machine was completed in 1955 and showed the influences both

---

Digital Computer Newsletter which was reprinted by ACM. See also, A. van Wijngaarden, "Moderne Rechenautomaten in den Niederlanden," *Nachrichtentechnische Fachberichte* (Braunschweig), vol. 4 (1956), pp. 60–61. Fortunately a big and important symposium on electronic computers was held in Darmstadt on 25–27 October 1955. At this meeting virtually all European machine developers presented papers. The proceedings appeared in the issue of *Nachrichtentechnische Fachberichte* just cited; hereafter referred to as *N.F.*

[11] W. L. van der Poel, "The Essential Types of Operations in an Automatic Computer," *N.F.*, pp. 144–145; B. J. Loopstra, "Processing of Formulas by Machines," *N.F.*, pp. 146–147; C. S. Scholten, "Transfer Facilities between Memories of Different Types," *N.F.*, pp. 118–119.

[12] See W. de Beauclair, *Rechnen mit Maschinen*, pp. 184–185, for a discussion of other French machines in the post-1956 era.

[13] The list includes, among others, Peter Henrici, Urs Hochstrasser, Werner Leutert, Werner Liniger, Hans Maehly, Heinz Rutishauser, and Andreas Schopf.

of Zuse's Z4 and Aiken's Mark IV.[14] Speiser went from there to distinguish himself further by becoming the first head of IBM's Research Laboratory in Zurich and then Director of Research for the well-known Swiss engineering firm of Brown, Boveri, where he is now.

From our point of view one of the most important figures in Zurich was Heinz Rutishauser, who played a key role in the development of automatic programming from its earliest days on through ALGOL 60 (above, p. 337).

GERMANY. In West Germany concern for electronic computers developed quite rapidly and in several places. Just as in the United States, World War II triggered off considerable interest in Germany in computing in general and in electronic computers in particular. There are two engineers who filed patent applications which show they had at a very early date the concept of using vacuum tubes to expedite computation. Walter Hündorf of Munich conceived of his *Elektrische Rechenzelle* on 6 April 1959. Helmut Schreyer, a colleague of Konrad Zuse (above, p. 250), conceived of his device on 11 June 1943, although he apparently started developing his ideas as early as 1937. Hündorf's ideas seem to have centered around special computing vacuum tubes and appear to me to be similar to comparable inventions of J. A. Rajchman, Richard Snyder, and others at RCA made during the wartime era. All these special tubes, in the event, disappeared in favor of simple, general-purpose tubes linked together by external circuitry. The special-purpose tubes were extremely difficult to make with the glass seal technologies of that time, and because of their specialized nature — hence low production volume — they were very expensive.

Schreyer's work was done at the Technische Hochschule in Berlin as his thesis topic. A small piece was constructed, but the machine itself was shelved by the Nazis as being *"völlig irreal und unwichtig."* [15] I suspect — although I do not know — that de Beauclair's claims for Schreyer's machine are overenthusiastic.

Perhaps the first significant developments in Germany arose at the Max Planck Institut für Physik in Göttingen. This work was in large measure due to Heinz Billing in collaboration with Ludwig F. B. Biermann, the astrophysicist. Their first machine, G1, was started in 1950, finished during 1951, and put in regular operation

---

[14] A. P. Speiser, "Eingangs-und Ausgangsorgane sowie Schaltpulkte der ERMETH," *N.F.*, pp. 87–89.

[15] de Beauclair, *Rechnen mit Maschinen,* p. 206.

in 1952.[16] This machine was more or less a test model for G2 (there was also a G1a) — which was in normal operation by December 1954 and not 1959 as stated by de Beauclair. Both G1 and G2 were serial machines. They were followed by the G3, a parallel machine.[17] (We at the Institute for Advanced Study were fortunate in having Dr. Elenore Trefftz of Billing's staff with us during the academic year 1952–53 and Billing in about 1956.) Later, in 1958, the Max Planck Institut was moved by its Director Werner Heisenberg to Munich, and it expanded first into the Max Planck Institut für Physik und Astrophysik and eventually into two separate institutes, one for physics under Heisenberg and one for astrophysics under Biermann. During this era a number of extremely able young astrophysicists from there such as R. H. F. Lust and A. Schlüter came to Princeton University to study with Martin Schwarzschild. While in Princeton they did a number of interesting and important calculations on the computer at the Institute for Advanced Study.

The development of computers in Germany was not confined to the Institut für Astrophysik. Already in 1952 Professor Hans Piloty of the Technische Hochschule of Munich, together with his son Robert, now Professor at the Technische Hochschule of Darmstadt and then a *privatdozent* who had recently received his doctorate at MIT, were very busy developing their machine. Others in this project should also be mentioned since they also have become important figures in the computer field. They are Walter Proebster and Hans-Otto Leilich of Piloty's Institut für Elektrische Nachrichtentechnik und Masstechnik as well as Friedrich L. Bauer and Klaus Samelson of the Mathematische Institut.[18]

This machine was known as PERM, Programmgesteurte Elektronische Rechenanlage München, and was very fast; it had both a core memory of 2,048 words and a magnetic drum of 8,192 words. Piloty very charmingly — from our point of view — described his machine in these words: "We took the famous work of Professor v. Neumann and Professor Goldstine as model and guide for our

[16] L. Biermann, "Überblick über die Göttinger Entwicklungen, insbesondere die Anwendung der Maschinen G1 und G2," *N.F.*, pp. 36–39; H. Öhlmann, "Bericht über die Fertigstellung der G2," *N.F.*, pp. 97–98; and K. Pisula, "Die Weiterentwicklung des Befehlscodes der G2," *N.F.*, pp. 165–167.

[17] A. Schlüter, "Das Göttinger Projekt einer Schnellen Elektronischen Rechenmaschine (G3)," *N.F.*, pp. 99–101.

[18] H. Piloty, "Die Entwicklung der Perm," *N.F.*, pp. 40–45; W. E. Proebster, "Dezimal-Binär-Konvertierung mit Gleitendem Komma," *N.F.*, pp. 120–122; F. L. Bauer, "Interationsverfahren der linearen Algebra von Bernoullischen und Graeffeschen Korvergenztyp," *N.F.*, pp. 171–176; and R. Piloty, "Betrachtungen über das Problem der Datenverarbeitung," *N.F.*, pp. 5–8.

first considerations. Since Dr. Goldstine is present here on our symposium, I am very glad of being able to express my deep gratitude to our teachers before him personally." [19]

Another early development in West Germany took place in Darmstadt where Professor Alwin Walther (1898–1967) headed the Institut für Praktische Mathematik at the Technische Hochschule. He was engaged in ballistic work for the military in Germany, and in 1943 or 1944 already began to conceive of a digital computer to aid his work. Then in 1951, with the help of his colleagues Hermann Bottenbush, Hans-Joachim Dreyer, Walter Hoffmann, Walter Schütte, Heinz Unger, and others, Walther began to build his DERA, Darmstädter elektronische Rechenautomat. It was not completed however until 1959.[20] The machine designer was much influenced by the ideas of Aiken, whom Walther admired greatly. In speed the machine was rather slow, a multiplication taking about 12 milliseconds and an addition about 0.8 milliseconds in contrast to the PERM which did an addition in about 9 microseconds.

These developments in the West German universities plus those of Zuse in industry served to stimulate a considerable interest in the computer field. At the present, this is so great that the Federal Government is putting substantial amounts of money into the Länder operated universities in order to encourage the creation of computer science departments.

In East Germany there was considerably less progress in electronic machines except at Dresden, where N. Joachim Lehmann built the D1 (completed in 1956) and its extension, the D2, at the Institute of Technology.[21]

AUSTRIA. During the 1950s the Institut für Niederfrequenztechnik of the Institute of Technology in Vienna was as active as its finances permitted. Under the direction of Heinz Zemanek, a distinguished leader of computer science in Europe, this Institute

[19] H. Piloty, op. cit.
[20] H. J. Dreyer, "Der Darmstädter elektronische Rechenautomat," N.F., pp. 51–55; W. Schütte, "Einige technische Besonderheiten von DERA, N.F., pp. 126–128; H. Unger, "Arbeiten der Darmstädter mathematischen Rechenautomat," N.F., pp. 157–160; H. Bottenbusch, "Unterprogramme für DERA," N.F., pp. 165–167.
[21] N. J. Lehmann, "Stand und Ziel des Dresdender Rechengeräte-Entwicklung," N.F., pp. 46–50; Lehmann, "Bericht über den Entwurf eines kleinen Rechenautomaten an der Technischen Hochschule Dresden," Ber. Math. Tagung, Humboldt-Universität Berlin (Berlin, 1953), pp. 262–270; also Lehmann, "Bermerkungen zur Automatisierung der Programmfertigung für Rechenautomaten (Zusammefassung)," N.F., p. 154.

used an IBM 604 to solve problems; it built a very small—about 700 relays—relay computer called URR-1, Universalrechenma- schine 1; a small relay machine to solve logistical problems LRR-1; and in the period 1955–1958 a transistorized machine with a magnetic drum.[22] Its multiplication time was around 0.4 milli- seconds, and it was known as a little May breeze, *Mailüfterl*. In an address on the machine, Zemanek with his usual Viennese charm compared the Austrian efforts with those of the Americans and said: ". . . wenn die Amerikaner einen 'Whirlwind' besitzen dann müsste man dieser Maschine auf gut wienerisch den Namen 'Mailüfterl' geben."

ITALY. The Italians did not undertake the development of an elec- tronic computer with the speed of some other European countries, but they nonetheless set up a group at Pisa with the idea of learn- ing the field. This resulted in the CEP, Calcolatrice Elettronica Pisana, an asychronous machine with cores and drums, produced after the time our history covers. The Institute at Pisa was known as Centro Studi Calcolatrici Elettroniche. Its machine was in some ways similar to the Institute for Advanced Study one in logical design. The International Center used commercially available equipment as did Professor Mauro Picone's famous Istituto Nazion- ale per le Applicazioni de Calcolo. This institute was then (1957) about 25 years old and had trained an extremely large group of applied mathematicians.

BELGIUM. Professor C. Manneback of Louvain University in Bel- gium was one of several scientists who was, from an early time, in- terested in the development of computers. In part, his interest caused the Institut pour l'Encouragement de la Recherche Scien- tifique dans l'Industrie et l'Agriculture and the Fonds National de la Recherche Scientifique to commission the Bell Telephone Man- ufacturing Co. of Antwerp to build a machine for Belgium. The construction started in 1951 and was finished in 1954. It was a magnetic drum machine with a 16-millisecond multiplication time and was used at the Centre d'Étude et d'Exploitation des Cal-

---

[22] H. Zemanek, "Die Arbeiten an elektronischen Rechenmaschinen und Informa- tionsbearbeitungsmaschinen am Institut für Niederfrequenztechnik der Tech- nischen Hochschule Wien," *N.F.*, pp. 56–59. See also Zemanek, "'Mailüfterl,' ein dezimaler Volltransistor-Rechenautomat," *Elektrontechnik u. Maschinenbau*, vol. 75 (1958), pp. 453–463.

culatrices Electronique in Brussels.[23] The design reflected
the profound influence Aiken had on European computer design.[24]

RUSSIA. The Russians have been very concerned about electronic
computers from the late 1940s on. I had on a number of occasions
received requests from a Russian trading company for the reports
by von Neumann and myself on electronic computing instruments.
By 1953 the BESM, Bystrodeistwujuschtschaja Elektronnajastschet-
naja Machina, one of the first Russian machines, was completed.
In 1955 it had a Williams tube memory of 1,024 words and a magnetic drum of 5,120 words arranged in five groups of 1,024 each. It
also had a small—376 words—germanium diode memory. The
operating times were quite good—between 77 and 182 microseconds for addition and 270 for multiplication. Later the Williams
tubes were replaced by magnetic cores.

This machine was designed under the direction of Academician
Sergei A. Lebedev of the Academy of Sciences of the USSR, Moscow. It was announced to the western world at the Darmstadt conference in 1955.[25] This Russian computer has a 39-binary-digit
word together with a three-address code and uses a floating point.
It was developed at the Institute of Precise Machines and Computing Techniques.

At the same conference I. I. Basilewski announced and discussed
the URAL, a magnetic drum computer completed in 1955. It was
built at the Scientific Research Institute of the Ministry of Machine
and Instrument Construction under the direction of B. I. Rameev.
It has 36 bit words, a 1,024-word drum, and a multiplying speed of
about 10 milliseconds. This machine is said to be the prototype for
a series of over 300 copies.[26]

The STRELA was another large Williams tube machine with a capacity of 1,023 words each of 43 binary digits. It, like BESM, was a
floating point machine and like it has magnetic tapes, which can
hold about 200,000 words. This machine was constructed in 1953

[23] J. L. F. de Kerf, "A Survey of European Digital Computers," Parts I, II, III,
Computers and Automation, vol. 9 (1960).

[24] M. Linsman and W. Pouliart, "Principales Caractéristiques dans la Machine
Mathématique IRSIA-FNRS," *N.F.*, pp. 66–68, and V. Belevitch, "Le Trafic des Nom-
bres et des Ordres dans la Machine IRSIA-FNRS," *N.F.*, pp. 69–71.

[25] S. A. Lebedev, "BESM, eine schnellaufende elektronische Rechenmaschine
der Akademie der Wissenschaften der USSR," *N.F.*, pp. 76–79.

[26] I. I. Basilewski, "Die universelle Elektronen-Rechenmaschine URAL für
ingenieur-technische Untersuchungen," *N.F.*, pp. 80–86.

under Basilewski's direction.[27] Later the Williams tubes were replaced by cores and about 15 copies were made by 1960. Its multiplication time was about 500 microseconds.

In addition to these three a number of other computers were built in the USSR during the same period: the PAGODA, the M1 and M2, MESM, KRISTALL, N.12. Moreover, since 1957, our cutoff date, a number of others have been started. The interested reader may wish to consult a paper by Carr, Perlis, Robertson, and Scott describing their fortnight in the USSR in the late summer of 1958.[28] They stated in their conclusions the following germane comparisons: "As to the equipment seen, the British EDSAC II (Cambridge) and Danish DASK were of later design than the BESM I. All three are of comparable speed."

CZECHOSLOVAKIA. In this country Antonin Svoboda was a pioneer. As early as 1952 he had been involved in building various kinds of calculating devices, and at the Darmstadt conference of 1955 he described his ARITMA as having "interesting similarities with the DERA computer." [29] (ARITMA is apparently a nationalized company making punch card equipment.) In addition, Svoboda and his colleagues at the Institute of Mathematical Machines of the Academy of Science built several small machines and a larger relay machine SAPO, Samočinný Počítač, with a magnetic drum.[30]

Svoboda is quite well known for his work on so-called residual classes — a very interesting and novel way to do arithmetic. This work was pioneered by M. Valach, who discovered the procedure, and Svoboda.

JUGOSLAVIA. The first machine in Jugoslavia was to be a magnetic core machine and was not due to be completed during the span of this account.

POLAND. In Poland three machines were made, two of them at the Instytut Maszyn Mathematysznych in Warsaw and one at the Research Institute for Electronic Computers of the Polish Academy of Sciences.[31]

[27] A. I. Kitov and N. A. Krinitskii, *Electronic Computers*, translated from the Russian by R. P. Froom (London, 1962).

[28] J. W. Carr III, A. J. Perlis, J. E. Robertson, and N. R. Scott, "A Visit to Computation Centers in the Soviet Union," *Comm. of ACM*, vol. 2, no. 6 (1959), pp. 8–20. This article also contains an excellent bibliography.

[29] A. Svoboda, "ARITMA Calculating Punch," *N.F.*, p. 72.

[30] J. Oblonsky, "Some Features of the Czechoslovak Relay Computer SAPO," *N.F.*, pp. 73–75.

[31] de Beauclair, *Rechnen mit Maschinen*, p. 103.

JAPAN. The Japanese were deeply concerned about computers and very early recognized their importance to their country. About 1943 an analogue computer was built to solve systems of linear equations. This work was done in the Electrotechnical Laboratory of the Ministry of Communications. In the same period a Bush-type differential analyser was installed in the Aeronautical Laboratory of Tokyo University. Then in 1944 the Japanese Scientific Research Council set up a research committee on electrical computing machines with Hideo Yamashita as chairman. This committee seems to have been active in encouraging the use of differential analysers, since six were installed in the period 1944–1952, one of which used electronic techniques.[32]

In 1951 the Japanese Permanent Delegate to UNESCO, Mr. Hagiwara, was already quite anxious to have a sub-center of the proposed International Computational Centre established in Japan. He and Yamashita met for a number of discussions with me in Paris in November 1951, and they followed this with a visit of some of their colleagues a few years later.

In one of the articles mentioned in note 32 there is a reference to the TAC, Tokyo Automatic Computer. It says that the device was to be completed in 1952 — then, in handwriting, "(if finances be obtained)." Apparently they were eventually, and the machine was completed in 1956 for the Research Committee for Electric Computers of the National Research Council of Japan in Tokyo. It used an electrostatic storage tube memory of 512 words and a magnetic drum of 1,536 words. It used binary numbers and a single-address code. It did a multiplication in about 8 milliseconds.

This was not the first machine the Japanese built. That was the so-called E.T.L. Mark I completed in 1952, which was followed by the E.T.L. Mark II in November 1955.[33] They were built under the direction of Motinori Goto with Yasuo Komamiya as chief designer.[34] These are relay machines. (Later, in 1957, Goto was to

---

[32] Memorandum, The recent development of computing devices in Japan. This is in my files but is undated and unsigned but has attached to it reprints of two articles in Japanese describing an A. C. Network Analyzer and a Differential Analyzer. Both of these are dated October 1951. They were probably given me by Yamashita and his associates.

[33] M. Goto et al., "Theory and Structure of the Automatic Relay Computer, E.T.L. Mark II," *Researches of the Electrotechnical Laboratory* (Tokyo, 1956).

[34] W. Hoffmann, *Digitale Informationswandler* (Braunschweig, 1962). In particular, see M. Goto and Y. Komamiya, "The Relay Computer E.T.L. Mark II," pp. 580–594. This book is an excellent compendium of material on computers worldwide prior to 1962 and has truly remarkable bibliographies in it.

go to Harvard to study for a year in Aiken's laboratory under a
United Nations Scholarship.) Then came E.T.L. Mark III and IV,
both of which were fully transistorized computers.[35] It is really
remarkable that the Japanese could have finished a relay computer
as late as November 1955 and a transistorized one in November
1957.

In 1954 Eiichi Goto invented an ingenious device called a parametron. At the same time, von Neumann independently conceived
of the same device. "However von Neumann's proposal has not
been followed up by the computer designers nor brought into
practice until the successful Japanese developments in this field
became known." [36]

ISRAEL. The Israelis were also active in our field from an early
stage. Chaim L. Pekeris, who has headed the applied mathematical
work at the Weizmann Institute of Science since its inception, was
at the Institute for Advanced Study on a number of occasions; his
first visit was from 1946 through 1948 when he became very
interested in having an electronic computer at his institute. He sent
E. Frei there and later J. Gillis and also persuaded Gerald Estrin to
take a leave of absence from the Institute for Advanced Study to
build the WEIZAC.[37] It was started in June 1954 and tests were to
begin in March 1955. The machine used a drum memory. It was a
member of the Institute for Advanced Study family.

RUMANIA. The Institute of Physics of the Rumanian Academy of
Sciences built several machines for itself and the University of
Bucharest known as CIFA-1, 2, 3.[38]

INDIA. Finally, the well-known Tata Institute of Fundamental
Research in Bombay, India, started early in 1955 the development

[35] *Ibid.*, pp. 575–649, "Digital Computer Development in Japan." This section of
Hoffmann's *Digitale Informationswandler* is a series of articles, with bibliography,
under the editorship of H. Yamashita. The authors are M. Goto and Y. Komamiya,
H. Takahasi and E. Goto, S. Takahashi and H. Nishino, T. Motooka, N. Kuroyanagi.
The topics covered are the Mark II, the Parametron, Memory Systems for Parametron
Computers, Mark IV, Magnetic Core Circuits, the Esaki Diode, and High-Speed
Arithmetic Systems.

[36] *Ibid.*, p. 595.

[37] Office of Naval Research, Digital Computer Newsletter, reprinted in *Jour. of
ACM*, vol. 2 (1955), p. 135.

[38] V. Toma, "CIFA-1, the Electronic Computer of the Institute of Physics of the
Academy of the Rumanian People's Republic," *Inter. Math-Koll., Dresden, 22–27
November 1957.*

of a pilot of an electronic machine of the Institute for Advanced
Study type. It was to have a core memory of 256 words and was in
November 1956 doing a number of test calculations.[39]

[39] Office of Naval Research, Digital Computer Newsletter, in *Jour. of ACM*, vol. 3
(1956), p. 110.

Index

abacus, 40

Aberdeen Proving Ground, Md., *see* Ballistic Research Laboratory

ACE / Automatic Computing Engine, 217-19

ACM / Association for Computing Machinery, 337-38

Adams, Charles W., 337n, 338-39

Adams, John Couch, 14, 77, 100; co-discoverer of Neptune, 30, 123

Adams-Moulton method, 100

Aiken, Howard Hathaway, 84, 119, 143, 219, 251, 275, 340, 350, 354, 356, 358, 361; carries out Babbage's program, 106-07, 111-16, 118; Mark I through IV, 118-19, 137; *see also* ASCC, Harvard

Airy, George Biddell, Astronomer Royal, 14, 29

Albert, Prince, 23, 33

Alexander the Great, 12

Alexander, J., 337n

Alexander, James Waddell, professor at the Institute for Advanced Study, 78, 80-81, 211, 241

Alexander, Samuel N., 315

ALGOL / Algorithmic Language, formulated by committees of ACM and GAMM, 338, 341, 354

al-Kāshī, Jamshīd ben Mas'ūd Mahmūd Ghiāth ed Dīn, Iranian astronomer, 5-6

Allison, Samuel King, directs Fermi Institute, 232

Alt, Franz Leopold, 115-16, 137, 145, 316n

American Association for the Advancement of Sciences, 77, 84

Amsterdam Mathematisch Centrum, 352

analog devices, 5-6, 39-44, 47-52, 54-56, 88-89, 351; cinema integraph, 98-99; continuous integraph, 89; differential analyzers, 49-51, 92-102, 134, 140-42, 153, 165-66; harmonic analyzers, 43-44, 47-49, 52-59; integrators and planimeters, 40-43, 54, 92-94; Kelvin on, 47-52, 102; network analyzers, 90-91, 97-98; tidal analyzers, 41, 44, 48-49

analytical engine, 11, 13, 19-22, 25, 266; *see also* Babbage, Charles

analyzer, *see* analog devices

Andrade, Edward Neville da Costa, 85

Andrew, Merle M., 243

Andrews, Ernest G., 106, 115, 118

APEC / All Purpose Electronic Computer, 349, 352

Applied Mathematics Panel, part of NDRC, 211-12, 216, 288-89

ARC / Automatic Relay Calculator, 349

Archibald, Raymond Clare, 8n, 16-17, 106-08, 219

Archimedes, 5, 12, 286

Arenberg, David Lewis, 248

Argonne National Laboratory, AVIDAC and GEORGE, 307

Aristotle, 174

ARMAC / Automatische Rekenmachine Mathematisch Centrum, 352

Armer, Paul, 256

Artin, Emil, 297-98

ASCC / Automatic Sequence Controlled Calculator, 112; *see also* Aiken, IBM machines

Ash, R., 340n

Astrahan, Morton M., 328n

astronomy, 5-6, 10; *see also* lunar theory

Atanasoff, John Vincent, 123-26, 148, 208

Atomic Energy Commission, 243, 307, 332; von Neumann appointed commissioner, 277; *see also* Los Alamos

Auerbach, A. A., 246n

Augustus, Gaius Julius Caesar Octavianus, 12

Australia, SILLIAC at Sydney, 307n

Austrian developments, URR-1, LRR-1 and Mailufterl, 356-57; *see also* Zemenak

automata, 271, 285; definition of, 274;